Clinical Biochemical and Hematological Reference Values in Normal Experimental Animals and Normal Humans

Second Edition

Clinical Biochemical and Hematological Reference Values in Normal Experimental Animals and Normal Humans

Second Edition

BRIJ M. MITRUKA
Director, Clinical Laboratory
Quality Control, Research and Development
BHP, Inc., West Chester, Pennsylvania

Formerly Associate Professor of Research Medicine
University of Pennsylvania, Philadelphia, Pennsylvania

Formerly Assistant Professor of Laboratory Medicine and
Laboratory Animal Sciences, Yale School of Medicine
New Haven, Connecticut

HOWARD M. RAWNSLEY
Professor and Chairman, Department of Pathology
Dartmouth-Hitchcock Medical Center
Hanover, New Hampshire

Formerly Professor of Pathology and Director of
Pepper Laboratories, University of Pennsylvania
Philadelphia, Pennsylvania

MASSON Publishing USA, Inc.
New York • Paris • Barcelona • Mexico City • Milan • Rio de Janeiro

Library of Congress Cataloging in Publication Data

Mitruka, Brij M.
   Clinical biochemical and hematological reference
values in normal experimental animals and normal
humans.

   Bibliography: p.
   Includes index.
   1. Chemistry, Clinical—Tables. 2. Hematology—
Tables. 3. Physiology, Comparative—Tables.
4. Laboratory animals—Physiology—Tables.
I. Rawnsley, Howard M. II. Title.
RB40.M43   1981   619′.0212   81-17157
ISBN 0-89352-163-9   AACR2

ISBN 0-89352-163-9

Library of Congress Catalog Card Number: 81-17157

Printed in the United States of America

DEDICATION

This book is dedicated to Anjuli, Kiren, Vicky and Eileen with affection

PREFACE FOR THE FIRST EDITION

The purpose of this book is to make available to the research worker in medicine and biology a collection of reference chemical, physiological, and hematological values in commonly used experimental animals.

The book emphasizes the major factors which influence the establishing of normal or reference values. Factors including collection, processing, and storage of specimens, methods and procedures, units of measurement, effects of drugs and toxicological agents, and effects of pathological conditions are discussed in detail. Clinical biochemical and hematological data on 14 species of experimental animals are presented and compared with those values reported in humans.

We hope this book will be a useful source of information on clinical laboratory values in these experimental animals.

Philadelphia, Pennsylvania   Brij M. Mitruka
Hanover, New Hampshire   Howard M. Rawnsley
July, 1977

PREFACE FOR SECOND EDITION

The main objective of this edition, as of the first edition, is to provide medical researchers, biologists and laboratorians a collection of biochemical and hematological reference values in commonly used experimental animals. Comparative values in hematology and blood chemistry of various age groups of humans are included to aid in the interpretation of experimental data.

Although the basic style and format of presentation are not changed significantly in this edition, a large volume of comparative animal data, published in the last 4 years, are added together with up-to-date original journal references and textbook references. In recent years much more emphasis has been placed on automation in clinical laboratory services. Therefore, in this volume methodology sections are expanded and automated hematological and biochemical methods are described in detail. A new section on interpretation of laboratory data in view of reference values is added in order to reemphasize the factors which influence the establishing of normal or reference values. Not only the factors such as collection, processing, preservation of specimens and effects of drugs are important on test results, other factors such as pathological condition of the patient or animal subject and methodology used to obtain the test results are often overlooked, thereby causing serious misinterpretations. The latter two factors are elaborated in greater detail in this edition. Selected references on these and other subjects relating to reference values are included with the aid of Med-Line and Med-Lar computer surveys of literature.

We hope this volume, like its first edition, will be well received by the experimental animal user as a valuable source of information on clinical laboratory values in these experimental animals.

West Chester, Pennsylvania   Brij M. Mitruka
Hanover, New Hampshire   Howard M. Rawnsley
June, 1981

ACKNOWLEDGMENT FOR THE FIRST EDITION

For excellent typing and proofreading, we thank Hazel Harber, Lily Chen, Jo Ann Doven, Barbara Grosso, Dorothea Adams and Robert Abbott. We are grateful to Jerry Knee and Irene Soroka for preparing illustrations included in this book. Our special thanks are due to Dr. D. V. Vadehra, Surendra Mitruka, Ravendra Mitruka and Vicky Abbott for their help in reviewing portions of the manuscript and checking references cited in this book.

It is a pleasure to acknowledge the courtesy of many authors, scientists and publishers who provided us with invaluable information and data included in this book. Previously published material is included with permission and sources are indicated.

The cooperation of the staff of Masson Publishing USA, Inc., and especially of Alan Frankenfield and Jeanne Rich, is also acknowledged gratefully.

ACKNOWLEDGMENT FOR THE SECOND EDITION

We thank Lilia Aragoncillo and Sharon Davidson for typing the manuscript. Without the enthusiasm and untiring efforts of Lilia Aragoncillo the preparation of this volume would have been very difficult. The cooperation and help of the staff of Medical library at Dartmouth Medical School and especially of Elaine A. Bent for Medlar search is also acknowledged gratefully.

CONTENTS

SECTION VI

SECTION VII

SECTION VIII

List of Tables

Introduction

The main objective of this book is to present reliable and quality controlled data on clinical biochemical and hematological values of a wide variety of normal experimental animals used in biological and medical research. A compendium of normal values on several well-defined species and breeds of experimental animals should be quite valuable in providing reference data of commonly used diagnostic tests in hematology and clinical chemistry laboratories. Reference values of normal laboratory animals are useful in clinical and biological investigations in which experimental animals are used to study human diseases. The ranges of normal values for laboratory animals can provide criteria for the selection of appropriate experimental animal species, not only for the study of pathological conditions, but also for toxicological studies and for the understanding of mechanisms of disease processes.

Since a wide variety of experimental animals is currently used for medical research, it is rather difficult to establish a normal range of chemical constituents in blood, plasma, serum, and other body fluids of all the animal species. Normal values reported in the literature for a given species of animal often range so widely that the clinical investigator has no valid assessment of a particular animal's physiological state. In addition, the values reported are not comprehensive, and in most studies only a few unrelated biochemical and hematological data on some animal species have been reported.

A number of factors contribute to the problems of the establishment and use of normal values:

1. Arbitrary choice of population—lack of preselection and screening.
2. Lack of consideration of age, sex, breed, strain, nutritional and physiological status and environmental conditions of the experimental animals.
3. Lack of consideration of physiological and analytical variation.
4. Varying statistical interpretation of data—a) assumption of normal distribution of values for the population, b) use of empirical limits to establish normal range.

The data presented in this book have been collected with these criteria in mind. Analytical variations were minimized by using standard quality control methods, including the use of replicate standard sera and analysis of the same specimen with different test methods at different time intervals. Physiological variations were considered by defining and selecting animal test populations. Different age groups, sexes and breeds of animal species were used to determine the normal range of clinical biochemical and hematological values. The nutritional and environmental status of the animals were kept constant throughout these studies.

Standard methods of sample collection, storage, and preparation are described in detail in Section II, and the methods of hematological and biochemical determinations used throughout these studies are described in Sections III and V. To eliminate systematic errors and to improve the precision and specificity of our methods, a broad-range concept of quality control was applied. Items considered included standardization of tests, precision, specificity, recovery studies, interlaboratory comparisons, and long-range stability of laboratory performance (Levey and Jennings, 1950; Hoffman and Waid, 1963; Waid and Hoffman, 1955; Henry, 1959, 1974; Anido et al., 1979; Wirth and Thompson, 1965; Neumann, 1968; Martinek et al., 1968; Corley, 1970; Zieve, 1966; Caraway, 1971; Amador et al., 1968; Moss et al., 1971). Most values presented here are indicative of mean value for the population available for testing. In general, these mean values agree with those reported in the literature (Coffman, 1970; Coles, 1973; Jennings and Mulligan, 1953; Kanenko and Cornelius, 1970; Medway et al., 1969; Laird, 1972; Altman and Dittmer, 1971, 1974; Burns and DeLannor, Jr., 1966; Albritton, 1952). Further comparisons of values reported by us to those reported by other authors are given in this revised edition.

The currently held concept in human clinical medicine of applying two times the standard deviation as the normal ranges for a given group of animals was applied because the populations were carefully selected with regard to age, sex, breed, and nutritional conditions of animals showing good normal distribution curves (Section VI). It has been reported that qualitative values for a given animal from a given species are strongly influenced by age (Laird and Fox, 1970; Benirschke, 1978); sex (Fox et al., 1970); Laird et al., 1970); breed (Coles, 1973, 1980); strain (Fox et al., 1970; Laird et al., 1970); environment (Roussel et al., 1971); and the physiological status of the animal (Laird and Fox, 1970; Fox et al., 1970; Anonymous, 1980; McPherson et al., 1978; Cherian et al., 1978; Werner and Marsh, 1975; Tsay et al., 1979; Cembrowski et al., 1971; Grasbeck et al., 1979).

Although intraspecies physiological variation contributed to the range of biochemical and hematological values, it could be minimized by selecting genetically pure populations of animals that are kept under constant environmental and nutritional conditions. Interspecies physiological variations of the animals, on the other hand, show large ranges of normal values because of inherent physiological differences. The physiological and other biological characteristics of some animal species and man are compared in Tables 1–3. These values may be useful in selecting an animal species as a model for the study of human disease. The data may also be useful in choosing between laboratory tests in experimental animals.

In human medicine the purpose of hematology and clinical chemistry laboratory results is to aid the physician in making decisions in the diagnosis and treatment of patients. A critical aspect of the use of laboratory data is their interpretation. Therefore normal or reference values in various populations of subjects, or in pooled specimens from representative populations, should be established by every laboratory, keeping in view the many variable factors—analytical as well as biological.

Individual laboratory tests are subject to a number of variables, and the physician and clinical investigator should be aware of this variability in his interpretation of any

Table 1. Comparative Biological Data of Experimental Animals and Humans

Species	Average Life Span (years)	Age at Puberty	Chromosome No. (Diploid)	Average Estrus Cycle (days)	Average Gestation Period (days)	Litter Size
Mouse (Mus musculus)	4	35±5.0 days	40 s	4	25	1—12
Rat (Rattus norvegicus)	5	50±10.0 days	42 m	4.5	21	1—9
Hamster (Mesocricetus auratus)	1.9	45±11.0 days	44 m	4	18	2—12
Guinea pig (Cavia porcellus)	6	62±8.0 days	64 m	18	68	1—8
Rabbit (Oryctolagus cuniculus)	13	7.0±1.5 months	44 s	Induced Ovulation	30	1—13
Chicken (Gallus domesticus)	24	--	78 s, 770	--	--	--
Cat (Felis catus)	23	10.5±3.5 months	38 m	22	63	1—6
Dog (Canis familiaris)	22	7±1 months	78 m	9	61	1—12
Monkey (Rhesus) (Macaca mulatta)	29	2.5±1.5 years	42	28	168	1
Pig (Sus scrofa)	27	7±1.5 months	40 s	11	114	4—6
Goat (Capra hircus)	17	8±2 months	60 s	21	151	1—5

3

Table 1 (con't.). Comparative Biological Data of Experimental Animals and Humans

Species	Average Life Span (years)	Age at Puberty	Chromosome No. (Diploid)	Average Estrus Cycle (days)	Average Gestation Period (days)	Litter Size
Sheep (Ovis aries)	20	7.5±1.5 months	54 m	17	150	1—4
Cattle (Bos taurus)	30	8±2 months	60 m	18.5	280	1—2
Horse (Equus caballus)	50	10±2 months	64 m	24	330	1
Man (Homo sapiens)	73	13.5±2.5 years	46 s,o,m	28.4	278	1

S = Spermatogonium; O = oogonium; m = somatic cell.

Source: Hafez (1970, 1971); Altman and Dittmer (1974); Peplow et al. (1974); Altman and Katz (1977); Hawkey (1975).

Table 2. Physiological Characteristics of Experimental Animals and Humans

Species	Body Wt (kg)	Surface Area (cm²)	Energy Metabolism cal/kg/day	Energy Metabolism cal/m²/day	Heart Wt (g/100g)	Heart rate (beats/min)	Stroke Vol (ml/beat)	Cardiac output L/min	Cardiac index $L/m^2/m$	Arterial Blood Pressure (mm Hg) Systolic	Arterial Blood Pressure (mm Hg) Diastolic
Rat	0.1—0.5	0.03—0.06	120—140 (B)	760—905 (B)	0.24—0.58	250—400	1.3—2.0	0.015—0.079	1.6	88—184	58—145
Rabbit	1—4	0.23	47	310	0.19—0.36	123—330	1.3—3.8	0.25—0.75	1.7	95—130	60—90
Monkey	2—4	0.31	49 (B)	675	0.34—0.39	165—240	8.8	1.06	---	137—188	112—152
Dog	5—31	0.39—0.78	34—39 (B)	770—800 (B)	0.65—0.96	72—130	14—22	0.65—1.57	2.9	95—136	43—66
Man	54—94	1.65—1.83	23—26 (B)	790—910 (B)	0.45—0.65	41—108	62.8	5.6	3.3	92—150	53—90
Pig	100—250	2.9—3.2	14—17 (B)	1100—1360 (B)	0.25—0.40	55—86	39—43	5.4	4.8	144—185	98—120
Ox	500—800	4.2—8.0	15 (B)	1635 (B)	0.31—0.53	40—58	244	146	---	121—166	80—120
Horse	650—800	5.8—8.0	25 (R)	2710—2770 (R)	0.39—0.94	23—70	852	188	4.4	86—104	43—86

B = Basal; R = Resting

Source: Swenson et al. (1970); Altman and Dittmer (1971, 1974); Coles (1980); Phyillis (1976); Altman and Katz (1977).

Table 3. Certain Physiological Values of Experimental Animals and Humans

Species	Body temperature (°C)	Whole blood volume (ml/kg body wt)	Plasma volume (mg/kg body wt)	Plasma pH	Plasma CO_2 content (mM/L)	CO_2 pressure (mm Hg)
Mouse	36.5±0.70	74.5±17.0	48.8±17.0	7.40±0.06	22.5±4.50	40.0±5.40
Rat	37.3±1.40	58.0±14.0	31.3±12.0	7.35±0.09	24.0±4.70	42.0±5.70
Hamster	36.0±0.50	72.0±15.0	45.5±7.50	7.39±0.08	37.3±2.50	59.0±5.00
Guinea pig	37.9±0.95	74.0±7.00	38.8±4.50	7.35±0.09	22.0±6.60	40.0±9.80
Rabbit	38.8±0.65	69.4±12.0	43.5±9.10	7.32±0.03	22.8±8.60	40.0±11.5
Chicken	41.4±0.25	95.5±24.0	65.6±12.5	7.52±0.04	23.0±2.50	26.0±4.50
Cat	38.6±0.70	84.6±14.5	47.7±12.0	7.43±0.03	20.4±3.50	36.0±4.60
Dog	38.9±0.65	92.6±29.5	53.8±20.1	7.42±0.04	21.4±3.90	38.0±5.50
Monkey	38.8±0.80	75.0±14.0	44.7±13.0	7.46±0.06	29.3±3.8	44.0±4.8
Pig	39.3±0.30	69.4±11.5	41.9±8.90	7.40±0.08	30.2±2.5	43.0±5.60
Goat	39.5±0.60	71.0±14.0	55.5±13.0	7.41±0.09	25.2±2.8	50.0±9.40
Sheep	38.8±0.80	58.0±8.50	41.9±12.0	7.48±0.06	26.2±5.00	38.0±8.50
Cattle	38.6±0.30	57.4±5.00	38.8±2.50	7.38±0.05	31.0±3.0	48.0±4.80
Horse	37.8±0.25	72.0±15.0	51.5±12.0	7.42±0.03	28.0±4.00	47.0±8.50
Man	36.9±0.35	77.8±15.0	47.9±8.70	7.39±0.06	27.0±2.00	42.0±5.00

Source: Spector (1961); Medway et al. (1969); Altman and Dittmer (1971, 1974); Altman and Katz (1977); Bailey (1979); Duncan and Prasse (1977); Henry (1979).

changes which he may note. Physiological parameters may fluctuate considerably. Diurnal, annual, and postural variations in plasma constituents of experimental animals (and man) have been well documented (Whitehead, 1977). Effect of some of these factors and others are discussed in detail in Section VII.

Because of the complexities involved, laboratory data on experimental animals under clinical study may be interpreted with reference to the baseline values presented in this book. However, whenever possible, the researcher should determine the baseline values on each individual animal in the group and compare them with the normal values given here. Although similar in breed, sex, age, and nutritional and environmental conditions, the animals used in this study show small intraspecies variations. Therefore those animals in the group that have larger variations should not be included in the study of disease processes.

In recent years, the term reference values has been used instead of normal values (Sunderman, 1975). Reference values may be defined as a set of values of a certain type of quantity obtainable from a single individual or a group of individuals corresponding to a stated description. This description must be available in detail if others are to use the reference values. For each type of quantity, a series of reference groups will be necessary, taking into consideration age, sex, race, menstruation, previous diet and exercise, posture, etc. (Dybkaer, 1972). It should be emphasized that the term reference values denotes the entire set of test results derived from a reference population. Lumsden and Mullen (1978) discussed certain approaches to the establishment of reference values. Much of the statistical analysis (e.g. test, analysis of variance, etc.) of the biological data depends on the assumption that the data are Gaussian (or normal). If the data are not Gaussian, Holt et al. (1980) suggested methods of transformation that will render them Gaussian thereby subjecting the data for analysis with standard statistical techniques.

Sunderman (1975) and others (McNeil et al., 1975; Wilson, 1973; Jungner and Jungner, 1970; Lusted, 1968; Koller, 1965; Martin et al., 1975; Holland, 1975) have proposed substituting the term discrimination value for critical value. This value represents the result of a diagnostic test and serves as a criterion to distinguish between two subgroups of a population being tested, based upon (1) the clinical sensitivity of the diagnostic test, (2) the clinical specificity of the diagnostic test, (3) the relative distribution of individuals between the two groups, and (4) the clinical cost of misclassification. The concept of discrimination value has been derived from clinical applications of Baye's theorem for conditional probability. The basic requirements for using this Bayesian approach in interpreting a diagnostic test are (1) clear-cut definitions of the disease being sought; (2) demarcation of the stage in the pathogenesis of the disease by the test to be evaluated; (3) knowledge of the sensitivity (true positive results) and specificity (true negative results) of the diagnostic test; and (4) knowledge of the clinical consequence (clinical costs) of positive or negative diagnosis (Martin et al., 1975). Galen and Gambino (1975) discussed the predictive value of "positive" and "negative" laboratory test results from the point of view of the potential users of clinical laboratory results. Archer (1978) used Baye's formula to calculate predictive values for positive and negative clinical laboratory test results. They showed that Baye's formula could be transformed to a linear function through the use of odds rather than probabilities, thus facilitating conceptualization, understanding, and memory.

The use of laboratory animals in research has increased tremendously in recent years, with a greater number and a greater variety of animals and species being utilized. The size, origin, diet, physiology and bio-chemistry vary not only in each species, but in each breed as well. The need to establish reference values of clinically healthy and defined laboratory animals is obvious.

Therefore, this book presents a compendium of baseline parameters of selected laboratory animals that are commonly used in experimental biology and medicine.

Sample Collection, Preparation, and Preservation

Accuracy in the clinical diagnosis of a disease in experimental animals frequently depends on the results of laboratory examinations of specimens from the diseased animal. A prerequisite for a good laboratory test result is the proper collection and handling of the specimen. If the properly collected specimen cannot be sent to the laboratory quickly, it must be preserved or stored in such a manner that the necessary tests can be conducted. This section provides basic information on collecting, storing, and processing clinical specimens from various experimental animals that are to be submitted to a clinical chemistry or hematology laboratory.

Collecting Blood Samples

A representative sample from a laboratory animal is frequently difficult to obtain. Unless the animal's feeding and watering are carefully controlled, specimens can vary as a result of food intake. To establish normal physiological parameters, specimens should be collected from an experimental species at frequent intervals. The normal range of clinical biochemical and hematological values should be determined by selecting an animal in a good state of preservation. An animal in the advanced stages of the disease is most desirable for determinations of values of the pathological condition. If the disease is a flock or herd problem, specimens should be collected from more than one diseased animal. Likewise, in a group of animals in an experiment, samples should be taken from each individual animal. It is desirable to have duplicate samples from an animal, but it may not be practical when small animals are used and many tests are required. If possible, specimens should be taken from one or two animals that have died recently in a group, and from animals in various stages of illness. The researcher should collect the specimens that are characteristic of the spontaneous disease or the disease induced in the experimental animals. Every attempt should be made to avoid contamination with hair, dirt, chemicals, or microorganisms.

Blood samples for hematology should be collected directly into a test tube or vial containing an adequate amount of anticoagulant for the amount of blood desired. Usually 5 ml blood from a peripheral vein (e.g., an ear vein) is sufficient for most hematological examinations. An alternate technique in the large animal species is to draw blood from the caudal vein or artery with a clean dry syringe. An anticoagulant is allowed to coat the inside of the syringe and the excess is discharged.

Different chemical tests require different types of specimens—i.e., clotted blood or blood containing a particular anticoagulant or preservative. Venous blood is usually used. Sometimes small amounts of capillary blood from the extremities of the experimental animal may be satisfactory, and very rarely arterial blood may be required. In some tests a small amount of hemolysis may not be of great importance, whereas in other tests it must be avoided. The methods of preventing hemolysis are described later in this section.

Blood samples from monogastric animals for chemical analysis should be taken when the subject has been confined without food for 12 hours. This will minimize the effects of food on the concentration of various blood constituents. Chemical analysis should be started as soon as practical after withdrawal of the blood to prevent the loss of glucose or other components as a result of bacterial, oxidative, or enzymatic processes. If blood is allowed to stand at room temperature, as much as 10 mg/dl of glucose may be lost per hour. Other changes may occur, such as the breakdown of nitrogenous constituents, alterations in the electrolyte composition of serum resulting from change in red cell permeability, and increase in inorganic phosphate.

Certain preservatives may be added to blood samples at the time of collection. For blood sugar determinations in which there is to be some delay before analysis, it is best to add a preservative that will inhibit glycolysis. Thymol and fluoride have been used for this purpose. Fluoride is generally simpler to use. It has some anticoagulant properties in itself, but relatively larger amounts are required than for oxalate. A mixture of 15 mg of potassium oxalate and 25 mg of sodium fluoride is satisfactory. Because of their insolubility, these compounds may be added in dry, powder form (three parts of powdered potassium oxalate and five parts of sodium fluoride). Evacuated tubes containing both fluoride and oxalate are available commercially. Because fluoride inhibits the activity of many enzymes, it cannot be used when any such reaction is involved. Amylase and urease are not inhibited by fluoride in the amounts up to 2 mg/ml of blood. However, most enzymatic determinations are made on the serum to which no anticoagulants are added. If the glucose determination involves preparation of a protein-free filtrate, and the first step in this preparation is the dilution and laking of the blood, the determination may be delayed at this point. Glycolysis in the hemolyzed blood will be low if the blood has been diluted at least tenfold.

Although it is always preferable to use serum rather than whole blood or plasma for measurements of most chemical constituents, sometimes whole blood or plasma may be required for many tests. Not all anticoagulants are equally satisfactory for every test. Some may interfere with the chemical reactions of a given test. Heparin is an excellent anticoagulant but is more expensive than other chemicals used for this purpose. Potassium oxalate is commonly used as an anticoagulant in the blood samples when chemistries of whole blood or plasma are required. It may be added to the collection tubes as 0.1 ml of a 20% solution, which is then dried at a temperature not over 100°C. (High temperatures may decompose the oxalate.) Other anticoagulants described above may be used, but in certain chemical tests they may interfere with the reactions.

The technique and site of blood collection from various experimental animals vary with the animal species. The methods of specimen collection from laboratory animals are briefly described in the following sections.

The Mouse

A small amount of blood can be sampled from a mouse by puncturing of the paws or ear with a lancet, puncturing of the retro-ocular sinus via the inner cauthus of the eye, clipping or sequential amputation of the tail and by lacerating the tail artery. The tail should be immersed in water at 113–118.4°F (45–48°C) before amputation of the tail to ensure copious bleeding. The animal can be restrained by placing it in a small tube from which the tail emerges. A sufficient quantity of blood can be collected from a mouse by puncturing the venous plexus localized in the orbit behind the eyeball; the plexus communicates with blood vessels in the brain. The animal is held in the left hand and grasped by the back, the thumb and index finger enclosing the neck and exerting slight pressure. This causes the veins to fill and the eyeball to protrude slightly. A glass pipette drawn to a fine point is inserted in the medial angle of the eye between the eyeball and the orbit. It is rotated to drill through the conjunctiva in the direction of the larynx or the site of exit of the optic nerve. The plexus is reached at a depth of 4 to 5 mm; the blood flows into the capillary pipette. When enough blood has been withdrawn, the pressure on the neck is relieved and bleeding immediately stops. Blood can be sampled repeatedly from the old puncture site by this technique.

Blood can be sampled from mice most conveniently by cardiac puncture. However, this procedure is not without risk, especially in mice. A general anesthesia with ether, chloroform, or an injection of pervocton (5 mg/100 mg body weight) is required to sample blood by this procedure (Schermer, 1967). A 2½ ml syringe with a 22–gauge needle affixed is most useful for cardiac punctures (Archer, 1965; Jacobs

and Andreanssens, 1970). By heart puncture 0.7 ml of blood can be obtained from a 25 g mouse that has a total volume of 2–2.5 ml. The ratio of blood volume to body weight of the normal mouse is about 6 to 7 percent (Table 1).

The Rat

Blood samples from an adult rat can be obtained by puncture or incision of the tail vein, by snipping off the tip of the tail with scissors, by capillary collection from the orbital sinus or by cardiac puncture. From fetal and neonatal rats, about 0.7 ml blood can be obtained by decapitation or amputation of an extremity. Small amounts of repeated blood sampling can be accomplished by snipping off the tail of the rat by similar technique as described with the mouse. Hurwitz (1971) successfully collected 0.5 to 1.0 ml of multiple samples from midventral artery of the tail of the rat using a heparinized Caraway tube and a 21–gauge, Finch hypodermic needle. Serial samples may also be drawn by cardiac puncture.

As much as 10 ml blood can be drawn by syringe from the jugular or femoral veins, or from the heart from an adult rat. In our experience, a rat weighing 325 g produces about 9 ml of blood by cardiac puncture that is about 2.77% (1/36 of the body weight). The blood volume actually available from the rat may be 5–8% or about 1/10 of the body weight (Table 1).

The Hamster

Blood can be sampled from the hamster by venous plexus of the orbit method described for rats and mice. The dorsal or lateral sides of the tail or ventral area of the paw can be used for smaller amounts of repeated blood sampling. The jugular or femoral vein in the inside of the hind leg or the heart puncture method can be used to

Table 1.   Blood Sampling from Animals and Man

Animal	Blood Volume per Body Weight (%)	Maximum safe volume; one bleeding (ml/kg)	Practical volume for diagnostic use* from adults, not in extremis (ml)
Mouse	6.00-7.00	7.70	0.10
Rat	6.00-7.00	5.50	0.50
Hamster	6.00-9.00	5.50	1.00
Guinea pig	6.00-7.00	7.70	1.00
Rabbit	6.00-7.00	7.70	1.00
Chicken (M and F)	8.80-10.0	9.90	1.00
Cat	6.00-7.00	7.70	1.00
Dog	8.00-9.00	9.90	2.00
Monkey (Macaca)	6.00-7.00	6.60	2.00
Pig	5.00-6.00	6.60	10.0
Goat	6.00-7.00	6.60	20.0
Sheep	6.00-7.00	6.60	20.0
Cow (dry)	6.00-7.00	7.70	100.
Horse	6.00-11.0	8.80	100.
Man	7.00-7.50	8.00	5.00

*Estimated for hematologic procedures.

Source: Schalm (1961); Bentinick-Smith (1969); Sonnensirth and Jarett (1980).

collect large amounts of blood. However, the animal should be anesthetized before the procedures are performed. Using a hamster-holding device and a 25–gauge 1/8–inch needle with syringe up to 2 ml can be safely withdrawn from the heart of a 100–g animal. Heart puncture can yield a total amount of about 5 ml blood from a hamster approximately 1/17 to 1/20 of the body weight (Hoffman *et al.*, 1968).

The Guinea Pig

The veins on the dorsal side of the ear are the most suitable for withdrawing small quantities of blood. If larger quantities (about 5 ml) are collected, a suction bill can be used. Serial blood samples may be more easily obtained by cutting into the nailbed or by vacuum assisted bleeding of the lateral marginal vein of a hind limb.

Repeated small blood samples may be also obtained from the orbital venous plexus. Larger volumes of blood can be obtained by cardiac puncture with or without anesthesia. About 19 ml of blood (4% of the body weight) can be collected from a 450 g guinea pig. If the animal is to be exsanguinated, the yield of blood can be increased by using carbon dioxide as an anesthetic. The absolute blood volume in guinea pigs varies from 4.5 to 8.3% of the body weight (Wagner and Manning, 1976).

The Rabbit

Blood collection is commonly done from the large and readily accessible veins or arteries of the ears. The animal should be firmly secured with a restraining box or board. When the veins are clearly outlined, a 23 to 25–gauge needle is inserted. Bleeding can be increased by compressing the ear. Blood may be collected directly into an evacuated syringe or into a vacuum flask when large blood volumes (30 to 50 ml) are

needed. Long term, repeated sampling may be achieved by the surgical insertion of a heparin-lock shunt into the carotid artery of adult rabbits.

Blood can also be drawn from rabbits from the jugular vein after depilation and the application of a ligature to the neck. Heart puncture may also be used to collect large quantities of blood using an 18–gauge, 1½—inch needle. The site of puncture is the third left intercoastal space, about 4 mm from the sternum. Twenty milliliters of blood can be drawn from an adult rabbit without harm. By exsanguination of the carotid artery, 70 ml of blood can be obtained from a 3150 g rabbit (approximately ¼ of the body weight). Total blood volume in a rabbit is estimated at 4.5 to 8.1% of the body weight (Kaplan and Timmons, 1979; Weisbroth *et al.*, 1974).

The Chicken

Small quantities of blood are collected by puncturing the comb or by snipping off a comb tip with scissors. Hemostasis can be effected by searing the wound with a hot metal instrument. Large amounts of blood are drawn from the wing vein (vena cutanea ulnaris) on the inside of the elbow joint after a few feathers are removed from the site. Hemostasis is effected by pressure with slight displacement of the skin. Sampling is facilitated by lightly holding the chicken between the knees. The blood volume achieved by exsanguination varies considerably (2.5–6.5% of the body weight). The cause of these differences is not known. Chickens with a small blood volume do not seem to be clinically anemic.

The Cat

Small amount of blood sample can be obtained from a cat from the marginal ear veins via a lancet puncture wound and the blood is collected directly into a diluting vial or capillary pipette for hematologic determinations. The cat should be wrapped in a towel with its head protruding. The veins are clearly outlined after rubbing the ear with ether or xylol. In some cases, it may be necessary to remove the hair by shaving or with depilatory agent. Larger samples may be collected from the external jugular vein, the cephalic vein of the forelimb, or the external saphenous vein of the hind limb. The great saphenous vein, the dorsal of which extends on the anterior medial edge of the tibia, may be used for taking repeated blood samples. This site must be shaved before the animal is bled. An assistant should hold the leg (near the hock) firmly with both hands. The vein then becomes clearly visible (but on the medial side, unlike the vein in the dog) and can be punctured with a hypodermic needle. Large amounts of blood can be taken from the cat by puncturing the jugular vein or the heart. However, general anesthesia by ether inhalation is required to perform this procedure. A syringe with a 19 or 20 gauge, ¾ inch needle is more satisfactory for drawing blood than a vacuum container, which may collapse the vein in the cat. Jacobs and Andreanssens (1970) described a method using a carotid shunt in the cat when repeated large volume samples are needed or when hemolysis free specimens are essential for the samples being collected. The total blood volume in the cat is estimated from 4.7 to 9.6% of the body weight. A cat weighing 1800 g yielded 60 ml of blood by exsanguination from the carotid artery that is 1/30 or 3.3% of the body weight.

The Dog

Blood samples from the adult dog are most commonly collected from the external jugular vein. Alternately, blood may be drawn from the small saphenous vein, the dorsal branch of which extends on the lower one-third of the thigh on the lateral side from the speroposterior to the infero-anterior aspect. An assistant should hold the dog's thigh with both hands above the hock to ensure fixation of the leg and congestion of blood. The thigh should be shaved before taking the blood sample. Anesthesia is not required for this procedure. An antibrachial cephalic vein of the foreleg may also be used. It extends on the medial side of the lower one-third of the forearm obliquely from the supero-anterior to the infero-posterior aspect, and is clearly outlined when the limb is compressed near the elbow joint. A 19 or 20–gauge, ¾-inch needle with a disposable syringe is more satisfactory for use than a vacuum tube which may cause the vein to collapse. The jugular vein may be used to collect larger quantities of blood. The dog is made to sit, the lower jaw is held in one hand, and the head turned upward and slightly sideways. An elastic band or rubber tube is placed on the neck near the chest, whereupon the vein clearly emerges on the side from which the head is turned. The site of the sampling must be shaved especially with the long haired breeds. The veins are more clearly outlined when the site is rubbed with ether or alcohol. A 2 or 3 ml Vacutainer with the appropriate needle may be used to draw blood via jugular puncture. Generally, blood sampling is tolerated by dogs without anesthesia. If the animal is very excitable, anesthesia is required. In young animals, venous bleeding is difficult so an intra-cardiac technique using a 19–gauge, 1½–inch needle may be used. The total blood volume of the dog is about 9.5 ± 1.7 ml/100 g of body weight. Dogs weighing 27 kg or more are able to donate approximately 500 ml blood each 3 or 4 weeks (15 to 20% of the total blood volume collected in slightly over 8 minutes) without significant changes in hematological and biochemical parameters. An average of 3.85% or 1/26 of the body weight can be obtained by exsanguination of a dog.

The Monkey

The cubital vein in the fold of the elbow can be used, or blood can be sampled from the great saphenous vein which is on the medial side of the lower leg. Small amount of blood for hematology can be obtained from the marginal ear vein by lancet puncture or clipping with a razor or surgical scissors or from a lancet puncture of the finger or heel. Infants and juveniles of most monkey species, or adults of the small or medium sized species may be bled from the femoral vein under manual restraint (Vagtborg, 1967). In infants, an assistant restrains the torso, head, and neck in a supine position on a towel while the technician uses one hand to hold the legs in a fixed position with the ankles crossed. With the other hand, a needle can be inserted into the femoral vein. When blood collection is complete, pressure must be applied to prevent hemorrhage. In the adult macaques, the technique used is similar except that the assistant holds both of the animal's arms above the elbow, with the arms pressed against the back in order to prevent biting and struggling during blood drawing. Blood sampling of juvenile or adult apes or baboons requires the use of Sernalyn Phencyclidine Hydrochloride (Berchelmann et al., 1973; Valerio et al., 1969). In multisampling of infant nonhuman primates, lancet puncture of the heel or finger is recommended, whereas puncture of saphenous vein with a hypodermic needle may be used for daily or more frequent blood sampling in older animals. Anesthesia may be effected by ether inhalation. However, the monkey can be bled without anesthesia if the arms and legs are held and if the animal is laid on the table with the limbs extended. Monkeys have a blood content of 7.35–9.02% of the body weight. Exsanguination from the carotid artery yields less than half of this quantity, i.e., 1/30 to 1/33 of the body weight. Lang (1968) recommends that removal of blood from the squirrel monkey be limited to 1 ml per 100 g body weight at any one time and only once per month.

The Pig

The most commonly selected site for bleeding the pig is the anterior vena cava, although the jugular vein and cephalic vein are also sometimes used. Ear veins or tail veins may be used as bleeding sites for very small pigs. To collect blood from the tail vein, a rubber band is applied as a tourniquet to the base of the tail. When the tail is engorged, the tip is cut with a sharp blade, and blood is milked into the tube. A bleeding procedure for adapting the orbital sinus technique for use in swine has also been described. With this method, 10 ml of blood can be collected in 5 to 6 seconds (Huhn et al., 1969). It is usually necessary to restrain the animal with its head turned on the side to have a clear view of the site of bleeding. For this purpose the lower jaw may be grasped from below, and the head may be bent upward and slightly to one side. A number of restraining devices are described by Tumbleson et al. (1968); Bustad (1965, 1966); Mackellar (1976) and Huhn et al. (1969). Newborn miniature pigs can be easily restrained by hand for collection of blood samples. It is recommended that a needle and syringe technique be adopted for bleeding swine since their erythrocytes are especially prone to lysis and are, therefore, easily damaged in vacuum collection tubes. The syringe should be fitted with an 18–gauge, 1½ or 2½ inch needle for adult swine and a 21–gauge, 1½ inch needle for piglets.

Sheep, Goats, Cows, Horses, and other Large Animal Species

The usual procedure in large animals is to collect blood directly from the jugular vein into a test tube or vial. About five ml of blood is sufficient for most hematological examinations. An alternate technique in the cow is to utilize the caudal vein or artery as a source of peripheral blood, and in this instance a clean, dry syringe or Vacutainer is used. However, more consistent blood values are achieved when the blood is collected from the jugular vein. It is usually necessary to restrain the animal with its head turned on the side to have a clear view of the site of bleeding. For this purpose the lower jaw may be grasped from below, and the head may be bent upward and slightly to one side. With sheep, the wool should be removed from the site of the vein. Hair may be removed from other animal species, but it may not be necessary. The vein may be made prominent by twisting a cord around the neck immediately above the chest. The vein, which is then clearly visible, can be punctured with the hollow needle. Table 1 lists total blood volumes of the large animal species. Usually about 50% of the total blood volume can be collected by exsanguination of a large animal.

Humans

Usually the veins of the ventral aspect of the elbow (antecubital) are used. A tourniquet is placed around the arm above the elbow and tightened to produce a pressure higher than that in the vein but lower than the diastolic pressure. The radial pulse should be checked after the tourniquet has been applied. If the pulse cannot be felt, the tourniquet should be loosened. In selecting a vein, the deeper ones which can be palpated, but not seen, are usually more satisfactory than the more superficial ones. The latter tend to have thick coats and to roll away from the needle. The skin at the site of venipuncture is cleaned with 70% alcohol and allowed to dry. The needle is pushed through the skin and the point advanced until it penetrates the vein. When blood appears in the syringe tip, the angle of insertion is changed by depressing the syringe, and the needle is advanced a few millimeters into the lumen of the vein. The tourniquet is then released, and after 10 seconds the blood is drawn into the syringe. After enough blood has been withdrawn, the needle is quickly taken out and a sterile gauze sponge is placed on the venipuncture site. The patient immediately exerts pressure on the sterile gauze to prevent blood from oozing out of the opening. A 20-gauge needle is used for most venipuncture, unless the patient is a child. When only a drop or two of blood is needed for a test, such as in hematological determinations, it is usually obtained by pricking the finger or ear lobe. The blood oozes from capillaries and is therefore called capillary blood.

Important factors of biological origin that contribute to variations in results include preparation of the patient, time of venipuncture, and diurnal variations. To minimize these variations, the following recommendations are made:

1. Greater standardization in the venipuncture procedure by defining the postural and venous stasis limitation, which the phlebotomist should follow.
2. Recording the time of day, the posture of the patient (supine or sitting), the time of venous stasis (short, medium, or long), and whether an intravenous solution was being administered to the patient.
3. Increasing the scope of quality control to include the pre-instrumental sources as well as the instrumental analytic variables. "Paired blood duplicates" could be used in addition to commercial quality control sera.

Collecting Other Biological Fluids and Tissues from Experimental Animals

In the diagnosis and study of many pathologic processes in experimental animals, it may be necessary to examine urine, cerebrospinal fluid, synovial fluid, amniotic fluid, and pus or tissue material. The following section briefly describes the methods of collecting these materials from research animals.

Urine Specimens. It is rather difficult to collect urine specimens from small laboratory animals such as mice, rats, and hamsters. A fine catheter or syringe may be used to collect the urine specimen from the urinary bladder, although it can be easily contaminated by blood or tissues during the operation. In general, urine from most species of larger animals can be collected for analysis as the animal voids. However the external urinogenital organs should be cleaned of all contamination because a small quantity of fecal material may significantly affect the results of urine analysis. Containers used for urine specimens must be clean. Disposable plastic or paper containers are convenient for collecting urine specimens. In some instances a catheter is used to collect a specimen from the experimental animal. In this case care should be taken to avoid any chemical or microbiological contaminant that might interfere with urine analysis. The addition of a chemical sterilant or excessive quantities of petrolatum or another lubricant to a urine specimen may alter the sample enough to change many of its chemical and physical characteristics.

In the small laboratory animals, including dogs and cats, urine can be collected by bladder puncture. A sterile syringe and needle should be used after the skin is sterilized. The bladder may be readily tapped to remove specimens for urinalysis or bacteriological examination, or both. Care should be taken to avoid entering an intestine, and the procedure should be attempted only when the bladder can be readily palpated. If the bladder is turgid, gentle pressure will often result in micturation and allow collection of an adequate specimen.

This technique is particularly useful with the cat. The urine specimen should be at least about 4 ounces, although small quantities can be utilized in many of the analyses of the modern clinical chemistry and microbiology laboratories.

If quantitative analyses of urine are required, a 24-hour specimen should be collected. In the larger animals 24-hour specimens can be collected by strapping a female collecting urinal onto a male animal. A collection cage can be used to collect 24-hour specimens of dog and cat urine. In collecting 24-hour specimens a preservative such as toluene or thymol should be added to the container in which the urine is to be collected.

Cerebrospinal Fluid (CSF). Cerebrospinal fluid is the product of the secretory activity of the choroid plexus. In man it circulates from the lateral ventricles through the third and fourth ventricles into the subarachnoid space. About four-fifths flow over the brain and one-fifth flows down around the spinal cord. It is absorbed through the arachnoid villi into the large sinuses and through the capillaries into the bloodstream. Some of the constituents of the spinal fluid are quantitatively dependent upon the concentration of these substances in the blood plasma. However the composition of spinal fluid differs materially from that of the blood plasma.

Although in human medicine examination of CSF has become essential for the diagnosis of diseases of the nervous system, in veterinary medicine this technique has not been extensively utilized. Only a few reports of examination of CSF in diseases of animals have appeared in the literature.

Cerebrospinal fluid may be removed from the cow either by suboccipal or lumbar puncture (Frankhauser, 1953). The area should be carefully clipped, shaved, and disinfected before the skin is punctured. The puncture is made with a sterile 5-inch,

16- to 18-gauge needle with a stylet. The principal problem encountered in entering the subarachnoid space at this point is the thickness of the skin and the struggling of the animal. The operator should make the puncture directly on the mainline. If, upon entering the subarachnoid space, the fluid does not flow readily, 5 ml may be withdrawn by gentle suction with a sterile syringe. CSF can also be collected from the lumbar area by inserting the needle in the soft depression located between the dorsal process of the last lumbar vertebra and the anterior end of the median sacral crest.

In the bovine species CSF is readily obtained by a suboccipital approach with a 3- to 4-inch, 16-gauge needle complete with stylet, with the animal standing or preferably in lateral recumbency. In sheep and goats the suboccipital cerebrospinal puncture is the most satisfactory technique. However the suboccipital cerebrospinal puncture in swine is very difficult, and usually a general anesthesia is required. Lumbosacral puncture is the method of choice in pigs weighing over 600 kg. As with sheep the puncture and aspiration is easily performed with the pig in a sitting position. In the horse the suboccipital puncture is the method of choice. Punctures made between the atlas and the axis have been described but are less satisfactory; a general anesthetic is usually necessary.

Cerebrospinal fluid from the dog is removed at the cisterna magna at the atlantooccipital articulation because lumbar puncture is extremely difficult (McGrath, 1960). Light anesthesia should be used because the cisterna magna puncture is in a vital area. In the cat the suboccipital puncture is recommended for removal of CSF. The technique is similar to that used in the dog. Not over 0.5 to 1 ml of fluid should be removed from the cat, because cats are quite susceptible to meningeal hemorrhage if too much fluid is withdrawn. Not over 10 to 20 drops of CSF should be removed from kittens. It is difficult to collect CSF from rodents and rabbits. Suboccipital puncture may be used to collect a few drops of CSF from these animals, or it may be collected directly from the ventricles of the brain. Light anesthesia is required.

Synovial Fluid. Synovial fluid is a modified connective tissue fluid found in joints and tendon spaces. It is characterized by a high mucin content which gives it its viscid nature. It is a product of the secretory activity of the synovial lining. Other components of the fluid are dialysates of the plasma such as glucose, electrolytes, and nonprotein nitrogen, in the same concentration as in plasma. Analysis of the synovial fluid in laboratory animals is valuable in the study of human arthritis. Synovial fluids from dogs and pigs may be obtained by aspirating the fluid with a sterile needle of the proper length and gauge. It is preferable to use at least an 18-gauge needle with a short bevel and equipped with a stylet. Five to 10 ml of the synovial fluid can be collected from certain joints of small domestic animals.

Amniotic Fluid and Genital Fluids. The study of amniotic fluid in animals is sometimes required to diagnose parasitic infections such as bovine trichomoniasis. The vaginal and cervical fluids in animals are examined to detect the proper time for breeding, to reveal reproductive disorders, and to detect early pregnancies. The laboratory examination of semen in animals is of considerable importance in determining the ability of males to produce viable spermatozoa in a high quality semen, especially in cattle, horses, and other domesticated animals. For small experimental animals these specimens are not routinely examined. The reader is referred to the articles of Fuchs (1966), Bone (1954), Foote (1958), Mann (1959, 1961), Rosenkrantz et al. (1961), and Coles (1980) for further information on this subject.

Tissue. Tissue specimens of laboratory animals are often examined for microbiology and for tumors or other diseases. A sample may be taken by biopsy, during surgery, or at necropsy. If the tissue specimen is small, mince it with sterile forceps and scissors and the entire sample can be used for microbiologic or clinical chemistry tests. If the specimen is medium in amount, grind it with a sterile pestle in a sterile mortar containing sterile sand or alundum. If a large amount of tissue is collected, remove several small representative portions from the desired areas with sterile instruments and transfer them to a sterile mortar for maceration. Store the remaining tissue in the deep freeze.

An ample quantity of liver, kidney, and stomach contents should be collected and submitted under refrigeration for the toxicologic examinations (e.g., the heavy metals mercury, lead, bismuth, arsenic, etc.). For examination of strychnine poisoning, refrigerated samples of stomach contents, liver, kidney, and urine should be submitted. For the analysis of insecticides, fatty tissues, liver, stomach contents, and blood samples preserved with an anticoagulant provide the best material for assay. Liver, stomach contents, and sometimes brain samples may be the best sites for demonstration of alkaloids. Occasionally the large animals may ingest hydrocyanic acid from certain plants in a field. To examine for this toxic chemical, plant material from the suspected field, stomach (rumen) contents, blood, and liver may be tested. Because hydrocyanic acid is a volatile chemical and identification of the toxic material is difficult, the specimens submitted to the laboratory should be fresh. Samples should be forwarded from the material suspected of containing nitrate; in addition stomach contents and blood may be tested.

For microbiological examinations, the tissue samples or organs of laboratory animals should be aseptically collected in individual containers. Coles (1980) described the collection of specimens from domestic animals for certain specific diseases. Similar methods of collecting specimens are applicable to other experimental animals, except that small amounts are sufficient with rodents, rabbits, chickens, and other small laboratory animals.

Test results can be rendered worthless by the improper collection and processing of the specimen.

Preparing Serum Specimens

The following general guidelines are recommended for processing serum, whole blood, and plasma samples:

1. Check laboratory requirements for the quantity of serum.

2. Allow the sample to clot for 30 minutes at room temperature; do not delay separation beyond 45 minutes because certain test results are not reliable when performed on specimens not separated within 30 to 45 minutes (Table 2). Laessig et al. (1976) examined changes in serum chemical values as a result of prolonged contact with the clot for as long as 48 hours. Glucose, lactic dehydrogenase, the transaminases, potassium, alkaline phosphatase, and iron showed significant changes. Other common tests including electrophoretic separations were not affected. The authors recommended that serum-erythrocyte (or clot) separation should be made as rapidly as possible even though the test ordered may not be critically affected.

3. Ring clot gently with a wooden applicator.

4. Centrifuge the specimen.

5. Transfer serum to another tube.

6. The sample should be stored in the refrigerator or in the freezer compartment ($< -20°C$) if the test cannot be performed immediately. Many chemicals are unstable at room temperature and some are unstable even when the sample is frozen (Table 3).

Table 2   Processing of Clinical Specimens

Results may be erroneous when *		
Separation is Delayed	Specimen is Hemolyzed	Specimen is Lipemic
Acid Phosphatase	Acid Phosphatase	Amylase
Carbon Dioxide	Aspartate Amino-transferase	Bilirubin
Coagulation studies		Blood Urea Nitrogen
Glucose	Bilirubin	
Lactate Dehydrogenase	Calcium	Calcium
Potassium	Hematological studies	Electrolytes
Uric Acid	Lactate Dehydrogenase	Glucose
	Magnesium	Hemoglobin
	Phosphorus	Total Protein
	Potassium	
	Protein Electrophoresis	
	Total Protein	

Source: Ono et al (1981), Ladenson (1980), Frank et al (1978)

* Varies with time of delay, conditions of storage, degree of
  hemolysis or lipemic and with the method used.

20

Table 3. Stability of Specimens for Clinical Biochemical and Hematological Tests

Test	SERUM At Room Temperature	Refrigerated	Frozen	Whole Blood with Additive
Acid phosphatase	1 hour; stable with disodium citrate	3-4 hours	Stable	
Alkaline phosphatase	2 days	7 days	Stable	No preservation available
Amylase	7 days plus	Stable	Stable	Untreated; 2 days
Bilirubin	2 days (in dark)	4-7 days (in dark)	90 days	EDTA; Heparin: 3 days (results not valid if hemolyzed)
BSP (Bromsulphalein test)	Stable	Stable	Stable	
BUN (Blood urea nitrogen)	3-5 days	7 days	Stable	Fluoride; fluoride-thymol: 5 days untreated: 3 days
Calcium	7 days	Stable	Stable	Heparin: 3 days
Cholesterol	7 days	7 days plus	Stable	Heparin; EDTA; Thymol: 1 day
Creatinine	1 day	1 day	Stable	EDTA: 3 days
Chloride	7 days	7 days	Stable	Ammonium oxalate: separate plasma within 1 hour
CO_2	4 days (combining power)	7 days	Stable	Heparin: 5 hours in ice; tube must be completely filled (CO_2 content)

Table 3 (con't.). Stability of Specimens for Clinical Biochemical and Hematological Tests

| Test | SERUM | | | Whole Blood with Additive |
	At Room Temperature	Refrigerated	Frozen	
Potassium	14 days	30 days	Stable	Ammonium oxalate: separate plasma within 1 hour
Sodium	14 days	30 days	Stable	Ammonium oxalate: separate plasma within 1 hour
CPK (Serum creatine phosphokinase)	Stable - 48 hours	Stable - 7 days	Stable	
LDH (Serum lactic dehydrogenase)	7 days	Unstable	Unstable	
SGOT (Serum glutamic oxaloacetic transaminase)	4 days	14 days	Stable	
Glucose	45 minutes; stable in sodium or potassium fluoride 10 mg/ml	2 days	Stable	2.5 mg potassium fluoride plus 5 mg sodium fluoride/ml of blood: stable
Hemoglobin				Oxalate; Heparin; EDTA: 7 days
Hematocrit				EDTA: 2-4 hours
Platelet count				EDTA: 2-4 hours
Red Cell count				EDTA: Heparin; Citrate oxalate; 2 days

22

Table 3 (con't.). Stability of Specimens for Clinical Biochemical and Hematological Tests

Test	SERUM			Whole Blood with Additive
	At Room Temperature	Refrigerated	Frozen	
Reticulocyte				EDTA: 1 day
Sedimentation Rate				EDTA: 4 hours
White cell count				EDTA: 3 days
Lipase	7 days	7 days plus	Stable	
Magnesium	7 days	7 days plus	Stable	
PBI (protein-bound iodine)	30 days	30 days	Stable	Untreated: 3 days plus
pH	30 minutes	5 hours	Unstable	Heparin in oiled syringe; kept cold in wet ice or Heparin completely filled vacuum tube; 5 hours
Phosphorus	4 days	7 days	Stable	
Albumin	4 days	30 days	Stable	
Globulin	4 days	30 days	Stable	
Total protein	7 days	30 days	Stable	
Prothrombin time	Unstable	Unstable	Unstable	Sodium oxalate: 1 hour; plasma refrigerated 4 hours
Triglycerides	4 days	7 days plus	Stable	
Uric acid	3 days	3-5 days	Stable	Heparin oxalate: 3 days; untreated: 3 days

23

Table 4 summarizes the stability of enzymes of diagnostic importance at various temperatures.

7. Check the serum for hemolysis (pink serum). Hemolysis occurs when the red blood cells are disrupted and the hemoglobin and other components spill into the serum rendering the specimen worthless for any test that measures the content of the red blood cell. Hemolysis can have an unpredictable effect on chemical determinations. Results can go up or down drastically. Potassium and magnesium are the most affected by hemolysis because their concentration is normally higher in erythrocytes than in the serum; therefore even minimal hemolysis can prevent accurate testing.

Other tests in which a hemolyzed specimen should be discarded are ACTH, acid phosphatase, BUN, bilirubin, bromide, calcium, LDH, SHBD, CC flocculation, icterus index, serum iron, lactic acid, lipase, NPN, pH, phosphorus, albumin, and pyruvic acid. Lipemic serum interferes with chemical determination, and it would be better to discard it for the following tests: albumin, bilirubin, thymol turbidity, platelet count, and uric acid. Since the regular discarding of serum specimens would be impractical, prevention is the key to controlling the effect of hemolysis and lipemia.

The common causes of hemolysis are (a) Use of improperly cleaned needles, syringes, or glassware. (b) Use of wet needles, syringes, or glassware. A drop of distilled water will hemolyze blood by itself by lowering osmotic pressure outside the RBC; then the cell will imbibe water and rupture. (c) Drawing the sample with too much pressure. If the syringe plunger is pulled back roughly or with too much pressure, currents are created in the blood which break up the RBCs. (d) Ejecting blood from the syringe too rapidly after removing the needle. (e) Putting the sample into chilled glassware. (f) Vigorous shaking

of the blood; too vigorous ringing of the blood clot before centrifuging; shaking rather than slow inversion of vacuum tubes to mix whole blood with the chemical reagents. (g) Extremes of temperature, the most common cause of hemolysis of chemically treated whole blood mailed to a laboratory. It is difficult to control this cause of hemolysis.

To prevent lipemia in tests in which it interferes with the chemical determination, it would be best to require a fasting specimen.

Preparing Whole Blood or Plasma Specimen

1. Check laboratory requirements for the quantity of sample.
2. Mix with the proper additive immediately. Many types of anticoagulants are used to prevent coagulation of blood. Table 5 lists the anticoagulants commonly used in hematology and clinical chemistry. The addition of oxalates, citrates, or EDTA (ethylenediamine tetra-acetate) removes calcium from the blood by forming insoluble or un-ionized calcium salts. Heparin inactivates thrombin and thromboplastin, and defibrination removes fibrinogen converted to fibrin. The correct choice and the correct amount of anticoagulant are important. An insufficient amount of anticoagulant may lead to partial clotting and the incorrect anticoagulant may lead to distortion of blood cells, whereas too much liquid anticoagulant dilutes the blood sample. When preservation of cells and cellular characteristics are important criteria, the dipotassium (or disodium) salts of ethylene diamine tetra-acetate (EDTA) are the anticoagulants of choice. These agents do not alter the size of cells, nor do they interfere with the normal staining characteristic of blood cells. EDTA may be used either in liquid or dry form. If used in the liquid form one drop of 10% solution is sufficient to prevent the coagula-

Table 4. Stability of Some of the Enzymes Used in Diagnosis

Enzyme	Body Fluid	25 C	4 C	-20 C	Anticoagulant
Acid phosphatase	Serum	4 hours	14 days	115 days	Inhibited by fluoride, oxalate
	Urine, undialyzed	7 days	15 days	Unstable	
Alkaline phosphatase	Serum	May increase 5-30% in 12 hours	Several days	20 months	Inhibited by EDTA, oxalate
	Urine, undialyzed	6 hours	6 hours	Unstable	
	Urine, dialyzed	24 hours	24 hours	Unstable	
Aldolase	Serum		Several days		
Amylase	Serum	27 days	7 months	2 months	Inhibited by citrate, oxalate
	Urine	2 days	2 months	2 months	
Catalase	Erythrocytes		4 days		Heparin
Glucose-6-phosphate dehydrogenase	Erythrocytes		7 days		Acid citrate dextrose, EDTA, oxalate, Heparin
Isocitric dehydrogenase	Serum	> 5 hours	Several days	21 days	
Lactic dehydrogenase	Serum	7 days	Unstable	Unstable	Inhibited by oxalate
	Urine, undialyzed	6 hours	12 hours	Unstable	
	Urine, dialyzed	24 hours	24 hours	Unstable	
	CSF	4 hours	14 days		

Table 4 (con't.). Stability of Some of the Enzymes Used in Diagnosis

Enzyme	Body Fluid	25 C	4 C	-20 C	Anticoagulant
Lipase	Serum	7 days	21 days		Inhibited by hemolysis
Phosphohexoseiso-merase	Serum	8 hours	14 days		
Ribonuclease	Serum		1 day		Inhibited by Heparin
	CSF			Several months	
Transaminase, serum aspartate	Serum	2 days	14 days	1 month	
	CSF	6 hours	14 days		
	Urine	30 minutes		2 days	

Source: Amador et al (1969).

Table 5. Common Anticoagulants for Blood Sampling

Anticoagulants	Composition	Amount re-quired ml/ ml blood	Useful for
Oxalates	Ammonium oxalate, 1.20 g	0.10	-Least distortion of cells
	Potassium oxalate, 0.80 g	(1 mg/ml dry form)	-Not recommended for plat-elet counts
	Distilled water, 100 ml		-Not recommended for nitro-gen determinations
Sodium citrate	Trisodium citrate, 3.80 g	0.15	-Coagulation tests (e.g., prothrombin time)
	Distilled water 100 ml		
ACD solution	Trisodium citrate, 2.20 g	0.15	-Platelet counts
	Citric acid, 0.80 g		-Blood banking and some isotope procedures
	Dextrose, 2.45 g		
EDTA	Potassium salt of ethy-linediamine tetraacetate, 1.00 g	0.10	-Platelet counts
	0.07% saline solution, 100 ml		-Preservation of cells and cellular characteristics
Heparin	Stock solution	Rinsed syringe enough for 5 ml blood specimen	-Osmotic fragility test
			-Not recommended for staining the blood smears
			-Effective for only 10-12 hours
Sodium Fluoride		2.5 mg/ml	-Preservation of glucose
Defibrination	Glass beads, 3-4 mm	20 beads/ 25 ml blood	-Serum preparation

tion of 5 ml of blood. EDTA is also a good preservative for the blood morphologic characteristics and stainability. Samples of blood may be stored in the refrigerator overnight. The EDTA is most compatible with saponinization of erythrocytes for use in the Coulter Electronic Cell Counter.

The anticoagulant oxalate is inexpensive and easy to prepare. If the sample is examined within an hour after adding oxalate anticoagulant, there is very little distortion of the cells. However the examination of blood later than that shows marked leukocytic degeneration because the ammonium will interfere with the results. Although the oxalate mixture produces the least distortion of the cell, it is not recommended for platelet counts.

Minute quantities of heparin which cannot be seen but which adhere to the barrel of the syringe after the heparin is aspirated and expelled, produce excellent anticoagulation. Heparin is the anticoagulant of choice for the osmotic fragility test because it does not affect the size of red blood cells, which when stained with Wright's stain have a light blue background. It is most commonly used in the liquid or dry form in a test tube. A syringe can be rinsed with a stock solution of heparin and will contain enough anticoagulating activity to preserve 5 ml of blood. If there is too much heparin, staining of the leukocytes becomes difficult and occasionally impossible. Also, heparin is the most expensive of the commonly used anticoagulants and it is effective for only 10–12 hours.

Sodium fluoride at a concentration of 10 mg/ml is used as an anticoagulant and preservative of glucose.

Defibrination with glass beads is accomplished by adding 25 ml of venous blood to a 125 ml Erlenmeyer flask containing 20 glass beads (3–4 mm). Rotate the flask until the beads are covered with fibrin and cease to make a rattling noise. Continue for another 2 minutes and decant the defibrinated blood. The fluid portion will be the serum, not plasma.

3. If using a vacuum tube, slowly invert the tube 8 to 10 times.

4. Fill the vacuum tubes as completely as possible.

5. If plasma is required, separate immediately.

Storing Clinical Specimens

It is desirable to perform chemistry, hematology, and microbiology tests on a specimen as soon as the sample is obtained from the experimental animal. However it is not practical to do all the tests on the same day that the specimen is received when multiple tests are required. Sometimes technical problems or the lack of sufficient personnel make it necessary to store the specimens. Occasionally the specimen may be sent to a distant laboratory for certain specialized tests. In these instances the specimens must be properly preserved and adequately packed so that the constituents are not significantly altered.

If specimens for hematology tests cannot be handled immediately, the test tube should be stoppered and placed in a refrigerator. For some tests (e.g., a plasma hemoglobin level) it is necessary to separate plasma from the red cells immediately after the specimen is obtained and before it is refrigerated. Blood specimens for platelet counts, white cell counts, sedimentation rate determination, and prothrombin time should not be stored longer than 2 hours. Before any examination is undertaken, stored anticoagulated blood must be thoroughly mixed on blood rotators for at least 2 minutes after it has reached room temperature. The specimens should be properly identified by describing the animal's species, age, and sex. Records should also include the investigator's experiment codes or the owner's name and address, the duration of the condition or outbreak, the mortality rate, clinical signs, number of animals affected, necropsy findings, treatment, history, and so on. It is also important to record the type of preservative or anticoagulant used on a specimen before it was stored for hematologic examination.

Blood taken for chemical analysis should be centrifuged to separate the serum, because the serum is most frequently used for most chemical tests. The concentration of many constituents of blood may change more or less rapidly on standing, particularly at room temperature. For example, the concentration of phosphates changes because of hydrolysis of organic phosphate esters, lipids change because of lipolysis, chlorides change because of shifts of chloride between cells and serum. For most tests, including some enzymes, serum or plasma may be stored for a few days in the refrigerator. For most enzymes the serum may be kept for several days if frozen.

Section III

Methods in Hematology

The methods used in hematologic determinations are standard and they are described in most textbooks and laboratory manuals on the subject (Hawkey, 1975; Wintrobe et al., 1974; Wintrobe, 1967; Williams et al., 1977; Blood et al., 1979). The methods used in this study to establish normal values are briefly described below, and different methods are discussed along with the interpretation of laboratory data.

A portion of each sample was centrifuged in a capillary tube for 5 minutes at 7500 × g in a microhematocrit centrifuge to determine the packed cell volume (PCV). Erythrocyte count (RBC), leukocyte count (WBC), PCV, and platelets were measured with a Model F Coulter Electronic Cell Counter (Coulter Electronic, Inc. Hialeah, Florida). A modified multipurpose reference reagent prepared from fresh human blood provided control values for RBC, WBC, hemoglobin, PCV, and MCV. The instrument was recalibrated specifically for animal blood. Settings on the Coulter Counter were: attenuation 0.707; aperture 8; erythrocyte threshold 15; and leukocyte threshold 30. Composite PCV values were also measured by the microhematocrit technique. A Bausch and Lomb Spectronic–20 (Bausch and Lomb, Inc., Rochester, New York) at a wavelength of 540 nm was used to determine the hemoglobin values by using a cynomethemoglobin standard. Differential leukocyte counts were determined by scanning Wright's-stained slides in the classic manner. In order to obtain more valid data, duplicate slides of each sample were made, each was counted by a different technologist, and the mean value was taken as the result. One hundred cells were identified and counted on each slide. Blood indices, mean corpuscular volume (MCV), mean corpuscular hemoglobin (MCH), and mean corpuscular hemoglobin concentration (MCHC) were calculated.

Experimental Animals

Clinically normal animals of various ages and both sexes were purchased from commercial sources in lots of 10 to 20. The details of sex, age, breed, and weights are presented with the normal value data in the next sections. The animals were allowed to equilibrate for at least 1 week after shipping to provide time for homeostatic return to physiological normality and to stabilize their nutritional and environmental conditions.

Small animals (mice, rats, hamsters, guinea pigs, rabbits, and chickens) were housed in appropriate cages in separate air-conditioned rooms with controlled relative humidity and air flow. Mice, rats, and chickens were housed 5–6 animals in a cage; rabbits and guinea pigs were housed in individual cages. Animals were provided with constant fresh water and laboratory chow supplied by a commercial animal feed supplier. Serum, whole blood, urine, or other body fluid from large animal species (horse, cow, pig, sheep, and goat) were acquired from a commercial supplier and also from animals maintained on a farm for experimental purposes. Cats, dogs, and monkeys were maintained in separate rooms in individual cages.

Animals were checked daily for clinical signs of illness, and apparently healthy animals were included in this study. Whole blood was drawn by heart puncture, from peripheral veins, or from the jugular vein (see Section II) and allowed to clot in centrifuge tubes. Small animals were bled in 4–5-day intervals. The blood of 4–5 mice was pooled whenever a larger quantity was required. The tubes were ringed and the serum was separated about ½ hour after clotting by centrifugation at 2500 rpm for 10 minutes. The serum was drawn off by pipette and immediately frozen. Serum samples were used for the determination of chemical constituents.

In general, samples from small animals (rats, mice, hamsters, and guinea pigs) were taken by cardiac puncture under light ether anesthesia; each animal provided one sample for chemistries or hematology. Large animals were bled from the jugular vein, usually on alternate days. Procedures for bleeding individual species of animal and their physiological constants are described in Section II.

Manual Methods of Cell Counting

Determination of the number of erythrocyte (RBC), leukocytes (WBC), and platelets (thromocytes) in the peripheral blood has long been a fundamental procedure in hematology. All cell counting methods are based on the dilution of capillary blood or well mixed correctly anticoagulated venous blood with special counting fluids in special counting pipettes. The dilution must be performed as accurately as possible and since the number of cells in blood is enormous, the dilution must be great (1:20 to 1:200). Therefore the volume of whole blood used is small and the errors of measurement are correspondingly large.

The diluting fluid is chosen to permit the cell to be counted selectively. Thus for erythrocyte counts the diluting fluid merely dilutes the blood while preserving erythrocytes morphology. In the methods for counting leukocytes the fluid lyses the erythrocytes while preserving the leukocytes. Most methods for platelet counting employ this same principle. For manual counting and some electronic methods, a counting chamber is used which permits examination of a precise volume of the diluted blood determined by a grid inscribed in the base of a chamber of exact depth. In other automated counting methods the cells in exact volume of cell suspension are counted. Counting may be performed visually or by automated means. Visual counting may be done under the bright-field or phase contrast microscope. The cells to be counted are identified visually and enumerated by the operator. Care must be exercised to count all the appropriate cells in the area defined by the grid. Manual and automated methods used in hematology for specific tests are described briefly in the following section:

Manual Methods

Erythrocytes. The normal physiologic functions of erythrocytes are gas exchange (the transport and release of oxygen and carbon dioxide), participation in the buffer system of the blood, and a role in the clotting mechanisms. The number of red blood cells (RBC) varies from species to species because of physiological variations or certain pathological conditions. The increase in the number of RBC's is called polycythemia. Secondary polycythemia (erythrocytosis) may be caused by dehydration (from fever, loss of fluid, etc.) or by anoxemia (high altitudes, pulmonary and cardiac disease methemoglobinemia). Progressive or primary polycythemia (polycythemia vera, erythremia) is an idiopathic disease of the erythropoietic tissue characterized by hyperplasia of the bone marrow and enlargement of the liver and spleen. A decrease in the RBC count is called anemia. Physiologically, adult women have a lower red cell count than men. In the circulation, mature red cells have the shape of a biconcave disc. Their diameter is approximately 8 μm and their thickness is about 2.5 μm at the periphery and 1 μm in the center. The surface is 160 μm^2 and the volume 90 μm^3 and the weight 30×10^{-12} g. Each of these values varies by 5–10% in normal humans. In the circulation in a test tube or in a thick preparation between slide and coverslip, red cells form rouleaux. The length of the rouleaux determines the speed of erythrocyte sedimentation (ESR) used clinically. Rouleaux formation is altered in pathologic conditions (e.g., infections, multiple myeloma, spherocytosis, erythroblastosis fetalis). In the presence of red cell antibody, red cells form clusters rather than rouleaux. The agglutinated red cells are difficult to disperse, particularly in the presence of high antibody titers. Normal red cells are easily deformable and return as easily to their biconcave shape. This deformability is physiologically very important since it allows a red cell with a diameter of 8 μm to pass through 3 μm capillaries.

The primary function of the red blood cell is the transport of respiratory gases and to achieve this, there must be sufficient circulating erythrocytes to meet the metabolic needs of the body. There is a compensatory increase in the number of peripheral red cells when there is a decrease in the amount of oxygen reaching the tissue cells of the body. If there is a surfeit of oxygen available, a decrease in circulating erythrocytes occur. Hemoglobin is formed by the developing erythrocyte in the bone marrow. The most rapid hemoglobin formation is in the class III (polychromatophilic) normoblast, when approximately 80% of the hemoglobin carried by the mature cell is formed. Under normal conditions the human body produces approximately 6.25 g of hemoglobin per day. The maximal effort of the body in the event of hemolytic disease has been calculated to be approximately 40 g per day. This means that the survival time of the red cell may be reduced to 18 or 19 days without the occurrence of anemia if the bone marrow functions at maximal capacity. If the destruction of the red cells exceed the maximal effort of the bone marrow, anemia develops.

Manual Red Blood Cell Counting. All counting methods are based on the dilution of capillary blood or well mixed, correctly anticoagulated venous blood with special counting fluids in special counting pipettes. The individual cells are counted in a counting chamber (hemocytometer).

To calculate the number of RBCs counted in a hemocytometer, the basic formula is:

No. of Cells/mm^3
Undiluted Blood

$$= \frac{\text{Total No. of Counted Cells in 5 Squares}}{\begin{array}{c}\text{Surface Area Counted} \times \text{Height} \\ \text{of Counting Chamber} \\ \times \text{Dilution of Blood}\end{array}}$$

The surface area counted can be calculated by the center square, which occupies 1 mm^2, and each of the 25 smaller squares that it contains, which occupy $1/25$ mm^2, and each of its 16 smallest squares, which occupy $1/400$ mm^2. Therefore a total of 80 (5×16) of these smallest squares are counted, and they occupy $80 \times 1/400$ mm^2 or $1/5$ mm^2.

The usual dilution of blood in the RBC cell pipette is 1:200. The depth of the counting chamber is $1/10$ mm. The number of erythrocytes per mm^3 of the body can be thus calculated as Counted Erythrocytes ($1/5 \times 1/10 \times 1/200$). Therefore the number of erythrocytes per mm^3 blood = counted erythrocytes $\times$ 10,000. The reproducibility of this method is $\pm 8\%$.

Platelets (Thrombocytes). Platelets are characterized by a tendency to agglutinate and to adhere to foreign surfaces. Therefore the blood specimens should be collected in a plastic syringe immediately mixed with an anticoagulant. There are several techniques available for counting platelets.

Indirect Methods. Platelets are counted in their relationship to red blood cells on fixed smears. This method is not reliable.

Direct Counting Methods. These are accurate methods and any one of the three methods may be used: (1) phase microscopy method, (2) Rees-Ecker method, or (3) Coulter Counter method.

Phase Microscopy Method of Brecher-Cronkite. (1) Use capillary blood or venous blood drawn by the 2–syringe method; the second syringe should be siliconized. (2) Empty the blood into a siliconized test tube and keep it in the refrigerator. (3) Draw the blood to the 0.5 mark in two red cell pipettes and dilute it with diluting fluid to the 101 mark (dilution 1:200). (4) Shake the pipettes for 3 minutes. (5) Expel the first four drops and fill both chambers of a hemocytometer. Use number 1 coverslips. (6) Place the hemocytometer in a closed Petri dish kept humid by moist filter paper

in the top half, and let it stand for 15 minutes. (7) Count the platelets by phase microscopy in the central square millimeter of each chamber. Platelets appear as round or oval bodies with a pink-purple hue. The number of platelets can be determined by the following formula:

No. of Platelets/mm^3 Blood

$$= \frac{\text{No. of Cells Counted} \times \text{Dilution}}{0.04 \times \text{No. of Squares Counted}}$$

This method is more accurate than the Rees-Ecker method (described below), and it is less difficult to perform because there are fewer particles that may mimic platelets. If the platelets clump, the procedure must be repeated. Ammonium oxalate (1%) in distilled water is used as diluting fluid in this method. The direct counting method should always be supplemented by a blood smear to confirm the counter number and to study platelet morphology.

Rees-Ecker Method. The diluting fluid consists of 3.8 g of sodium citrate, 0.2 ml of neutral formaldehyde (40%), 0.1 g of brilliant cresyl green, and 100 ml of distilled water. An erythrocyte diluting pipette is first rinsed with diluting fluid, and blood is drawn to the 0.5 mark; Rees-Ecker diluting fluid is then drawn to the 101 mark. The pipette is shaken for several minutes, several drops are discarded, and both sides of the counting chamber are filled. The hemocytometer should now be placed in a moist chamber, such as a Petri dish containing a piece of wet filter paper, and allowed to stand for 10 minutes. Count the platelets in the entire central ruled area in each side of the counting chamber. Thrombocytes are identified as lilac-colored, rod-shaped or oval shaped bodies approximately one-half the diameter of erythrocytes. The number of thrombocytes counted in the ruled areas is multiplied by 1000 to give the total thrombocytes per cubic millimeter.

This method is subject to the same sampling and technical errors as the red cell count. Because platelet agglutinates invalidate the count, all glasswares must be kept clean to prevent clumping.

Plate Count by Coulter Counter. (1) Aspirate blood into plastic sedimentation tube. (2) Allow sedimentation to occur. (3) Cut the tube either at the junction of the red cells and plasma or in the center of the red cell portion. (4) Count the platelet suspension on the Coulter Counter. (5) Correct the count by using a coincidence and dilution chart and the factor due to excess platelet concentration and hematocrit (Sipe and Cronkite, 1962).

Interpretation. Thrombocytosis or thrombocythemia is used to describe an increase in the number of platelets. The count may reach several million and the condition may be associated with a tendency to clotting as well as to bleeding. Pathologic thrombocytosis may occur in polycythemia vera, in malignant tumors, in chronic myelogenous leukemia, after splenectomy, and in splenic atrophy, myelosclerosis, and megakaryotic leukemia. Physiologic thrombocytosis occurs in pregnancy, during menstruation, following hemorrhage and exercise, and after adrenaline injection.

Decrease in platelets, thrombocytopenia, becomes critical when it reaches 50,000 platelets/mm³, at which level the prothrombin consumption test becomes shortened and clinicaly bleeding may follow minor trauma. At a level of 20,000 platelets/mm³, spontaneous bleeding occur. Thrombocytopenia may be caused by bone marrow damage, congenital absence of megakaryocytes, aplastic anemia due to marrow failure, or by drugs, infections, and other conditions.

Leukocytes. The normal physiologic functions of leukocytes are phagocytosis and antibody production thus defending the body against foreign material. The phagocyte system is comprised of the granulocyte (neutrophils, eosinophils, and basophils) and monocytes. The granulocytes are named according to the staining characteristics of their cytoplasmic granules. They are capable of phagocytizing small particles such as bacteria and therefore are also called microphages. Monocytes are capable of engulfing large particles such as cellular debris and they are called macrophages. The immunocyte system is not concerned with phagocytosis, but rather with production of immunoglobulins (antibody) and with cell mediated immunity. Cells of this system all belong to the lymphoid series. The two systems are related in that the macrophages are necessary to "process" antigens for the lymphocytes (B cells) that secrete the specific antibody. Also neutrophils are more efficient at phagocytizing bacteria that have been coated with antibody (Payne *et al.*, 1976).

Granulocytes are produced in the bone marrow through various stages of proliferation and maturation (myeloblast, promyelocyte, myelocyte, metamyelocyte, band and segmenter) (Bointon and Farquhar, 1966). The segmented granulocyte has prominent nuclear segmentation or lobulation. The number of lobes generally increases with the age of the cell. In man, dog and cat, the lobes are sistinct and connected by very fine nuclear filaments. In other animals, the nucleus has fewer pronounced constrictions so that only one lobe is present, but there is an uneven nuclear membrane. In most animals only mature granulocytes are seen in the blood. In some species, 1 to 2% "band forms" may be normal, but more than this is abnormal. The granules in the cytoplasm are membrane-bounded, electron-dense bodies and they serve a storage function. Azurophilic (primary) granules are formed on the inner face of the Golgi apparatus predominantly during the promyelocyte stage. They contain several enzymes, acid phosphatase, arylsulfatase, indosylesterase, myeloperoxidase, 5–nucleotidase, galactosidase, glucuronidase. Specific (secondary) granules are formed on the outer face of the Golgi apparatus. These granules are rich in alkaline phosphatase and deoxyribonuclease enzymes. Normal granulocytes contain glycogen which is stained red by the periodic Schiff (PAS) technique, a reaction which is negative after amylase digestion. They also contain lipid which stains black with the Sudan black B stain. The granulocytes are also positive for peroxidase and alkaline phosphatase activity.

White Blood Cell (Leukocyte) Count

Leukocytes may be enumerated by using a counting chamber made of a single piece of glass with two elevated platforms, each having etched on it the improved Neubauer ruling. Each platform on which the rulings are engraved is surrounded by a moat, and on each side of the platform there is an elevated glass support at a height that will allow for a distance of 0.1 mm between the bottom of the cover glass and the ruled area. The improved Neubauer ruling of the platform consists of a system of squares in which a square measuring 3×3 mm is divided into nine equal squares, each measuring 1 mm³. The four corner squares are divided into 16 intermediate squares measuring 0.25×0.25 mm. These four corners are used for the total leukocyte count. The center square millimeter is subdivided into 25 squares and each of these is subdivided into 16 small squares which measure 0.05×0.05 mm. As a rule, five of the middle-sized squares are used in making the total erythrocyte count.

To perform the total leukocyte count, draw blood to the 0.5 mark on a white cell diluting pipette, then draw 0.5% acetic acid or Turk diluting fluid to mark 11. Mix as for the red blood cell count. The dilution is 1:20. The dilution fluid (Turk diluting fluid) is made with 1 ml of glacial acetic acid, 1 ml of Gentian violet (1% aqueous), and 100 ml

of distilled water. The solution should be filtered before use. The acetic acid produces hemolysis of all the nonnucleated red cells, but nucleated red cells are counted together with the white cells. Discard one-third, wipe the tip and place a drop on one side of the double chamber. Allow 3 minutes for settling and count the four corner square millimeter areas on each of the two preparations. In each large square begin counting at the extreme upper left and move toward the right. Count all cells in the upper four quadrants. Then drop down to the lower line of squares and count all the cells, move back to the next row of squares and continue. Count the cells that touch dividing lines to the left and above and disregard cells touching the dividing lines to the right and below. The greatest variation should not exceed ±12 cells.

Calculation. A total of 8 squares are counted. The sum of the numbers of cells counted is multiplied by 25 to give the number of leukocytes per mm³ of blood.

No. of Leukocytes/mm³ Blood

$$= \frac{\text{No. of Cells Counted} \times \text{Dilution}}{\text{No. of Large Squares Counted} \times \text{Height of Chamber}}$$

If eight large squares are counted, multiply as follows:

If the dilution is 1:10: cells counted × 12.5 = leukocytes/mm³ blood
If the dilution is 1:200: cells counted × 250 = leukocytes/mm³ blood

The leukocyte count in normal humans by this method is 7,000 WBC/mm³ (5,000–11,000).

Interpretation. The leukocyte count may be falsely low in chronic lymphatic leukemia because of the fragility of lymphocytes. Also, counts over 100,000 WBC are probably inaccurate. Nucleated erythrocytes are enumerated along with leukocytes in the leukocyte count so that a correction must be made when nucleated erythrocytes are observed in the blood smear.

Corrected WBC Count =
Counted WBC − X

$$X = \frac{\text{No. of Nucleated RBC/100 WBC}}{100 + \text{No. of Nucleated RBC/100 WBC + WBC Count}}$$

The number of nucleated red blood cells/100 WBC is derived from the differential count.

Automated Methods in Hematology

Over the past several years there has been a steady increase in the use of automated methods for routine hematological determinations. Although most of this equipment is very expensive, the improved efficiency has tended to lower the overall costs of laboratory tests. Also, automated analysis can minimize random technical errors thus improving the precision of tests and reliability of laboratory data. Some of the more common and better accepted automated instruments used in hematology laboratory tests are briefly described below:

Types of Automated Systems

Coulter Counter. (Coulter Electronics Inc., Hialeah, Fla. 33010, and Mississauga, Ontario). This is the system used by the authors and will be described in detail.

SMA 7A (Technicon Corp., Tarrytown, N.Y. 10591). This instrument measures the same seven parameters as the Coulter S. The hemoglobin is measured as cyanmethemoglobin, while the hematocrit is measured indirectly by the measurement of whole blood electrical conductivity. Cells are counted using an optical system based on the principle that light scatters as a blood cell flows through a small illuminated area. Coincidence is negligible. The results are graphed on a strip chart.

Hemalog System (Technicon Corp.). This system measures those parameters already mentioned plus conductivity cell volume, conductivity cell volume packed cell volume ratio, prothrombin time, partial thromboplastin time, and platelet count. The system can be expanded to include leukocyte differential counts based on the cytochemical staining reactions of different white cells. The hematocrit is measured by automatic centrifugation. Cell counts are coincidence free and results can be displayed visually or on a digital printout.

Autocytometer (Fisher Scientific Co., Pittsburgh, Pa. 15219). This semiautomated machine measures white cells and red cells by the scattering of light that occurs as cells pass through a dark field.

Hemalyzer (Fisher Scientific Co.). An automated version of the autocytometer measures the same parameters as the Coulter S. As each cell flows through an 18 micron aperture created by laminar flow, a 20 micron laser beam is interrupted. The hematocrit is measured by the pulse amplitudes produced by narrow angle diffraction measurements. The hemoglobin is measured as cyanmethemoglobin.

The choice of instrument can be problematic. The multichanneled instruments are very attractive for large institutions. A laboratory can very quickly become dependent on these machines so that any breakdown is a disaster. Consequently, it is necessary for each laboratory to have a backup system. This can be readily provided by the semiautomated machines. Accuracy and reliability are probably the most important requisites for any automated system. It is useful, when contemplating the purchase of such a system, to communicate with laboratories in which that system is being used. An objective appraisal can thus be obtained from workers who use the system on a daily basis. Whatever system is chosen, it is essential that the manufacturer's instructions regarding cleaning, maintenance, and servicing be followed. Some of the more commonly used automatic blood cell counters are described in detail below.

The Coulter Counter (Coulter Electronics, Inc., Hialeah, Fla.). This device has been widely employed for blood cell counts (Brecher *et al.,* 1956; Richar and Breakell, 1959). It measures a suspension of blood cells in a metered volume of an electrolyte medium flowing through a small orifice. An electric current is applied between platinum electrodes on either side of the aperture. Since each cell is a relative nonconductor, there is a momentary decrease in conductance corresponding to the passage of each cell through the orifice. The result is a series of pulses which can be counted electronically. With appropriate standardization and correction for coincidence and dilution factor, the number of pulses can be expressed as a blood cell count. Also the magnitude of each pulse is proportional to the volume of displaced electrolytes. Electronic averaging of these pulse heights has been employed as a direct measurement of the mean corpuscular volume of erythrocytes (Dacie and Lewis, 1975). Red cells are counted in an isotonic medium containing both red and white cells. Since the number of red cells generally exceed that of white cells by a factor of approximately 500 or more, the error in red cell count due to the presence of white cells is usually negligible. However, this error progressively increases as the white cell count rises, and red cell volume measurements are significantly altered when the white cell count exceeds 50,000 per mm^3. Leukocyte counting is performed after lysis of the red cells by surface active agents. Platelet counting can be performed on platelet rich plasma after removal of the red cells by spontaneous sedimentation (Bull *et al.,* 1965; Sipe, 1962).

The Coulter Counter Model S. This is an elaborate device which simultaneously computes and prints results for the hemoglobin concentration, hematocrit, white cell count, red cell count, mean corpuscular volume (MCV), mean corpuscular hemoglobin (MCH), and mean corpuscular hemoglobin concentration (MCHC). The instrument contains six separate Coulter counters, three each for the enumeration of red cells, and white cells. Hemoglobin (Hb) is measured colorimetrically as cyanmethemoglobin. The instrument measures the white cell count, the MCV, and the Hb concentration directly. The other results are electronically derived. Computations are performed electronically by analog devices and subsequently digitized for output to a printer or alternatively, directly to the computer. Red cell and white cell counts are performed in triplicate, and electronic averaging and comparison are performed with automatic rejection of data from one or all channels if significant variation occurs. The hematocrit measurement is derived from the electronic multiplication of the red cell count and the MCV measurement. The precision of the red cell count and the MCV values are greatly improved over that attainable by manual methods. The reproducibility of the white cell count and the hematocrit is comparable to that of carefully performed manual techniques.

Coulter S–Plus Counter

The S–Plus generates the seven hematologic parameters that have been associated with the Coulter S from the start:

1. White Blood Cell (WBC) or Leukocyte Count.

2. Red Blood Cell (RBC) or Erythrocyte Count.
3. Hemoglobin (Hgb) Concentration.
4. Hematocrit (Hct).
5. Mean Corpuscular (Erythrocyte) volume (MCV).
6. Mean Corpuscular Hemoglobin (MCH).
7. Mean Corpuscular Hemoglobin Concentration (MCHC).

Parameters 1, 2, 3 and 5 are directly measured while others are automatically calculated from measured parameters. Another parameter added in S–Plus is the measurement of platelet (PLT) or thrombocyte count. There are also two additional parameters, the red cell distribution width (RDW) and the mean platelet volume (MPV). The RDW is derived from the distribution of red cell volume recorded during the RBC. The clinical significance of RDW depends on the recognized fact that red cells of patients with pernicious anemia tend to be larger than those from the normal population, and the red cells of patients with hemorrhage tend to be smaller than normal. In both cases the RDW indicates the associated increase in standard deviation of red cell volumes.

One ml of whole blood sample is aspirated into the instrument and it is then analyzed by a completely automated cycle. The sample is first split by a Y tube into two segments. The red cell and platelet counting portion is diluted 1:6250, while the white cell counting and hemoglobin measuring portion is diluted by 1:251. If only a capillary sample of whole blood is available for the CBC, 44.7 μl of this sample may be prediluted in 10 ml of Isoton II prior to testing by the S–Plus. The required volume of diluent may be dispensed by the instrument into the sample cup with the capillary sample by pressing the Diluent Dispense button. A separate prediluted sample aspirator tip is provided for this mode of testing. After predilution the sample is aspirated by pressing the Prediluted sample button. In this automated mode there is no further dilution in the WBC/Hgb portion of the system, while the RBC/PLT segment of the sample automatically is further diluted as it enters the RBC/PLT bath. Each test solution, RBC/PLT and WBC/Hgb, is pulled through three identical apertures for cell counting by the well known Coulter impedance change techniques. Also well established is the fact that the size of each pulse caused by a cell passing through an aperture is proportional to cell volume. Therefore, as a fixed volume of solution passes through each aperture, the analyzer collects data on the number of cells, the distribution of their volumes, and their average volume. Statistical correction for more than one cell coincidentally within an aperture region is automatically made. The electronic analysis of the range of red cell volumes leads to the RDW result. The reason for the three identical apertures for each determination is to improve the statistical accuracy of the results. Circuit logic "votes" out the results from one of the apertures if they significantly deviate from results from the other two. Results from at least two must agree within established statistical limits. The accepted results from either two or three apertures are electronically averaged for the printed answer.

In routine patient testing the entire automatic cycle is initiated by pressing either the Prediluted Sample or the Whole Blood button. Normal operating cycle times are 34 seconds in the whole blood sampling mode and 45 seconds in the prediluted sample mode. A platelet count less than 20,000 can cause these times to extend by up to 16 seconds, to insure a good statistical data base. The Plot button on the diluter front panel causes the X–Y recorder to plot the average platelet distribution curve. The platelet count is collected from data generated by the same three aperture used for the RBC at the same time as the RBC. In order to increase the sensitivity of these apertures to the smaller platelets, these aperture dimensions have been reduced from 100 μm diameter, 75 μm length (still used for WBC) to 50 μm diameter, 60 μm length. The reported platelet count is derived from a sophisticated calculation compared to the RBC and WBC. All cells passing the RBC apertures with volumes of 36 μm^3 or more are added for the RBC, whereas all lysed cells passing through the WBC apertures with volumes of 45 μm^3 or more are added for WBC. In the case of platelets, the microprocessor fits a thrombocyte log's normal distribution curve to the raw data. These raw data gathered between 2 and 20 μm^3, may include debris near the 2 μm^3 threshold or red cell fragments near the 20 μm^3 threshold. The raw data are tested against a set of mathematical criteria, and the log's normal distribution curve is fitted from 0 to 70 μm^3. The debris and the red cell fragments, if any are thereby eliminated enabling calculation of the total number of thrombocytes and their sizes from 0 to 70 μm^3. If the mathematical criteria are not met for a particular sample, a printed code for this no-fit condition alerts the operator who may then command an X–Y plot of the thrombocyte data.

There are a number of improvements made on the Coulter S–Plus over its predecessor Coulter S–Counter. These include automatic aperture cleaning with high current flow for each cycle, separate Hgb blank for each sample; simultaneous, not sequential RBC and WBC dilutions; automatic cycling of diluent and cleaning solution 5 times on start-up and shutdown; microprocessor monitors and checks on many electronic functions; error codes determined by microprocessor; printer more reliable; electronic tuning replaces cams; means of visually monitoring the clarity of Hemoglobin reagents and blank; bubble rate adjustments no longer needed; improved backwash; aerosols eliminated; system closed; few mechanisms; all accessible from the front; and ability to monitor each measured parameter in each aperture.

Coulter S6–Plus II Counter. This is the latest development in Coulter S–Counters in which two new parameters have been added—lymphocyte percent and absolute number of lymphocytes, bringing the total number of parameters for each sample to 12. Coulter has formulated **ISOTON PLUS** and **LYSE S** Plus for the S–Plus II to insure that volumetric distinctions between cells are carefully maintained. This enables the determination of both lymphocyte percent and lymphocyte number on the S–Plus II. The video data terminal on the S–Plus II automatically displays for each sample the results of the 12–parameter CBC and the size distribution curves for WBC, RBC and PLT. On command from the operator full scale curves for individual size distribution curves or selected apertures may be displayed. Other new features on the S–Plus II include the ability to test two samples at the same time, in addition to added flags and indicators as well as more built-in test capabilities.

SMA 7A. Another instrument that performs simultaneous analysis of Hb concentration, red cell count, white cell count, hematocrit, and red cell indices is the SMA 7A (Technicon Corporation, Ardsley, N.Y.). The instrument measures Hb colorimetrically as cyanmethemoglobin. The hematocrit is indirectly estimated by the correlated measurement of whole blood electrical conductivity. Counting of red cells and white cells is based on the scattering of light that occurs when a blood cell flows through a small illuminated area projected by an optical system. The sampling volume is optically defined and is very small. Errors of coincidence are negligible. Electronic computations are performed to derive the mean corpuscular volume, mean corpuscular hemoglobin, and mean corpuscular hemoglobin concentration and the results are expressed as a graphic tracing on a strip chart recorder. Alternatively the output can be digitized for transmission to a computer based reporting system. The reproducibility and accuracy of this system has been reported by Nelson (1969) and Lappin *et al.* (1969). The Hemalog system is closely related to the SMA 7A differing only by substitution of an automatic centrifuge for determining the packed cell volume (Saunders and Scott, 1974). This system provides data which agree closely with those obtained by manual methods (Ohringer, 1970).

Differential Count

The examination of a smear of peripheral blood is one of the most informative laboratory procedures in hematology. For differential counts, a blood smear is prepared on a slide and stained by one of the staining procedures. The stained smear is then examined under the microscope. The minimum number of leukocytes counted in arriving at a differential count is 100, but if there are abnormal cells present or if a blood dyscrasia is suspected, 200 or more cells should be enumerated. It has been suggested by some workers that 100 cells should be classified on the differential count for every 10,000 total leukocytes. This procedure increases the accuracy of the determination.

In the differential count it is important for the operator to observe the characteristics of the erythrocytes and the number and nature of the thrombocytes (platelets). A search should be made for the stippling of erythrocytes, diffuse basophilia, and other erythrocytic abnormalities.

Preparing Blood Smears. A medium-sized drop of blood is placed near one end of a clean slide. Capillary or fresh venous blood (without anticoagulant) should be used. A spreader slide is placed in front of a drop at an angle of about 45°, and blood is allowed to spread in the angle between the slides. Then, just before the blood has spread quite

to the edges, the spreader slide is pushed ahead of the drop of the blood. The smear should be air-dried at once. A stain of proper thickness will dry quickly and appear yellow.

Coverglass Method. Coverslip preparations are preferable if a careful examination of the cellular characteristics is required. They are superior because the leukocytes are better distributed, and thus they provide a more suitable field for examination and more leukocytes per field than the slide method. Cell definition is also generally better.

Coverslips should be carefully cleaned and stored in 95% alcohol and dried with a lintless soft cloth immediately before use. Square coverglasses measuring 22 × 22 mm and of No. 1 thickness are usually used. While holding the clean dry coverslip by its edge in one hand, place a small drop of blood on the center of the slide. The drop should be small enough so that on spreading it will not reach the edges of the coverglass. Place a second clean dry coverglass diagonally on the first to form an eight-pointed star; the blood will immediately spread by capillary action. Grasp the top coverglass by its corners and slide them apart with a smooth motion. Wave the coverglass in the air to enhance drying. The coverslips should be identified and stained with Wright stain. Anticoagulants distort white blood cells; therefore only fresh blood should be used for smears. For the best results, smears should be stained soon after they have dried.

Staining the Blood Smear. Several stains and staining techniques are used, but most are of the Romanowsky type, incorporating basic and acidic aniline dyes which bring out the contrasting colors red and blue. The best known and most widely used stains of this type are the Wright and Giemsa stains. Both of these stains are available in powder form from most laboratory supply companies.

Wright's Stain. This stain may be utilized in the conventional manner or it may be used in a dip technique. Apply Wright's stain for 3 minutes and cover the slide completely. The mixture consists of 0.1 g of powdered Wright stain and 60 ml of acetone-free alcohol. Let the mixture stand for about 1 week before using and shake it vigorously each day. Add buffer in about the same quantity as the stain and mix them by blowing gently on the surface of the stain. Leave for 3–6 minutes. Keep the slide horizontal and float off any scum with distilled water to prevent the precipitate from sticking to the slide. Then wash the stain with distilled water until the smear is pink and translucent, and allow the slide to dry by letting the edge rest on a blotter. Remove the stain on the back of the slide by cleaning with moistened gauze. The slide may be covered with Permount (Fisher Scientific Company, New York).

A well stained smear will appear pink macroscopically. Microscopically the red cells are stained pink and the white cell nuclei are blue. There should be no precipitate. If precipitate forms, it may be due to inadequate or incorrect washing, failure to float off the scum, a dirty slide, dust, or an overconcentrated stain. If the stain is too blue, it may be due to inadequate washing, too thick of a film, overstaining, or excessive alkalinity of water, buffer, and/or stain. If the smear is too red, the pH of the water, buffer, or stain may have been too acid. The nuclei and basophilic cytoplasmic components should stain blue, neutrophilic granules are lilac, eosinophilic granules are orange, mass cell granules are deep blue-violet, nucleoli are usually blue violet, red cells are pink, the cytoplasm of mature monocytes is pale gray blue, and the cytoplasm of neutrophils is pale pink.

Giemsa's Stain. Fix the smear in absolute methyl alcohol for 3–5 minutes. Wash it and place it in diluted Giemsa's stain for 15–30 minutes (3.8 g of Giemsa stain powder and 200 ml of glycerin at 60°C for 2 hours; 312 ml of absolute methyl alcohol). A 1% solution of sodium carbonate instead of water intensifies the stain. The solution above is diluted 1:10 to make diluted Giemsa stain. Wash the stain and allow it to dry.

The appearance of cells in Giemsa's stain is similar to Wright's stain. There are other specific stains for staining components of blood cells which may be used for specific purposes. Some of these are listed below.

Alkaline Phosphatase Stain. Alkaline Phosphatase is used to stain the granules in the blood cells. The granules of segmented and band granulocytes and of metamyelocytes are phosphate positive and stain from pale brown to black, with a grading from 0 to 4+ according to the number and color of the granules. In the normal blood smears, most neutrophils are Grade 0; a few are Grade 1+ to Grade 2+. In infections, myeloproliferative disorders, and pregnancy, the alkaline phosphatase content of the granules is increased. In acute and chronic myelogenous leukemia, the alkaline phosphatase within the neutrophils is decreased.

For the details of the staining method and application, the reader is referred to Kaplow, 1963.

Acid Phosphatase Stain. This stain is used to demonstrate the leukocytic acid phosphatase enzyme. The activity appears as round or rod-like dark blue granules within the cytoplasm of leukocytes, occasionally red cells, and platelets. Films must be examined under oil immersion (Roasles *et al.*, 1966; Kaplow and Burstone, 1964).

Peroxidase Stain. The peroxidase stain is used to distinguish myeloid cells because their granules contain the enzyme peroxidase. Benzidine or a benzidine derivative is used as an indicator of leukocyte peroxidase activity because, when oxidized, it precipitates in the form of brown to blue granules (Kaplow, 1965). The peroxidase-positive cells contain yellowish-green to bluish and brownish-green granules. All cells of the granulocytic series from the promyelocytic stage onward are peroxidase positive, increasing in degree as the cells mature. All other cell types are negative, except the monocytes which may show a weak reaction.

Periodic Acid-Schiff (PAS) Stain. Cells of granulocytic series from myelocytes become increasingly positive. Myeloblasts and promyelocytes are negative, whereas mature polymorphonuclear cells are strongly positive. Cells of the lymphocytic series occasionally show positive granules. Monocytes show fine to coarse granules, sometimes with a positive tinge in the cytoplasm. Megacaryocytes and platelets are strongly positive. Normal erythroblasts and mature red cells are positive. In erythemia and thalassemia, the young normoblasts (erythroblasts) are PAS positive (McManus, 1946).

For other stains for the determination of dehydrogenase and diaphorases and staining nucleic acids the reader is referred to Balogh and Cohen (1961); and Kurnick (1955).

Schilling Method of Differential Leukocyte Count. This method counts the percentile distribution of the various white blood cells. The percentage of each group is based on the total number of leukocytes counted, expressed as a percentage of 100 leukocytes as described previously.

The increase of young cells in the peripheral blood is called a shift to the left. It is the result of bone marrow stimulation and leads to the appearance of an increased number of band granulocytes, metamyelocytes, and even myelocytes, so that a leukemialike picture may be produced. If the cells show toxic granulation and other toxic changes, a degenerative shift, as seen in severe infection, is produced. If there is no evidence of toxic damage, it is a regenerative shift like that seen following hemorrhage. The hypersegmented polymorphs of pernicious anemia produce a shift to the right.

Neutrophils. The primary function of neutrophils is to defend the body against bacterial invaders. They migrate to and accumulate in sites of infection (inflammation) by chemotaxis process involving specific antibodies, complement components, fragments and complexes (C_3, C_5, C_{3a}, C_{5a}, $C-$), and several pseudoglobulins. Neutrophils are more adept at phagocytosis than either eosinophils or basophils. When opsonizing antibodies are present, the phagocytic ability of neutrophils is enhanced and the band cells and metamyelocytes are even more avidly phagocytic than the mature neutrophils. Phagocytic activity is also enhanced by fever. Neutrophils are not capable of producing antibodies.

In humans, neutrophils are produced in the bone marrow from mesenchymal reticuloendothelial stem cells. Their primary function is to defend the body against bacterial invaders. The percentage of neutrophils in man varies from 29 to 65%.

Leukocytosis is most commonly the result of an absolute increase in neutrophils, and the term neutrophilia is sometimes used synonymously with leukocytosis. Most frequently leukocytosis is seen in acute infections in which its degree depends upon the severity of the infection and patient's resistance. Other causes of neutrophilia are intoxications, including those produced by metabolic disturbances (uremia, diabetes, eclampsia) and those produced by certain chemicals (lead, mercury, etc.). Venoms also produce leukocytosis. Neutropenia is seen in association with rapidly growing tumors, with acute hemorrhage, particularly into one of the body cavities (thorax, peritoneum, joints, etc.), with sudden hemolysis of erythrocytes, and with leukemias and surgical trauma.

Eosinophils. Eosinophils are large, fragile cells found mainly in the tissues adjacent to mucus membranes, e.g., gastrointestinal, respiratory and urinary traits. Eosinophils mature in the same manner as the neutrophils, but most of the eosinophils in the normal blood contain nuclei with only two lobe masses. The mature granules stain red, and immature granules are often black to gray. The eosinophilic granules contain antihistaminic substances. Eosinophils are found in increased numbers in situations involving decomposition of body protein and at sites of antigen-antibody reaction. They disappear from the circulation during stress, e.g., during early bacterial infection and leave the blood for the tissues, returning during the stage of convalescence presumably when their defense job has been completed. In the differential count all the eosinophils are placed in one group, except for eosinophilic myelocytes. The myelocytes are counted separately because they have a greater significance, being found in leukemia or leukemoid blood pictures in the peripheral blood. Normally there are 0–7.2% eosinophils in adult humans, yielding an absolute count of 0–570 eosinophils/mm^3 of the blood. The normal eosinophils count is lowest in the morning and then rises from noon till after midnight.

An elevated eosinophils count (eosinophilia) is caused by diseases of blood-forming organs. Included in this category are myelocytic leukemia, eosinophilic leukemia (up to 85%) following splenectomy (19%), and erythremia. Eosinophilia is also seen in allergic conditions including asthma, hay fever (15–30% increase), urticaria, parasitism (10–30%), and especially the diseases due to Trichinella and Filaria and the most common cestodes, trematodes, and nematodes, skin diseases such as psoriasis, eczema, pemphigus, and

scabies, certain infectious diseases such as scarlet fever, erythema multiforme, and other pyogenic infections, etc.

In humans, eosinophils also appears to be a consequence of antigen-antibody interaction, but is much more common in atopic disease than in immune disease mediated by nonreaginic antibody. The antigen-antibody complex attracts eosinophils by activating complex. An eosinophil chemotactic factor has been isolated and appears to be very specific.

A decrease in eosinophil count (eosinopenia) is seen in acute infections and in marked intoxication producing uremia. The count is also decreased after the administration of ACTH and cortisone, in hyperadrenalism (Cushing's syndrome) and disseminated lupus, and with stress burns.

Basophils. Basophils are somewhat smaller than eosinophils, measuring 10–12 μ in diameter in adults. The granules often overlie the nuclei, but do not fill the cytoplasm as completely as the eosinophilic granules do. They are soluble in water and may therefore be washed out during the staining procedure, leaving a typical granulocyte nucleus and a vacuolated cytoplasm. They are peroxidase negative. The nucleus is indented in several places, giving rise to an overleaf pattern. The cytoplasm is pale blue to pale pink. Basophils are functionally related to their morphologically similar connective tissue counterparts, the tissue mast cells. Both have the function of storing and producing heparin, histamine, and serotonin.

Basophils are widely distributed in connective tissue, particularly that around blood vessels. They play a central role in acute systemic allergic reactions. In vitro exposure of basophils from a sensitized individual to the sensitizing antigen results in histamine release. Sensitized basophils have bound IgE (the reaginic antibody). Sensitization as well as histamine release from cultured mast cells has been accomplished in vitro using tissue derived from dog mast cell tumors. Basophils are increased in myelocytic leukemia, basophilic leukemia, myeloproliferative disorders (e.g., polycythemia vera), and hypersensitivity states. They decrease following steroid therapy, in stress (e.g., myocardial infarction and bleeding peptic ulcer), in anaphylactic shock), and in hyperthyroidism.

Lymphocytes. The lymphocytic series which includes cells with diverse functions is concerned with resistance to infection, antibody production, and tissue rejection. The cells have little motility. Knowledge of the origin and differentiation of cells in the development of the immune system has come from phylogenetic, ontogenic, and other experimental studies in animals, and from elucidation of the various developmental immunodeficient disorders of man. Lymphoid stem cells which are derived from pheripotential stem cells in the yolk sac migrate to the fetal liver, spleen and bone marrow. From these sites lymphoid stem cells migrate to the thymus where further differential occurs under the influence of the microenvironment of the thymus to produce circulating T–Cells (thymus dependent cells). Other lymphoid cells, under the influence of the bursa of Fabricius in birds or some mammalian equivalent (fetal liver, bone marrow, gut associated lymphoid tissue) differentiate to become B–Cells (bursa or bone marrow dependent cells). The stage of differentiation from lymphoid stem cells to T–Cells and B–cells is antigen-independent. Further differentiation of T–Cells to activated T–Cells and B–cells to antibody-producing plasma cells is antigen-dependant and involves a complex interaction of many cellular and noncellular elements of the immune system. Circulating T–cells and B–cells cannot be distinguished under the light microscope, but they are most readily separated by the nature of their membrane surface marker. The prop-

erty of human T–cells to form rosettes with sheep red blood cells is most often employed to identify these cells. T–cells also react with nonspecific mitogens such as a phytohemagglutin (PHA) and concanavallin A (Con A). On exposure to these agents T–Cells transform from a small lymphocyte into a large pyroninophilic lymphoblast. Approximately 70 to 80% of lymphocytes in the peripheral blood and 90% or more in the thoracic duct show T–cell characteristics. T–cells are also the predominant cell in the paracortical area of the lymphnodes, in the periarterial lymphatic sheath of the spleen, and in other thymus dependent areas of the peripheral lymphatic structures. The average life span (intermitotic period) is approximately 4 years, and same cells survive 20 years or more as compared to a life span of three days for short-lived lymphocytes, which are mostly B–cells. The mechanism of the initial cell-mediated immune response to antigen involves a complex series of events that is still poorly understood. B–cells are easily identified by the presence of membrane bound immunoglobulins. Monomeric IgM, IgD and IgG are the most predominant immunoglobulins on the surface; membrane bound IgA and IgE and found rarely. B–cells are also characterized by the presence of surface receptors for the third component of complement (C_3) and probably the Fc portion of immunoglobulins. Some 10 to 20% of peripheral blood lymphocytes have B–cell characteristics. When B–cells are exposed to antigen, a complex sequence of events is initiated which involves interaction between antigen, macrophages, T–Cells (''helper'' or ''suppressor'' cells), dentritic reticulum cells, and B–cells. The B–cells proliferate to form memory cells and to undergo terminal differentiation to immunoglobulin secreting plasma cells.

The completely transformed lymphocyte cells are 20 to 30 μm in diameter with a homogeneous markedly basophilic cytoplasm, often with a somewhat more lightly stained cytoplasm immediately adjacent to the nucleus. In the dry smears the diameter of small lymphocytes varies from 7 to 10 μm. When Wright's stain is used, the nucleus of the small lymphocyte which is usually round, but sometimes indented, appears almost to fill the entire cell. Lymphocytes at times show definite staining abnormalities in disease stages, particularly in virus infection. Normal human blood contains about 21–49% lymphocytes, yielding an absolute count of 1510 to 3980/mm³ of blood with only an occasional large lymphocyte, except in children.

An increase in lymphocytes (lymphocytosis) may be caused by one or more of the following conditions:

1. All conditions that have an associated neutropenia may have a relative lymphocytosis (over 40%), but there is seldom an absolute increase.
2. Lymphocytic leukemias are accompanied by a marked increase in lymphocytes.
3. During the recovery stages of certain infections, the total number of lymphocytes may increase.
4. Some chronic conditions are accompanied by lymphocytosis.
5. Adrenocortical insufficiency is often manifested by an increase in the absolute number of lymphocytes.
6. Lymphocytosis sometimes occurs following vaccination.
7. Hyperthyroidism has been reported to be accompanied by lymphocytosis.

A relative decrease in the number of lymphocytes (lymphopenia) occurs in myelocytic leukemia, in Hodgkin's disease, neutrophilic leukocytosis, lupus erythematosus, and early acute irradiation syndrome.

Monocytes. The monocytes are phagocytic cells related to the tissue macrophages. They function in the body defense against infection, particularly in the formation of granulomas and giant cells, and in the destruction of damaged tissue. They also play a role in immune reactions probably by processing the antigen before it provokes a response from the lymphocyte. The monohistiocytic series comprises a variety of phagocytic cells, related by origin and function which include the blood monocytes, alveolar macrophages, peritoneal macrophages, Kupffer cells, and free and fixed macrophages of the bone marrow and other organs. It is generally believed that all of these cells are derived from bone marrow monoblasts and promonocytes, enter the blood as monocytes and later enter the tissues to develop into macrophages. In peripheral blood smears large monocytes have a diameter of 30 to 40 μm, their nucleus is large, located centrally or eccentrically, usually irregular, kidney shaped or multilobed. Small monocytes measure 20 to 30 μm in diameter. The cytoplasm is dusky blue. Occasionally it contains a few azurophilic granules, its nucleus is round to oval. The life span of monocytes in the circulation has been estimated at 30 to 100 hours. The cells leave the blood randomly and regardless of their age, by diapedesis, after they have become adherent to the endothelium. In sites of inflammation, they accumulate very rapidly. The life span of macrophages (histiocytes) is long and can attain 75 days or more. The death of monocytes and histiocytes proceeds in an unknown manner, but it is known that the cells when damaged and particularly after intense, phagocytosis can in turn be phagocytized by other histiocytes which is frequently seen in the course of infection and severe immunologic reaction.

Normal values vary from 2.20 to 10.2% of the total leukocyte count. An increase in monocytes (monocytosis) is caused by hematologic disorders such as monocytic leukemia, myeloproliferative disorders, lymphomas, multiple myelomas, and bacterial infections such as tuberculosis, collagen diseases, ulcerative colitis, and regional enteritis.

Phagocytic monocytes (macrophages)

may be found in small numbers in the peripheral blood in many conditions, such as severe infections, lupus erythematosus, hemolytic anemias, agranulocytosis, and thrombocytopenic purpura. Phagocytic cells are normally found in the bone marrow.

Automated Leukocyte Differential Counters

Two main types of automated differential counter are available. The Hemolog D (Technicon, Tarrytown, N.Y. 10591), uses cytochemical stains to differentiate between the different forms of leukocytes. Large number of cells are counted thus increasing the accuracy of the differential. The main disadvantage in the cells are counted in a fluid medium, so that a peripheral blood smear is still necessary. Other counters, LARC (Distributed by Canadian Laboratory Supplies, Toronto, Ontario), Honeywell (Honeywell Inc., Hopkins, Minnesota 55343), Coulters (Coulter Electronics, Hialeah, Fla. 33010 and Mississanga, Ontario) use the peripheral blood smear. The morphologic characteristics of the leukocyte are programmed into a computer, and the machine recognizes them in turn as they are viewed through a scanner. Although these machines count less cells than the cytochemical method, a smear is available for examination.

Hemalog D System

The Technicon Hemalog D is a multichannel system designed for continuous differential counting of leukocytes, based on the distinction between cell population using cell size analysis, two cytoenzymatic reactions and a cytochemical reaction. Analysis of variation in cell size distinguishes several populations from background noise, lymphocytes, eosinophils, and other elements present in blood including neutrophils and basophils, monocytes and large cells with no peroxidase activity. The peroxidase reaction also separates three populations: elements with negative peroxidase reaction (background noise, lymphocytes, large peroxidase negative cells), elements with medium peroxidase activity (monocytes, neutrophils, and basophils), and elements with strong peroxidase activity (eosinophils). The combination of these two parameters in an oscilloscope results in identification of basophils in one channel, monocytes in a second channel, and a combination of lymphocytes, neutrophils, eosinophils, "large unclassified cells," and "high peroxidase cells" in a third channel. Basophils are detected after staining with Alcian blue dye, using a technique (low pH, quarternary ammonium salts, and lanthanium ions) which minimized nuclear staining with that dye. Monocytes are identified as large cells which stain intensely for nonspecific esterase activity measured with α-napthylbutyrate substrate and coupling of the α-naphthol product with diazotized basic fuchsin. Inhibitors of neutrophil and plasma esterases are added to the reaction mixture. Neutrophils and eosinophils are identified after peroxidase staining utilizing a mixture of 4–chloro–1–naphthol and Hydrogen peroxide at pH 3.2. Cells are fixed with formaldehyde prior to staining in order to ensure localization of the enzyme. Eosinophils are differentiated from neutrophils by their more intense peroxidase stain. A small proportion of neutrophils stain more intensely than the remainder ("high peroxidase cells"), but are differentiated from eosinophils by light scattering. Monocytes, typical lymphocytes and immature granulocytes do not stain with the peroxidase reagents and are detected as "large unstained cells." Lymphocytes are identified as small unstained cells. This system permits classification of 10,000 cells per minute in each channel. The results agree with manual differential counts of 200 or more cells (Pierre and O'Sullivan, 1974; Simmons et al., 1975).

LARC (Leukocyte Automatic Recognition Computer). It is a pattern recognizing system which employs blood films prepared on glass slides by a "spinner" method (Megla, 1973). This technique produces even distribution of cells on the slide by the centrifugal force applied by a rapid rotation of the slide after a blood sample has been placed on its surface. The stained cells are examined by digital image processing, utilizing a television microscope system interfaced with a computer. The slide is automatically moved on a mechanical stage and cells are centered in the image field. Color differences are measured by consecutive scans of the cell using color filters selected as appropriate for the stain used. The LARC system incorporates a computerized stage which records the coordinates of each cell counted so that unclassified cells can be reviewed visually by the operator. The classification determined by the operator is thus incorporated into the final count. The LARC classifier has been subjected to limited field testing and appears to give results comparable to standard procedures, except that band neutrophil counts are higher and segmented neutrophil counts lower than obtained with manual methods. However, the LARC and manual methods are equivalent in detecting a "left shift." Abnormal cells are reliably detected by the automatic system. The technician can scan the slide for erythrocyte morphology and estimation of the platelet count.

Hematrak. This system also utilizes computerized pattern recognition method to identify nucleated cells on Wright's stained blood films prepared by the "wedge" technique (Dutcher *et al.*, 1974). The technician reviews the slide for quality of preparation and staining, evaluates red cell morphology and platelet count, selects an area of the slide suitable for differential leukocyte counting, and activates the electronic system. The machine can be instructed to stop for any unidentified cells so

that the technician can provide visual identification. The system recognizes segmented neutrophils, bands, lymphocytes, monocytes, eosinophils and basophils. Satisfactory agreement between manual differential counts and machine counts have been obtained in preliminary trials in clinical laboratory (Benzel *et al.*, 1974; Egan *et al.*, 1974; Christopher, 1974).

Other Hematological Methods

Hemoglobin. Hemoglobin (Hb) is an iron-containing conjugated protein (heme + globin) which has the physiologic function of transporting oxygen and carbon dioxide. Each molecule of hemoglobin consists of 1 molecule of globin linked to 4 heme molecules, and each is able to reversibly bind 4 molecules of oxygen to form oxyhemoglobin. In man about 55% of every red blood cell is hemoglobin.

There are many methods available for the determination of Hb, but they all vary in accuracy, ease of performance, and reproducibility. Some of these methods are briefly described below.

Colorimetric Methods. A large percentage of error results from these methods; therefore they are not recommended for Hb determination.

Direct Matching. This method is used in the Tallqvist and Dare methods. These methods are inaccurate because of the difficulties encountered in matching Hb with artificial standards.

Acid and Alkaline Hematin Production Methods. These methods, which also have a large percentage of error, are used in the methods of Haden and Hausser, Sahli, and Newcomber. The errors can be attributed to the difficulty in matching acid or alkaline hematin with colored glass, and also to incomplete solubility. Complete transformation of the hemoglobin into acid hematin requires a longer time; therefore these tests have to be read at a standardized period. Acid hematin is not stable and its color is affected by plasma color. Fetal hemoglobin is acid-resistant or alkaline resistant.

Gasometric Methods. These methods are accurate and can be used to standardize the technique used routinely. They are based on the amount of gas first bound by hemoglobin and then released and measured.

Oxygen Capacity Method. This method measures oxyhemoglobin, leaving out such hemoglobin derivatives as carboxyhemoglobin, methemoglobin, and sulfhemoglobin. When fully converted to oxyhemoglobin by a weak alkaline solution (such as 0.05% ammonium hydroxide), 1 g of hemoglobin combines with 1.36 ml of oxygen.

Carbon Monoxide Saturation. Carbon monoxide dissolved in the plasma may cause inaccurate results with this method. Certain compounds that readily absorb the gas, such as sulfhemoglobin and methemoglobin, may also cause inaccuracies.

Spectrophotometric Methods. These methods are based upon the measurement of hemoglobin derivatives at specific wavelengths.

Oxyhemoglobin Methods. Hemoglobin is transformed into oxyhemoglobin by diluting it with a weak alkaline solution and measuring the maximal absorption at 540 nm.

One of the simplest ways to measure oxyhemoglobin is with the Spencer Hemoglobinometer. The instrument measures oxyhemoglobin by light absorption through a green filter. A drop of blood is placed on a glass plate and cells are laked by a hemolytic agent such as saponin. The depth of blood is controlled by placing a second glass plate over the one containing laked blood and pressing them together. The glass chamber is inserted into the hemoglobinometer and the green color is matched with the standard.

Cyanmethemoglobin Method. This technique is the most commonly used and accepted method for determining Hb concentration. With this method, Hb is converted into cyanmethemoglobin by the addition of a potassium ferricyanide solution. All hemoglobin derivatives are measured. The pigment cyanmethemoglobin and reagents are stable and can be accurately standardized. The absorption band of cyanmethemoglobin in the region of 540 nm is broad, and its solution can be used in both filter photometers and narrow-band spectrophotometers.

Exactly 5 ml of diluent (cyanmethemoglobin reagent) is placed into a clean cuvette. The reagent is made of 200 mg of potassium ferricyanide, 50 mg of potassium cyanide, 140 mg of potassium phosphate (monobasic), 0.5 ml of Sterox SE (apolythiol compound), and up to 1000 ml of distilled water. The solution is also commercially available as diluent pellets. To prepare the diluent 1 pellet is dissolved in 250 ml of distilled water. A 0.02 ml potion of the blood specimen is added to the diluent solution in a Sahli or equivalent pipette. The pipette should be rinsed several times with the diluted blood. The dilution is 1:251. After blood is added, the tube should be stopped and inverted two or three times. The mixture of blood and reagent is allowed to stand for approximately 10 minutes for maximum conversion of Hb to cyanmethemoglobin. The optical density or transmittance of the unknown is determined at 540 nm (using the diluent as blank) and compared with a standard solution. A standard solution of cyanmethemoglobin can be purchased commercially. The concentration of cyanmethemoglobin is given in mg/dl and is usually in the order of 60 mg hemoglobin/dl. (This corresponds to 15 g

Hb/dl undiluted blood). The equivalence of this standard can be calculated by the following formula:

Equivalence of Standard
  in g/dl of Hb

$$= \frac{\text{Concentration Hb in Standard} \times 251}{1000}$$

The contents of a tube containing a known quantity of cyanmethemoglobin are placed in a matched cuvette. The tube containing the standard is placed into a previously warmed and zeroed spectrophotometer and the percentage of transmission is recorded. This reading represents the known quantity of cyanmethemoglobin in the standard solution. Similar readings can be obtained on other concentrations of a standard solution, and a standard curve can be made by utilizing a single-cycle semilog graph paper. Actual hemoglobin values on blood samples are determined by measuring the percentage of transmission and comparing it to the standard curve.

In addition to the methods described above, there are various other methods that may be used to determine Hb concentration. These methods include the iron content method of Wong, which measures the amount of iron in a measured amount of whole blood, and the specific gravity method. Hemoglobin and hemoglobin compounds such as carboxyhemoglobin, methemoglobin, and sulfhemoglobin can be measured spectroscopically on the basis of the number and position of their absorption bands. Since differences in the absorbence of the different hemoglobin derivatives are usually not large, and since the position of their absorption peaks does not differ greatly, a spectrophotometer that can be read to 0.001 optical density unit and that has a very narrow band width is required for their determination. The Hb values in experimental animals reported in this section were determined by the cyan-methemoglobin method described previously.

Sources of Error in Hemoglobin Determination

Each step in the estimation of Hb and preparation of a standard curve is a potential source of error. Unclean (moist) syringes and pipettes, incorrect pipetting, incorrectly calibrated pipettes, inadequate rinsing of pipettes, and blood adhering to the outside of pipettes may contribute to erroneous results. Errors in Hb determinations may also be caused by inaccurate calibration of an instrument with incorrect standards, by using nonmatched cuvettes, and by variation in the photocells. Errors may also arise from incorrect sampling. For example, capillary blood may be altered by squeezing, etc., and venous blood may be altered by prolonged stasis, inadequate mixing of the blood with anticoagulated blood before each test. Also, the method chosen may be inadequate to give accurate, reproducible Hb values.

Hematocrit—Volume of Packed Red Cells (PCV)

Hematocrit measures the proportion of red blood cells to plasma in the peripheral blood, but not in the entire circulation. The body hematocrit gives the ratio of total erythrocyte mass to total blood volume. Upon centrifugation, blood is separated into three distinct parts. The mass of erythrocyte at the bottom, which is referred to as packed red cell volume or PCV, and a white or gray layer of leukocytes and thrombocytes (platelets) immediately above the red cell mass, which is most commonly referred to as the buffy coat, represent about 10,000 WBC/mm³ of blood. In leukopenia and thrombocytopenia the layer will be thin, whereas in leukocytosis or thrombocytosis will be thicker. The hematocrit (Ht) reading is recorded as the number of millimeters of packed red cells/ 100 mm blood, including the volume (%) of packed red cells/dl blood.

The hematocrit is readily measured by the Wintrobe macrohematocrit method. A Wintrobe tube is a test of 3 mm uniform bore which is doubly calibrated with 10 cm scales with millimeter divisions. To measure Ht, fill the Wintrobe tube to the 10 mark with venous blood and anticoagulate by adding 1 drop of Versine to 5 ml of blood by means of a hematocrit pipette. The tube is centrifuged at 3000 rpm for 30 minutes in a standardized centrifuge. The level of packed cell is read directly from a centimeter and millimeter scale, and this figure is multiplied by 10 to give the volume of packed cells/dl blood. In reading the Wintrobe hematocrit tube, the exact level of the erythrocytes immediately beneath the buffy coat should be recorded, as should the thickness of the layer of leukocytes. Errors in reading must be avoided. The type and amount of anticoagulant are critical in this method. The average error of the Wintrobe method is ±1.13%. It is therefore one of the most reliable hematologic procedures.

Microhematocrit Method. A capillary hematocrit tube approximately 7 cm in length and having a bore of about 1.0 mm is used. These capillary tubes can be purchased with an anticoagulant already in them to facilitate direct filling from a venous or capillary puncture. Plain tubes are available for use with blood containing an anticoagulant. Capillary tubes are filled with capillary action, the outside is carefully dried with a piece of gauze, and the opposite end of the tube is sealed. These sealed tubes are then placed in a special high-speed centrifuge with the sealed end near the outside rim of the centrifuge. The tube is centrifuged for 2 minutes at 12,000 rpm

after which the volume of packed cells is ready from a scale held against the capillary tube in such a way that the top of the plasma column coincides with the 100% line and the bottom of the packed red cells falls on the zero line.

The most significant source of error inherent in packed cell volume determination by this method is centrifugation. In the process of centrifugation, platelets, plasma, and leukocytes are trapped between the red cells. The proportion of packed cell volume attributable to trapped plasma is somewhat greater than that attributable to the presence of either platelets or leukocytes. This error is increased by a lack of proper centrifugal force and by an incorrect length of time for centrifugation. It is important to standardize the time of centrifugation. In samples that are properly centrifuged, the error is between 2.25 and 8.0%. However the microhematocrit method has certain advantages over the Wintrobe method: (1) The amount of blood required can be obtained from finger puncture with children or from the extremities of experimental animals, and therefore is useful in pediatric practice and with small laboratory animals. (2) The time required for the entire process is less. (3) The capillary tubes are disposable. The main disadvantages of this method are that it requires a special reader, that it is impossible to determine the erythrocyte sedimentation rate, and that it is difficult to evaluate the depth of buffy coat.

Purpose of Hematocrit Determination. The hematocrit is used for calculation of blood constants (see Section IV) and as a check on the red blood cell count because it measures the concentration of red blood cells. Although it is not an accurate measure of blood volume, the degree of hemoconcentration in shock associated with surgery, trauma, and burns can be judged by the hematocrit. In conditions associated with hydremia, such as cardiac decompensation, pregnancy, and excessive administration of fluids, the hematocrit values will be below normal. The PCV is decreased in anemia and it is increased in primary and secondary polycythemia. Under normal circumstances hemoglobin and erythrocyte counts can be estimated from the macrohematocrit reading—1 mm on the hematocrit scale corresponds to 0.34 g hemoglobin and to 107,000 RBC/mm³ of blood.

Blood Indices and Corpuscular or Blood Constants

On the basis of the erythrocyte count and hemoglobin and hematocrit values, it is possible to express characteristics of the individual red blood cells as the erythrocyte constants and indices. The most commonly used constants are helpful in designating characteristics of red cells such as size, hemoglobin content, and hemoglobin concentration. These values are of particular importance in determining the morphological type of anemia, and they may also be of assistance in selecting therapy.

Mean Corpuscular Volume (MCV). The MCV is the volume of the average red blood cell of a given blood sample and it is calculated by the formula:

$$\frac{\text{Hematocrit} \times 10}{\text{RBC in Millions/mm}^3} = \text{MCV in } \mu^3$$

The usual normal MCV in man is 87.0 (80–94) μ^3 in absolute units. An MCV below 80 μ^3 indicates microcytosis, whereas a value of over 94 μ^3 indicates macrocytosis. The MCV allows classification of the red cells into normocytes, microcytes, and macrocytes if the values are appraised by the examination of well-prepared blood film in which variations in the size of red cells and the general trend in their size are readily perceptible.

Mean Corpuscular Hemoglobin (MCH). MCH, the amount of hemoglobin by weight in the average red cell of the sample of blood, is calculated by the formula:

$$\frac{\text{Hb in g/100 ml Blood} \times 10}{\text{RBC in Millions/mm}^3} = \text{MCH}$$

in micromicrograms ($\mu\mu$g)

The normal average in man is 29.5 (27–32) $\mu\mu$g. In newborn babies and infants and in macrocytic anemias the MCH is higher than normal. In iron deficiency anemia it is lower.

Mean Corpuscular Hemoglobin Concentration (MCHC). The MCHC is the proportion of hemoglobin (W/V) contained in the average red cell of the sample of blood. It is calculated by the following formula:

$$\frac{\text{Hb in g/100 ml Blood} \times 10}{\text{Hematocrit}}$$

$$= \text{MCHC in } \%$$

The normal average MCHC in man is 35% (33–38%). In hypochromasia the MCHC is lower than normal. Higher than normal concentrations (hyperchromasia) are not possible because the normal red cell contains the maximal amount of hemoglobin. Red cells can therefore be classified on the basis of MCHC as normochromic or hypochromic.

Interpretations of Mean Corpuscular Values

Mean corpuscular values should be preferred to indices such as column index, color index, and saturation index, because mean corpuscular values utilize absolute figures, whereas the indices are computed by means of relationships that are arbitrarily selected to represent 100% as being normal. As mentioned previously, the mean corpuscular values are utilized to classify anemias morphologically. However the morphological classification has little reference to the cause of anemias; it represents only an estimation of alterations in the size and hemoglobin concentration of individual red cells. In most anemic conditions, alterations in the average size of red cells (MCV) are paralleled by similar changes in MCH or in the weight of Hb in red cells. With microcytic cells the MCH is also decreased, and this is referred to as a microcytic hypochromic anemia. Such alterations are specific for iron deficiency or failure to utilize iron properly in the formation of Hb. This includes anemias due to chronic blood loss, copper, and pyridoxine deficiencies. Normocytic anemias are characterized by having a normal MCV and MCH; they are detected only by a decreased number of erythrocytes or by a low packed cell volume. Such anemias usually occur when there is a depression of erythrogenesis, such as in chronic infectious diseases and nephritis, in irradiated animals, in malignant neoplasms, or in any conditions in which marrow becomes hypoplastic. This is the most common type of anemia found in domestic animals and it often indicates that other diseases are present.

Sedimentation Rate of Red Blood Cells (ESR)

The ESR is a measurement in millimeters of the rate of settling of erythrocytes in anticoagulated blood in 1 hour. ESR values are affected by the shape and size of erythrocytes, concentration of erythrocytes, composition of blood plasma, physical attributes of the test tubes such as positioning, height, and width, anticoagulant used, and temperature. In man, the normal erythrocytes sediment slowly, in contrast to agglutinated red cells which have a high velocity because of their increased mass. Therefore any factor that leads to clumping

or to rouleaux formation of erythrocytes will increase the ESR, whereas factors preventing clumping will retard the rate. The ESR is relatively low in animals and is increased in inflammatory diseases and diseases in which there is tissue necrosis and degeneration. However in animals there is a great variation in the ESR values. The shape and size of the red cells influence the sedimentation rate only slightly. Severe spherocytosis or severe sickling retards the sedimentation rate. Macrocytes fall more rapidly than microcytes. Increased red cell volume (polycythemia) slows the sedimentation rate, whereas in anemia it is accelerated. Variations in quantity and distribution of plasma protein fractions greatly influence the sedimentation rate. Fibrinogen exerts the greatest influence. In disease associated with hypofibrinogenemia, such as liver disease, the rate is slow. Hydremia and increases in plasma lipids retard the sedimentation rate, whereas heme concentration accelerates it. Too short a tube (column of blood) and too narrow a bore lower the sedimentation rate. Alignment of the tube at an angle as little as 4–6° accelerates the rate. The temperature under which the test is performed affects the rate between and 15–25°C. The rate tends to accelerate as the temperature rises.

Determining Sedimentation Rate. The ESR is easily determined by filling a Wintrobe hematocrit tube or by using disposable sedimentation tubes. A 5 ml blood specimen is placed in a test tube with dried potassium and ammonium oxalate or 1 drop of Veresene. The test is set up within 2 hours by filling the Wintrobe tube to the 10 cm mark in a vertical position, and the fall of the red cell column is read at 1 hour in millimeters. The test should be made at room temperature between 22 and 27°C. After the sedimentation test has been completed at 1 hour, centrifuge the cells to a constant volume for 30 minutes at 3000 rpm to pack the cells and to determine their volume.

Normal values for the Wintrobe sedimentation rate in men equal 0–9 mm, in women 0–15 mm, and in children 0–13 mm.

Westergren Method of ESR Determination. The Westergren ESR determination is performed in a disposable glass tube 300 × 2.8 mm I.D. calibrated at 2 mm intervals over the lower 200 mm of the tube. Whole blood anticoagulated with EDTA is diluted 4:1 with a 3.8% solution of sodium citrate. After thorough mixing of the diluted blood, the Westergren tube is filled, placed in a rack to maintain a vertical position, and read in the usual manner after 1 hour at 27°C. The Westergren method has been endorsed by the standards committee in Europe and the United States, although the superiority of the Westergren technique has never been clearly demonstrated. The Westergren method provides an extended scale and the appearance of increased sensitivity. The greater length of the Westergren tube has the further advantage of delaying the packing effect. However the Westergren technique has the disadvantage that the citrate dilution lowers the concentration of asymmetric macromolecules, affects sensitivity, and decreases the ESR. At the same time dilution tends to increase the ESR by lowering the hematocrit. The net effect of dilution is a slight relative lowering of the ESR in the Westergren method compared with the Wintrobe method. Other important considerations of ESR methods are summarized below (Koepke, 1981).

Interpreting the ESR Test. The test has high sensitivity and low specificity, and is therefore difficult to interpret. Since concentrations of fibrinogen and globulins are commonly elevated in inflammatory disease, the sedimentation rate is useful as a screening test to detect the presence of infection, particularly chronic infection, or to monitor the progress of chronic inflammatory processes such as rheumatoid arthritis. An accelerated rate first of all points to a possible change in the distribution and quantitation of plasma proteins and demands further investigation. Acute and chronic inflammatory processes, as well as necrotizing tumors and lymphomas, leads to an accelerated rate.

The sedimentation rate increases with age, averaging 0.85 mm/hr/5 years in men. In women there is similar increase up to age 50 (menopause), after which the increase is much greater (2.8 mm/hr/5 years).

The ESR is decreased in polycythemia, hypofibrinogenemia, increased blood viscosity, hereditary spherocytosis, hypochromic microcytic anemia, sickle cell anemia, congestive failure, administration of salicylates, cortisone, and ACTH (adrenocorticotropic hormone), high blood sugar level, and in hemoglobinopathies.

Method	Blood Volume (ml)	Time (min)	Convenience	Effect of Hct	Screen	Monitor Disease	Clinical Acceptance
Wintrobe	1.5	60	good	minimal	excellent	fair	good
Westergren	2.0	60	fair	moderate	good	excellent	good
ZSR	0.5	5	very good	none	excellent	fair	poor

Table 1. Conversion of Hematological Values from Traditional Units into SI Units

Constituent	Traditional Units	Multiplication Factor	SI Units
Clotting time	Minutes	0.06	ks
Prothrombin time	Seconds	1.0	arb. unit
Hematocrit (Erythrocytes, volume fraction)	%	0.01	1
Hemoglobin	g/100 ml	0.6205	m mol/l
Leukocyte count (Leukocytes, number concentration)	per mm^3	10^6	10^9/l
Erythrocyte count (Erythrocytes, number concentration)	million per mm^3	10^6	10^{12}/l
Mean corpuscular volume (MCV)	μ^3	1.0	fl
Mean corpuscular hemoglobin (MCH) (Erc-Hemoglobin, amount of substance)	pg	0.06205	fmol
Mean corpuscular hemoglobin concentration (MCHC) (Erc-Hemoglobin, substance concentration)	%	0.6205	m mol/l
Erythrocyte sedimentation rate	mm/hour	1.0	arb. unit
Platelet count (Blood platelets, number concentration)	mm^3	10^6	10^9/l
Reticulocyte count (Erc-Reticulocytes, number fraction)	% red cells	0.01	1

Source: Young (1975).

Section IV

Hematological Reference Values

Blood is an important index of physiological and pathological changes in the organism. Such changes can be evaluated only if the normal values are known. A vast amount of data has accumulated over the years relating to the hematological aspects of animals used in experimental medicine. Unfortunately, much of this information consists of a mass of apparently unrelated facts because it has been derived from experiments primarily designed to elucidate some point of human physiology or pathology rather than to deepen the knowledge of the animal hematology.

However it has become increasingly apparent that a more fundamental approach to the actual functioning of the laboratory animals is necessary if such experiments are to be correctly evaluated in terms of any human or more general biological application. It is also important to have comparative values for these laboratory animals and humans so that the animals can be better utilized in attempts to understand the disease processes. This chapter describes the findings of different investigators on the normal hematological values of various strains of laboratory animals. Whenever necessary, important references are cited for further information. The comparative hematological values of 14 species of animals and man are presented as a readily available source of information for the medical investigator.

The necessity of establishing a basis for comparison has led to the hematological investigation of numerous breeds and species of apparently healthy animals. There are some variations in the normal values within a species of animal, but much of this variation can be explained by the lack of a standard method, and by differences in age, sex, number, and strain. It should be emphasized that unless the experimental animals are kept under carefully controlled conditions, a range of normal hematological values may vary widely because of spontaneous or latent diseases, genetic factors, diet, and other environmental conditions. Valid comparisons cannot be made between healthy and diseased stock or between animals of different strains. In all experimental work animals in different test groups should be littermates of the same strain kept free from disease and under identical environmental conditions.

Many of the methods used in hematologic determinations are standard (used in clinical hematology laboratories) and they are described in various textbooks and laboratory manuals (see Section III). The purpose of this section is not to describe the methods but to present the comparative values in normal laboratory animals and their interpretations in the study of human disease.

Mice

Table 1 lists the hematological values in male and female mice of ICR strain. The erythrocyte count is high (6.9–11.7 million/mm^3). Our data correlated well with those reported in literature (Table 2). Polychromatophilic erythrocytes are common and some normoblasts occur. The differences in values between male and female mice are within the margin of error. Differences in red blood corpuscle values were found to be primarily caused by maintenance factors such as the nutritional, environmental, and pathological conditions of the animals, rather than by the sex, breed, or strain of the animal species.

The animal's age is important in the study of erythrocyte counts; the erythrocyte count is lower in young animals than in adults. Kunze (1954) has reported an average of 3.7 million erythrocytes at the time of birth, 4 million after 4 days, 5.18 million after 10 days, 5.9 million after 14 days, and 9.34 million at 2–3 months of age. It is interesting that the spleen of the mouse contains erythroblasts and megakaryocytes and thus is a hemopoietic organ.

The hemoglobin values in mice that have been reported in the literature vary widely because of the method of determination. No important sex- or age-linked differences have been found. However during the first 2 months of life the hemoglobin value is much lower, which is in agreement with the smaller erythrocyte count. The mean corpuscular volume (MCV) and hemoglobin concentration data (MCH, MCHC) do not vary widely as a result of differences in sex, age, or breed.

White blood cell counts in mice do vary widely. Variations of as much as 30 times can be within physiological limits. One rea-

son for this variation may be the technique used to collect blood sample from the tail of the animal. A slight compression of the tail causes admixture of tissue fluid to tail blood. The leukocyte count is lower in the first few weeks of life than in adult mice: 4.5 on the day of birth, 7.8 at the age of 5 days, 8.7 at age of 5 months, and greater than 9.0 at 1 year. We found 12,000–20,000 leukocytes/mm^3 in the animals a little over 1 year of age.

The blood of the mouse has a high percentage of lymphocytes—rarely less than 60% and in some animals as high as 90% of the total WBC. Neutrophils range from 12 to 22%; some animals have as few as 5%. Basophils usually vary between 0.1 and 1%, eosinophils between 0.2 to 0.5%, and monocytes between 1 and 2%; there is a small percentage (2–6%) of "transitional" forms. A large number of neutrophils of the mouse blood show the "ring" type of nucleus, a "doughnut" shape rather than the multilobulated form seen in man. Most eosinophils also appear to have the ring type of nucleus.

Comparison of the data from different strains within a species of mice shows that the differences in hematological values are within the physiological limits of variation (Table 3).

Rats

The erythrocyte count of the blood of the rat varies widely. The red cell count is 6–10 million/mm^3. The normal values of adult animals are attained at 4 months of age; until then they are lower. No significant sex differences are found in the RBC counts of rats. However, slightly lower erythrocyte and hemoglobin values have been reported in females than in males: In males the RBC is 7.5 million with a hemoglobin of 17.9 g/100 ml, whereas in females the RBC is 7.6 million with a hemoglobin content 18.2 g/100 ml (Kleinburger, 1927). A charac-

Table 1. Hematological Values of the Normal Albino Mouse*

Test	Unit	Male (145 animals)			Female (128 animals)		
		Mean	S.D.	Range	Mean	S.D.	Range
Erythrocytes (RBC)	$(\times 10^6/mm^3)$	9.30	1.20	6.90-11.7	9.10	1.12	6.86-11.3
Hemoglobin	(g/dl)	11.3	0.10	11.1-11.5	10.9	0.10	10.7-11.1
MCV	(μ^3)	49.0	0.75	47.5-50.5	49.5	1.25	47.0-52.0
MCH	$(\mu\mu g)$	12.2	0.25	11.7-12.7	11.9	0.42	11.1-12.7
MCHC	(%)	27.2	2.00	23.2-31.2	25.9	1.80	22.3-29.5
Hematocrit (PCV)	(ml%)	41.5	4.20	33.1-49.9	42.1	1.20	39.7-44.5
Sedimentation rate	(mm/hr)	0.45	0.23	0.00-0.91	0.43	0.22	0.00-0.87
Platelets	$(\times 10^3/mm^3)$	232.	80.0	157.-412.	250.	77.0	170.-410.
Leukocytes (WBC)	$(\times 10^3/mm^3)$	14.2	0.87	12.5-15.9	12.9	0.42	12.1-13.7
Neutrophils	$(\times 10^3/mm^3)$	2.16	0.15	1.87-2.46	2.20	0.07	2.07-2.34
	(%)	17.4	2.10	13.2-21.6	17.1	0.70	15.7-18.5
Eosinophils	$(\times 10^3/mm^3)$	0.38	0.02	0.34-0.41	0.31	0.01	0.29-0.33
	(%)	2.09	0.36	1.37-2.81	2.41	0.18	2.05-2.77
Basophils	$(\times 10^3/mm^3)$	0.09	0.01	0.08-0.10	0.06	0.01	0.06-0.07
	(%)	0.52	0.15	0.22-0.82	0.49	0.18	0.13-0.85
Lymphocytes	$(\times 10^3/mm^3)$	11.2	0.60	10.0-12.4	9.28	0.29	8.70-9.85
	(%)	72.6	5.10	62.4-82.8	71.9	2.98	65.9-77.9
Monocytes	$(\times 10^3/mm^3)$	0.45	0.05	0.35-0.55	0.40	0.05	0.30-0.52
	(%)	2.35	0.06	2.22-2.47	2.04	0.03	0.98-1.11

*Charles River Strain (ICR), Caesarian derived, weighing 20-25 g.

Table 2.   Hematological Values of the Normal Mouse

Test	Unit	Benirschke[1] et al (1978)		Harrison[2] et al (1978)		Melby and[1] Altman (1976)		Mitruka &[3] Rawnsley (1977)
		Mean	S.D.	Mean	S.D.	Mean	% C.V.	Range
Erythrocytes (RBC)	$(\times 10^6/mm^3)$	7.85	2.12	6.74	0.47	8.60	10.0	6.86-11.7
Hemoglobin	(g/dl)	11.9	0.94	14.5	0.80	14.2	6.00	10.7-11.5
MCV	(μ^3)	–	–	–	–	51.0	–	47.0-52.0
MCH	$(\mu\mu g)$	–	–	–	–	17.0	–	11.1-12.7
MCHC	(%)	–	–	–	–	33.0	–	22.3-31.2
Hematocrit (PCV)	(%)	41.0	2.13	42.0	3.00	45.0	10.0	33.1-49.9
Sedimentation rate	(mm/hr)	–	–	–	–	–	–	0.00-0.91
Platelets	$(\times 10^3/mm^3)$	–	–	421	179	232	63.0	157 - 412
Leukocytes (WBC)	$(\times 10^3/mm^3)$	–	–	9.90	2.50	9.20	34.0	12.1-15.9
Neutrophils	$(\times 10^3/mm^3)$	–	–	2.20	1.20	1.80	–	1.87-2.46
	(%)	–	–	–	–	20.0	–	13.2-21.6
Eosinophils	$(\times 10^3/mm^3)$	–	–	–	–	0.08	–	0.29-0.41
	(%)	–	–	–	–	0.90	–	1.37-2.77
Basophils	$(\times 10^3/mm^3)$	–	–	–	–	0.00	–	0.06-0.10
	(%)	–	–	–	–	0.00	–	0.13-0.85
Lymphocytes	$(\times 10^3/mm^3)$	–	–	7.40	2.00	7.30	39.0	8.70-12.4
	(%)	–	–	–	–	80.0	–	62.4-82.8
Monocytes	$(\times 10^3/mm^3)$	–	–	–	–	0.02	–	0.30-0.55
	(%)	–	–	–	–	0.20	–	0.98-2.47

Source: [1] Derived from Literature

[2] Values from male $B_6D_2F_1$ mice. Cell counts were made with a Model ZF Coulter Counter (Coulter Electronics, Inc., Hialeh, Florida)

[3] Data from Normal Albino Mice, Charles River strain (ICR), Caesarian Derived weighing 20-25g, 128 females and 145 males. See text for methodology.

Table 3. Comparative Hematological Values of Different Strains of Mice

| Test | Unit | Multi-mammate mouse Promys (Mastomys) natalensis | Deer Mouse (Peromyscus maniculatus) | | | HA/ICR strain | Inbred strain ICR Charles River |
			Wild type (+/+)	Hetero-zygous (+/sp)	Sphero-cytic (sp/sp)		
Erythrocytes (RBC)	$(x10^6/mm^3)$	7.05-8.73	10.2-12.0	10.9-12.7	11.1-13.2	8.31-10.5	7.10-9.80
Hemoglobin	(g/dl)	11.0-14.8	11.8-15.4	11.4-14.8	11.4-14.7	11.3-14.4	14.5-16.6
MCV	$(\mu 3)$	46.6-60.3	34.8-45.7	32.9-44.1	31.0-42.6	36.5-47.8	42.5-58.4
MCH	$(\mu\mu g)$	15.3-20.2	11.4-12.8	10.5-11.9	9.8-11.6	10.1-11.5	12.0-14.5
MCHC	(%)	31.3-35.3	29.1-32.1	27.9-31.1	28.5-32.0	29.5-33.9	30.0-34.5
Hematocrit (PCV)	(ml%)	35.1-49.0	43.3-53.3	40.2-50.3	41.5-51.6	37.6-47.8	39.0-48.7
Platelets	$(x10^3/mm^3)$	208.-446.	190.-340.	188.-325.	200.-350.	160.-275.	210.-380.
Leukocytes	$(x10^3/mm^3)$	5.40-10.6	6.91-12.9	7.38-12.2	7.10-12.6	6.11-11.4	7.62-12.4
Neutrophils	(%)	8.00-48.0	8.25-40.8	10.5-38.1	11.6-40.8	10.9-39.4	12.0-35.1
Basophils	(%)	0.00-0.50	0.00-0.50	0.00-0.60	0.00-0.54	0.00-0.55	0.00-0.35
Eosinophils	(%)	0.00-0.50	0.10-0.93	0.00-0.80	0.00-0.90	0.00-0.75	0.00-1.00
Lymphocytes	(%)	57.0-93.0	62.0-90.0	64.0-87.0	61.0-82.5	62.0-83.5	65.0-84.0
Monocytes	(%)	0.00-7.00	0.00-5.40	0.00-6.58	0.00-5.20	0.00-5.50	0.00-5.00

Source: Russell et al (1951, 1966); Brown and Dougherty (1956); Burns and DeLannoy (1966); Chapman (1968);
Dieterich (1972); Finch and Foster (1973).

teristic feature of rat erythrocyte is a striking anisocytosis, the extent of which varies from rat to rat. There are erythrocytes whose diameter reaches only one-third of the normal diameter of 5.7–7 μ (average 6.2 μ).

Normoblast forms constitute about 0.12–2% of the red cells, and polychromatophilic erythroblasts constitute 1–18%. The large number of polychromatic erythrocytes is accompanied by a large number of reticulocytes—2–5% in adults and 10–20% in young animals (Schermer, 1967). Hemoglobin values vary considerably in accordance with the wide range of variation in the erythrocyte count. In the animal species we found 13.4–15.8 g/100 ml in males and 11.5–16.1 g/100 ml in females (Table 4). However, Albritton (1961) reported a range of 12 to 17.5 g/100 ml (average 14.8), with hematocrit values ranging from 39 to 53%.

As in the mouse, the spleen of the rat has hemopoietic activity, especially in younger animals. Both erythrocytes and granulocytes are formed here, while megakaryocytes lie close to the sinusoids and bud off the platelets into them. The normal values of blood platelets were found to range from 150 to 460,000/mm³ in our studies, compared to the literature values of from 121,000 to 460,000/mm³.

Our data show that leukocytes in rat blood range from 6.6 to 12.6 ($\times$ 10³/mm³), whereas the literature indicates a much wider range (4,800–30,000) in the leukocyte count. One of the reasons for this wide variation in literature values may be that the various authors used different species of rats in evaluating WBC. Another reason for the wide variability in the rat's leukocyte count is that the animal is highly excitable during blood sampling, and this may cause leukocytosis; as much as a 12% increase in lymphocytes has been reported as a result of this excitement (Farris and Griffith, 1963). Other factors such as age, sex, castration of male rats, and even slight

changes in nutritional conditions do not cause major changes in leukocyte counts.

The blood of the rat also has a high lymphocyte count. In the literature the figures in differential counts are lymphocytes 62–75%, large lymphocytes 5–8%, basophils 0–0.5%, monocytes 0–1%, and "transitionals" 0–2%. Plasma cells are also occasionally present. Our findings show the following differential counts in the SPF albino rat: lymphocytes 50–84.5%, neutrophils 4.40–49.2%, eosinophils 0–2%, basophils 0–0.6%, and monocytes 0–2% (Table 4). Hematological values reported in literature are in agreement with our data (Table 5). Table 6 presents the comparative hematological data of twelve strains of rats. In general the values reported reveal that species differences in blood counts are within physiological limits of variation. No large differences in the erythrocyte counts, hemoglobin, hematocrit, and differential WBC counts are caused by interspecies or strain differences.

Hamsters

Hematological values in hamsters vary considerably, but differences in values between males and females are not considerable (Table 7). Literature values fall within the ranges reported in this book (Table 8). Erythrocytes range between 4 and 10 million with an average of 7 million. The erythrocytes have a diameter of 5–7 μ (average 6 μ). A small proportion of erythrocytes shows polychromasia. Nucleated or basophilic cells are usually rare, but reticulocytes are found up to 3%. The hemoglobin (Hb) concentration ranges from 14.5 to 19.2 g/dl in males and from 13.1 to 18.9 in females. Higher Hb values are found in starving hamsters and in 2- to 3-week-old animals.

The leukocyte count in hamsters is between 5,000 and 10,000/mm³, of which about 75% are lymphocytes. During hiber-

Table 4. Hematological Values of the Normal Albino Rat*

Test	Unit	Male (160 animals)			Females (140 animals)		
		Mean	S.D.	Range	Mean	S.D.	Range
Erythrocytes (RBC)	$(\times 10^6/mm^3)$	8.95	0.40	8.15-9.75	7.98	0.61	6.76-9.20
Hemoglobin	(g/dl)	14.6	0.58	13.4-15.8	13.8	1.15	11.5-16.1
MCV	(μ^3)	53.8	2.00	49.8-57.8	58.2	3.65	50.9-65.5
MCH	$(\mu\mu g)$	16.3	1.00	14.3-18.3	17.3	0.85	15.6-19.0
MCHC	(%)	30.8	2.30	26.2-35.4	31.3	2.40	26.5-36.1
Hematocrit (PCV)	(ml%)	47.4	1.50	44.4-50.4	44.1	3.26	37.6-50.6
Sedimentation rate	(mm/hr)	1.22	0.27	0.68-1.76	1.10	0.26	0.58-1.62
Platelets	$(\times 10^3/mm^3)$	340.	70.0	150.-450.	319.	75.0	160.-460.
Leukocytes (WBC)	$(\times 10^3/mm^3)$	9.92	0.96	8.00-11.8	9.60	1.50	6.60-12.6
Neutrophils	$(\times 10^3/mm^3)$	2.42	0.23	1.95-2.88	2.57	0.40	1.77-3.38
	(%)	24.4	9.10	6.20-42.6	26.8	11.4	4.40-49.2
Eosinophils	$(\times 10^3/mm^3)$	0.03	0.01	0.03-0.04	0.06	0.01	0.04-0.08
	(%)	0.35	0.14	0.09-0.63	0.60	0.68	0.00-1.96
Basophils	$(\times 10^3/mm^3)$	0.02	0.01	0.01-0.03	0.01	0.01	0.00-0.03
	(%)	0.20	0.20	0.00-0.60	0.00	0.20	0.00-0.40
Lymphocytes	$(\times 10^3/mm^3)$	7.48	0.72	6.03-8.90	6.95	1.08	4.78-9.12
	(%)	75.4	8.90	57.6-83.2	72.4	11.1	50.2-84.5
Monocytes	$(\times 10^3/mm^3)$	0.02	0.01	0.01-0.04	0.03	0.01	0.02-0.04
	(%)	0.23	0.21	0.00-0.65	0.35	0.73	0.00-1.81

*Wistar strain, Caesarian derived rats, weighing 180-250 g.

Table 5.   Hematological Values of the Normal Rat

Test	Unit	Benirschke[1] et al (1978) Mean	S.D.	Melby and[1] Altman (1976) Mean	% C.V.	Mitruka and[2] Rawnsley (1977) Range
Erythrocytes (RBC)	$(x10^6/mm^3)$	7.83	0.62	7.30	10.0	6.76-9.75
Hemoglobin	(g/dl)	14.8	0.80	15.2	8.00	11.5-16.1
MCV	(μ^3)	59.0	3.20	62.0	5.00	49.8-65.5
MCH	$(\mu\mu g)$	18.9	1.20	21.0	8.00	14.3-19.0
MCHC	(%)	32.5	1.10	34.0	5.00	26.2-36.1
Hematocrit (PCV)	(%)	46.1	2.50	45.0	8.00	37.2-50.6
Sedimentation rate	(mm/hr)	-	-	-	-	0.58-1.76
Platelets	$(x10^3/mm^3)$	-	-	-	-	150 - 460
Leukocytes WBC)	$(x10^3/mm^3)$	9.98	2168	9.80	25.0	6.60-12.6
Neutrophils	$(x10^3/mm^3)$	-	-	1.90	54.0	1.77-3.38
	(%)	24.5	8.00	19.0	39.0	4.40-49.2
Eosinophils	$(x10^3/mm^3)$	-	-	0.16	94.0	0.03-0.08
	(%)	1.70	1.30	1.60	89.0	0.00-1.96
Basophils	$(x10^3/mm^3)$	-	-	-	-	0.00-0.03
	(%)	0.08	0.28	-	-	0.00-0.60
Lymphocytes	$(x10^3/mm^3)$	-	-	7.40	28.0	4.78-9.12
	(%)	71.1	8.70	76.0	10.0	50.2-84.5
Monocytes	$(x10^3/mm^3)$	-	-	0.26	58.0	0.01-0.04
	(%)	2.50	2.0	2.70	94	0.00-1.81

Source: [1] Derived from Literature
[2] Data from 140 females and 160 males Wistar, Albino Caesarian derived rats weighing 130-250 g.

Table 6. Hematological Values of Certain Strains of Rats

Test	Unit	Long-Evans (BLU:(LE))	Wistar/Lewis Albino	Osborne-Mendel	Fisher Inbred Strain 344/Cr	African White-tailed (Mastromys albecaudua)
Erythrocytes (RBC)	$(\times 10^6/mm^3)$	5.98-8.30	7.20-9.60	6.26-8.96	6.68-9.15	4.95-11.4
Hemoglobin	(g/dl)	13.1-16.7	12.0-17.5	14.3-17.7	13.4-17.2	10.0-16.8
MCV	(μ^3)	52.0-69.0	57.0-65.0	52.0-66.0	54.0-67.5	45.1-60.0
MCH	$(\mu\mu g)$	18.5-23.5	14.6-21.3	18.8-23.3	17.0-21.8	12.2-20.5
MCHC	(%)	32.0-38.5	26.0-38.0	32.0-42.0	26.0-35.5	22.0-40.0
Hematocrit (PCV)	(ml%)	39.0-48.0	42.5-49.4	39.4-46.2	46.0-52.5	37.0-49.0
Leukocyte (WBC)	$(\times 10^3/mm^3)$	3.30-7.90	5.00-8.96	6.23-12.6	5.35-11.2	4.0-12.2
Neutrophils	(%)	5.50-35.5	9.0-34.0	4.50-23.5	11.5-41.6	17.5-29.0
Basophils	(%)	0	0.00-1.50	0	0	0.00-2.0
Lymphocytes	(%)	60.0-93.5	65.0-84.5	72.0-94.0	43.0-79.5	62.5-80.0
Monocytes	(%)	0.00-5.50	0.00-5.00	0.50-3.50	0.00-2.00	0.50-8.00
Eosinophils	(%)	0.00-1.50	0.00-2.50	0.00-1.00	0.00-4.00	0.00-3.50

Source: Godwin et al. (1964); Burns and DeLannoy (1966); Hall et al. (1967); Schermer (1967); Kozma et al. (1969); Burns et al. (1971); Strett (1971); Benirschke et al. (1978); Melby and Altman (1976).

Table 6 . (continued) Hematological Values of Certain Strains of Rats

Test	Unit	Long-Evans	Wistar (caesarian)	Cotton rat (Sigmadon hispidus)	Rice rat (Oryzomys palustris)
Erythrocytes (RBC)	$(\times 10^6/mm^3)$	6.96-9.56	4.95-11.4	7.70-7.82	8.50-8.78
Hemoglobin	(g/dl)	13.6-16.8	10.0-16.8	15.1-15.9	15.3-15.9
MCV	(μ^3)	46.1-68.5	-	-	-
MCH	$(\mu\mu g)$	15.9-21.1	-	-	-
MCHC	(%)	29.8-34.2	-	-	-
Hematocrit (PCV)	(%)	43.0-51.8	37.0-49.0	44.5-46.9	45.1-47.1
Leukocyte (WBC)	$(10^3/mm^3)$	5.94-10.7	4.00-24.5	5.97-6.61	4.57-5.41
Neutrophils	(%)	9.70-40.1	-	-	-
Basophils	(%)	0.00-0.28	-	-	-
Lymphocytes	(%)	50.1-86.5	-	-	-
Monocytes	(%)	0.70-5.30	-	-	-
Eosinophils	(%)	0.50-6.70	-	-	-

Table 6. (continued) Hematological Values of Certain Strains of Rats

Test	Unit	SPF/Sprague-Dawley (C.F. New York)	SPF/Wistar hooded Rattus norvegicus	Sprague-Dawley
Erythrocytes (RBC)	$(\times 10^6/mm^3)$	6.26-8.96	5.98-8.30	7.21-8.45
Hemoglobin	(g/dl)	13.4-16.4	13.1-16.7	14.0-15.6
MCV	(μ^3)	52.0-66.0	52.0-69.0	52.6-65.4
MCH	$(\mu\mu g)$	-	-	16.5-21.3
MCHC	(%)	32.0-33.5	32.0-35.5	30.3-34.7
Hematocrit (PCV)	(%)	40.0-49.0	39.0-48.0	41.1-51.1
Leukocyte (WBC)	$(\times 10^3/mm^3)$	9.40-14.9	3.30-7.90	7.30-12.7
Neutrophils	(%)	4.50-23.5	5.50-35.5	8.50-40.5
Basophils	(%)	-	-	0.00-0.36
Lymphocytes	(%)	72.0-94.5	60.0-93.5	53.7-88.5
Monocytes	(%)	0.50-3.50	0.00-5.50	0.50-4.50
Eosinophils	(%)	-	-	0.40-3.00

Table 7. Hematological Values of the Normal Hamster*

Test	Unit	Male (84 animals)			Female (80 animals)		
		Mean	S.D.	Range	Mean	S.D.	Range
Erythrocytes (RBC)	$(\times 10^6/\text{mm}^3)$	7.50	1.40	4.70-10.3	6.96	1.50	3.96-9.96
Hemoglobin	(g/dl)	16.8	1.20	14.4-19.2	16.0	1.45	13.1-18.9
MCV	(μ^3)	70.0	3.19	64.8-77.6	70.0	3.00	64.0-76.0
MCH	$(\mu\mu\text{g})$	22.4	1.27	19.9-24.9	23.0	1.40	20.2-25.8
MCHC	(%)	32.0	2.23	27.5-36.5	32.6	2.40	27.8-37.4
Hematocrit (PCV)	(ml%)	52.5	2.30	47.9-57.1	49.0	4.90	39.2-58.8
Sedimentation rate	(mm/hr)	0.64	0.16	0.32-0.96	0.50	0.10	0.30-0.70
Platelets	$(\times 10^3/\text{mm}^3)$	410.	75.0	367.-573.	360.	121.	300.-490.
Leukocytes (WBC)	$(\times 10^3/\text{mm}^3)$	7.62	1.30	5.02-10.2	8.56	1.54	6.48-10.6
Neutrophils	$(\times 10^3/\text{mm}^3)$	1.68	0.29	1.11-2.25	2.48	0.30	1.88-3.08
	(%)	22.1	2.50	17.1-27.1	29.0	3.12	22.8-35.2
Eosinophils	$(\times 10^3/\text{mm}^3)$	0.07	0.02	0.04-0.12	0.06	0.01	0.04-0.08
	(%)	0.90	0.32	0.26-1.54	0.70	0.24	0.22-1.18
Basophils	$(\times 10^3/\text{mm}^3)$	0.08	0.01	0.05-0.10	0.04	0.01	0.03-0.05
	(%)	1.00	2.00	0.00-5.00	0.50	0.70	0.00-2.10
Lymphocytes	$(\times 10^3/\text{mm}^3)$	5.60	0.85	3.69-7.51	5.81	0.70	4.41-7.20
	(%)	73.5	9.40	54.7-92.3	67.9	8.52	50.9-84.9
Monocytes	$(\times 10^3/\text{mm}^3)$	0.19	0.03	0.12-0.26	0.20	0.02	0.16-0.25
	(%)	2.50	0.80	0.90-4.10	2.40	1.00	0.40-4.40

*Syrian, golden hamsters, weighing 75-100 g.

Table 8. Hematological Values of the Normal Hamster

Test	Unit	Benirschke et al (1978)[1] Mean	S.D.	Melby and Altman (1976)[1] Mean	% C.V.	Mitruka and Rawnsley (1977)[2] Range
Erythrocytes (RBC)	$(10^6/mm^3)$	7.50	2.40	7.20	12.00	3.96-10.3
Hemoglobin	(g/dl)	16.8	1.20	14.8	12.00	13.1-19.2
MCV	(μ^3)	71.2	3.19	64.0	-	64.0-77.6
MCH	(μμg)	22.3	1.27	21.0	-	19.9-25.8
MCHC	(%)	32.0	2.23	32.0	-	27.8-37.4
Hematocrit (PCV)	(%)	52.5	2.30	46.0	9.00	32.9-58.8
Sedimentation rate	(mm/hr)	-	-	-	-	0.30-0.96
Platelets	$(x10^3/mm^3)$	310	62.8	300	-	300 - 573
Leukocytes (WBC)	$(x10^3/mm^3)$	7.62	1.30	6.30	31.00	5.02-10.6
Neutrophils	$(x10^3/mm^3)$	-	-	1.70	-	1.11-3.08
	(%)	21.9	5.50	27.0	-	17.1-35.2
Eosinophils	$(x10^3/mm^3)$	-	-	0.06	-	0.04-0.12
	(%)	1.10	0.02	1.10	-	0.22-1.54
Basophils	$(x10^3/mm^3)$	-	-	-	-	0.03-0.10
	(%)	-	-	0.00	0.00	0.00-5.00
Lymphocytes	$(x10^3/mm^3)$	-	-	4.30	-	3.69-7.51
	(%)	73.5	9.40	68.0	13.0	50.9-92.3
Monocytes	$(x10^3/mm^3)$	-	-	0.19	-	0.12-0.26
	(%)	2.50	0.80	2.90	30.0	0.40-4.40

Source: [1]Derived from Literature
[2]Data from 80 female and 84 male Syrian, Golden Hamsters weighing 75-100 g.
See text for methodology.

nation, leukocyte counts have been reported to be low (Schermer, 1967), averaging about 2,500/mm³ with a neutrophil:lymphocyte ratio of 45%:45%. When the animal is awake (during the night), the neutrophil count varies between 17 and 35%. Neutrophils in the hamster have a diameter of 10–12 μ, annular (or less regular) pyknotic nuclei, and a cytoplasm with moderately dense pink granulation of round or rod forms (like pseudoeosinophils or specially granulated cells). Basophils and eosinophils are rare and monocytes vary from 0.5 to 5%.

In general, the leukocyte picture of the hamster resembles that of other rodents. Comparison of the hematological values of European hamsters, Chinese hamsters, and Mongolian gerbils reveals a marked similarity of nearly all values in females and males (Table 9). However the total number of leukocytes in gerbils is higher (6.51–21.6 × 10³/mm³) compared to that of hamsters (Ruhren, 1965).

Guinea Pigs

Hematological values in guinea pigs show considerable physiological variations, but factors such as sampling at different times of day, feeding, different breeds and sexes seem to have many significant effects on the normal blood picture. Table 10 presents the basic hematological parameters of normal albino guinea pigs. Erythrocytes are in the range of 3.5 to 7 million/mm³; females have slightly fewer erythrocytes than males. The erythrocytes are moderately anisocytotic, with a corpuscle diameter of 6.6–7.9 μ. Polychromatic erythrocytes amount to about 1.5% in adult animals, about 4.5% in young animals, and about 25% in neonates (Schermer, 1967; Albritton, 1958). Hemoglobin values vary from 11.0 to 17.0 g/dl. White blood cells show considerable physiological variations; however, as in other rodent species, lymphocytes predominate

(45–80%). Many small lymphocytes are found to be similar in size to erythrocytes.

Kurloff bodies are distinctive structures found in large lymphocytes, particularly in the splenic sinusoids, but also in small numbers in the peripheral blood of guinea pigs. The Kurloff bodies or Kurloff corpuscles (KC) are always found in the cells of striking similarity, all of which resemble large lymphocytes. The nucleus is invariably elongated, only slightly pyknotic, and it often seems depressed and pushed aside by the KC. The KC varies in size, but the typical KC occupies the larger part of the cellular body of which only a small zone resembling a signet ring remains visible. It is always seen as an unstained round or oval vacuole in which deep dark violet cocci and short rod-like structures are visible at varying densities. The rod-like structure often resembles the azure granules of the large lymphocytes in form and color. Further details on KCs can be found in the literature on this subject (Andrew, 1965; Schermer, 1967).

Neutrophils, as in rabbits, are markedly granulated and are therefore often referred to as pseudoeosinophils. They vary from 20 to 55%, which is in agreement with the values reported by Dumas (1953) and Albritton (1958). Schermer (1967) found that neutrophils in guinea pigs range from 18 to 35%. Eosinophils vary from 1 to 7% and are larger in diameter (10–15 μ) than neutrophils; pseudoeosinophils have a diameter of 10–12 μ. The nuclei of eosinophils are less pyknotic and less segmented, but they usually show many indentations.

Basophils in guinea pigs' blood are characteristic because the form and thickness of their granulation are seldom found in other animals. They are the size of neutrophils, sometimes even larger, and have a lobulated rather homogeneously stained nucleus. Monocytes vary from 1 to 6% and are generally larger than the large lymphocytes. Comparative hematological values of the normal guinea pigs are presented in Table 11.

Table 9. Comparative Hematological Values of Certain Strains of Hamsters and Gerbils

Test	Unit	European hamster (Cricetus cricetus)	Chinese hamster (Cricetulus griseus)	Mongolian gerbil (Meriones unguiculatus)
Erythrocytes (RBC)	$(x10^6/mm^3)$	6.04-9.10	4.40-9.10	7.87-9.97
Hemoglobin	(g/dl)	13.4-15.5	10.7-14.1	15.2-16.8
MCV	(μ^3)	58.7-71.4	53.6-65.2	46.6-60.0
MCH	$(\mu\mu g)$	18.6-22.5	15.5-19.1	16.1-19.4
MCHC	(%)	26.4-32.5	27.0-32.0	30.6-33.3
Hematocrit (PCV)	(ml%)	44.0-49.0	36.5-47.7	46.0-52.0
Leukocytes (WBC)	$(x10^3/mm^3)$	3.40-7.60	2.70-9.60	6.51-21.6
Neutrophils	(%)	3.50-41.6	14.8-23.6	2.00-23.0
Basophils	(%)	0.00-0.20	0.00-0.50	0.00-1.00
Eosinophils	(%)	0.00-2.10	0.30-3.10	0.00-4.00
Lymphocytes	(%)	50.0-95.0	68.1-84.8	73.0-97.0
Monocytes	(%)	0.00-1.00	0.00-2.40	0.00-3.00

71

Source: Ruhren (1965); Burns and DeLannoy (1966); Moore (1966); Schermer (1967); Mays (1969); Dontewill (1974); Dillon and Glomski (1975).

Table 10. Hematological Values of the Normal Albino Guinea Pig*

Test	Unit	Male (110 animals)			Female (95 animals)		
		Mean	S.D.	Range	Mean	S.D.	Range
Erythrocytes (RBC)	$(\times 10^6/mm^3)$	5.60	0.62	4.36-6.84	4.75	1.20	3.35-6.15
Hemoglobin	(g/dl)	14.4	1.38	11.6-17.2	14.2	1.42	11.4-17.0
MCV	(μ^3)	77.0	3.00	71.0-83.0	91.0	2.45	86.1-95.9
MCH	$(\mu\mu g)$	25.7	0.75	24.2-27.2	29.7	0.80	23.1-26.3
MCHC	(%)	34.3	2.28	29.7-38.9	31.3	1.55	28.2-34.4
Hematocrit (PCV)	(ml%)	42.0	2.50	37.0-47.0	45.4	2.25	40.9-49.9
Sedimentation rate	(mm/hr)	1.50	0.20	1.10-1.90	1.70	0.25	1.20-2.20
Platelets	$(\times 10^3/mm^3)$	500.	120.	260.-740.	450.	92.0	266.-634.
Leukocytes (WBC)	$(\times 10^3/mm^3)$	11.5	3.00	5.50-17.5	10.8	2.80	5.20-16.4
Neutrophils	$(\times 10^3/mm^3)$	4.83	1.26	2.31-7.35	3.36	0.87	1.62-5.10
	(%)	42.0	7.00	28.0-56.0	31.1	5.40	20.3-41.9
Eosinophils	$(\times 10^3/mm^3)$	0.44	0.11	0.22-0.70	0.38	0.10	0.18-0.57
	(%)	4.00	1.50	1.00-7.00	3.50	1.75	0.00-7.00
Basophils	$(\times 10^3/mm^3)$	0.07	0.02	0.04-0.12	0.02	0.01	0.01-0.03
	(%)	0.70	0.50	0.00-1.70	0.20	0.30	0.00-0.80
Lymphocytes	$(\times 10^3/mm^3)$	5.64	1.47	2.70-8.58	6.85	1.78	3.30-10.4
	(%)	49.0	6.75	40.0-62.5	63.4	8.50	46.4-80.4
Monocytes	$(\times 10^3/mm^3)$	0.49	0.12	0.24-0.75	0.19	0.05	0.09-0.30
	(%)	4.30	0.50	3.30-5.30	1.80	0.40	1.00-2.60

*Hartley strain, ranging in weight from 500 to 800 g.

Table 11. Hematological Values of the Normal Guinea Pig

Test	Unit	Coles[1] (1980) Range	Benirschke[1] et al. (1978) Mean	S.D.	Benjamin[1] (1978) Range	Melby and[1] Altman (1976) Mean	% C.V.	Mitruka &[2] Rawnsley (1977) Range
Erythrocytes (RBC)	($\times 10^6$/mm^3)	5.00-8.00	4.92	0.54	5.00-8.00	5.40	12.0	5.49-8.69
Hemoglobin	(g/dl)	10.0-16.0	12.4	1.30	10.0-16.0	13.4	12.0	11.4-13.5
MCV	(μ^3)	50.0-67.0	84.1	4.50	50.0-68.0	81.0	-	54.6-62.0
MCH	($\mu\mu$g)	-	-	-	16.6-22.0	25.0	-	16.5-18.8
MCHC	(%)	30.0-34.0	30.1	1.20	30.0-34.0	30.0	-	26.1-34.1
Hematocrit (PCV)	(%)	32.0-50.0	41.2	3.60	32.0-50.0	43.0	10.0	37.5-44.2
Sedimentation rate	(mm/hr)	1.00-14.0	-	-	-	-	-	2.30-8.10
Platelets	($\times 10^3$/mm^3)	296 - 616	530	149	-	-	-	161 - 368
Leukocytes (WBC)	($\times 10^3$/mm^3)	10.0-14.0	11.2	2.85	11.0-22.0	9.90	30.0	7.80-20.7
Neutrophils	($\times 10^3$/mm^3)	0.00-5.00	2.70	13.0	0.00-4.00	3.90	-	2.26-8.07
	(%)	28.0-47.0	-	-	28.0-47.0	38.0	-	21.7-47.7
Eosinophils	($\times 10^3$/mm^3)	-	-	-	-	0.38	-	0.39-1.03
	(%)	1.00-11.0	2.20	2.20	0.00-11.0	3.50	-	2.10-7.80
Basophils	($\times 10^3$/mm^3)	-	-	-	-	0.00	-	0.12-0.41
	(%)	0.00-2.00	0.21	0.28	0.00-2.00	0.30	-	0.60-2.70
Lymphocytes	($\times 10^3$/mm^3)	-	-	-	-	5.40	-	4.59-12.1
	(%)	39.0-60.0	67.8	14.4	39.0-52.0	55.0	-	41.3-68.5
Monocytes	($\times 10^3$/mm^3)	-	-	-	-	0.30	-	0.34-1.07
	(%)	2.00-10.0	3.10	3.10	2.00-10.0	2.70	-	2.46-5.84

Source: [1]Derived from literature
[2]Data from 95 female and 110 male
Hartley Albino guinea pigs weighing 50 to 500 g. See text for methodology.

Rabbits

The erythrocyte count in rabbits varies between 5 and 8 million/mm^3 and shows marked anisocytosis. The diameter of the rabbit erythrocyte is 6.5 to 7.5 μ (Albritton, 1952). A characteristic of rabbit blood is the occurrence of numerous thorn apple forms in the smear, and polychromasia is found in 1–2% of erythrocytes. The erythrocytes include a large number of reticulocytes (1–7%) in adult animals and higher numbers in the young. Erythrocytes and hemoglobin concentration are slightly higher in male rabbits than in females.

The leukocyte count in rabbits fluctuates considerably because of rhythmic diurnal fluctuations, nutritional variation, and differences in age, sex, and breed. The differential count of one normal rabbit may fluctuate considerably when counted over a period of 1 month. Our data show leukocyte counts of 5,000–12,500/mm^3 in males, and 5,200–10,600/mm^3 in females, with lymphocytes 31–52%, neutrophils 36–50%, eosinophils 0.5–5%, basophils 2.5–7.5%, and monocytes 4–13.5% (Table 12). Many authors have reported a large variation, mainly in lymphocytes, due to the age and nutritional condition of the animal (Benirschke *et al.*, 1978; Melby and Altman, 1976; Albritton, 1958; Schalm, 1965; Schermer, 1967) (Table 13). The neutrophils in the rabbit are analogues of the neutrophils in man and are granulated (pseudoeosinophils). Many other mammals and birds also have cells with such granulations in a more or less pronounced form.

Pelger's nuclear anomaly of leukocytes, which is characterized by a lack of the segmentation normally seen during the maturation of leukocytes, is most apparent in neutrophils (Pelger, 1943). As a result the peripheral blood contains almost exclusively round, single, or poorly segmented nuclei without even a depression. These animals more frequently suffer severe skeletal deformities and a higher mortality.

Data on hematological values of different breeds of rabbits are compared in Table 14. In general, variations due to breed are not as pronounced as those due to the individual physiology, nutrition, age, and sex of the animal.

Opossums

Table 15 lists the normal hematological values of three species of opossums. Rothstein and Hunsaker (1972) found no significant differences between immature and mature opossums in any blood parameter. The normal hematologic values of one species of opossums compare fairly closely with those for related marsupial species and for other mammalian species including humans. A greater difference in individual hematologic values in opossums is attributable to physiological stress and to bacterial and parasitic infestations. Timmons and Marques (1969) identified various internal parasites including *Trichuris* species, *Capilleria* species, *Physaloptera* species and *Entamoeba coli* (cysts) in Virginia opossums. Therefore it is difficult to define a normal opossum. Parasitic infestation may cause increased platelet, lymphocyte, and eosinophil counts. In general, blood cell morphology of the opossum has a striking resemblance to that of humans and other mammals, especially to that of a variety of monkeys described by Schermer (1967). The erythrocytes of the opossum, for example, are almost identical in dimension and form to those of humans and monkeys, and they exhibit almost the same degree of anisocytosis and poikilocytosis.

Other Small Mammalian Species

Table 16 presents the hematologic values of a few mammalian species that have recently been reported to be useful for medical research. A few nucleated red cells

Table 12. Hematological Values of the Normal Rabbit*

Test	Unit	Male (120 animals)			Female (80 animals)		
		Mean	S.D.	Range	Mean	S.D.	Range
Erythrocytes (RBC)	$(\times 10^6/mm^3)$	6.70	0.62	5.46-7.94	6.31	0.60	5.11-6.51
Hemoglobin	(g/dl)	13.9	1.75	10.4-17.4	12.8	1.50	9.80-15.8
MCV	(μ^3)	62.5	2.00	58.5-66.5	63.1	1.92	57.8-65.4
MCH	$(\mu\mu g)$	20.7	1.00	18.7-22.7	20.3	1.60	17.1-23.5
MCHC	(%)	33.5	1.85	30.0-37.0	32.2	1.74	28.7-35.7
Hematocrit (PCV)	(ml%)	41.5	4.25	33.0-50.0	39.8	4.40	31.0-48.6
Sedimentation rate	(mm/hr)	2.00	0.50	1.00-3.00	1.75	0.40	0.95-2.55
Platelets	$(\times 10^3/mm^3)$	480.	88.0	304.-656.	450.	90.0	270.-630.
Leukocytes (WBC)	$(\times 10^3/mm^3)$	9.00	1.75	5.50-12.5	7.90	1.35	5.20-10.6
Neutrophils	$(\times 10^3/mm^3)$	4.14	0.82	2.53-5.75	3.43	0.58	2.26-4.60
	(%)	46.0	4.00	38.0-54.0	43.4	3.50	36.4-50.4
Eosinophils	$(\times 10^3/mm^3)$	0.18	0.04	0.11-0.25	0.16	0.04	0.10-0.21
	(%)	2.00	0.75	0.50-3.50	2.00	0.60	0.80-3.20
Basophils	$(\times 10^3/mm^3)$	0.45	0.08	0.28-0.62	0.34	0.06	0.22-0.46
	(%)	5.00	1.25	2.50-7.50	4.30	0.95	2.40-6.20
Lymphocytes	$(\times 10^3/mm^3)$	3.51	0.68	2.14-4.88	3.30	0.57	2.17-4.43
	(%)	39.0	5.50	28.0-50.0	41.8	5.15	31.5-52.1
Monocytes	$(\times 10^3/mm^3)$	0.72	0.14	0.44-1.00	0.71	0.12	0.47-0.95
	(%)	8.00	2.00	4.00-12.0	9.00	2.20	6.60-13.4

*New Zealand young adult white rabbits weighing 3.00-5.00 kg.

75

Table 13. Hematological Values of the Normal Rabbit

Test	Unit	Benirschke et al.[1] (1978)				Melby&Altman[1] (1976)		Mitruka&Rawnsley[2] (1977)
		1 Year		Adult				
		Mean	S.D.	Mean	S.D.	Mean	%/C.V.	Range
Erythrocytes (RBC)	(x10^6/mm^3)	7.73	0.78	7.79	0.51	6.2	9	5.11–7.94
Hemoglobin	(g/dl)	15.97	1.30	15.95	1.18	13.4	17	9.80–17.4
MCV	(μ^3)	63.62	2.47	61.08	2.45	60.0	6	57.8–65.4
MCH	(μμg)	20.70	1.07	20.48	1.10	23	8	17.1–23.5
MCHC	(%)	32.52	1.04	33.54	1.20	35	6	28.7–37.0
Hematocrit (PCV)	(%)	49.08	3.98	47.58	2.89	39	8	37.0–50.0
Sedimentation rate	(mm/hr)	–	–	–	–	–	–	0.95–3.00
Platelets	(x10^3/mm^3)	447	215	449	179	650	25	270–656
Leukocytes (WBC)	(x10^3/mm^3)	4.91	2.19	7.46	3.15	8.1	23	5.20–12.5
Neutrophils	(%)	4.5	–	–	–	32	34	36.4–54.0
	(x10^3/mm^3)	–	–	–	–	2.7	43	2.26–5.75
Eosinophils	(%)	4.5	3.6	2.2	2.0	1.30	126	0.50–3.50
	(x10^3/mm^3)	–	–	–	–	0.10	101	0.10–0.25
Basophils	(%)	0.4	0.6	0.4	0.5	2.40	195	2.40–7.50
	(x10^3/mm^3)	–	–	–	–	0.19	54	0.22–0.62
Lymphocytes	(%)	54.2	11.6	42.7	19.8	63	23	28.0–52.10
	(x10^3/mm^3)	–	–	–	–	5.20	39	2.14–4.88
Monocytes	(%)	6.3	3.5	4.3	3.0	4.10	142	4.00–13.40
	(x10^3/mm^3)	–	–	–	–	0.29	53	0.44–1.00

Source: [1] Derived from literature.
[2] Data from 80 female and 120 male young adult rabbits. Weighing 3–5 kg. New Zealand white. See text for methodology.

Table 14. Hematological Values of Different Species of Rabbits

Test	Unit	Inbred NZW (Oryctolagus cuniculus)	Dutch belted (Lepus europaeus)	Polish white (Lepus europaeus)	Black-tailed jack rabbit (Lepus californicus californicus)
Erythrocytes (RBC)	($\times 10^6$/mm^3)	4.33-6.82	4.80-6.30	4.60-6.32	6.59-8.56
Hemoglobin	(g/dl)	9.37-13.9	12.2-16.3	11.9-16.3	13.7-17.5
MCV	(μ^3)	57.1-90.0	62.7-88.1	67.7-80.3	57.6-70.0
MCH	($\mu\mu$g)	16.8-24.9	22.0-29.4	22.0-30.1	18.1-23.1
MCHC	(%)	22.9-34.1	28.5-38.1	29.7-40.6	28.8-36.8
Hematocrit (PCV)	(ml%)	31.6-50.0	34.8-48.9	36.7-43.5	42.0-53.0
Platelets	($\times 10^3$/mm^3)	180.-630.	126.-490.	210.-750.	170.-798.
Leukocytes (WBC)	($\times 10^3$/mm^3)	6.00-13.0	4.00-13.0	7.45-13.3	2.20-14.7
Neutrophils	(%)	36.0-52.0	30.0-50.0	16.4-49.9	13.0-81.5
Eosinophils	(%)	0.50-3.50	0.50-5.0	0.13-1.63	0.00-8.00
Basophils	(%)	2.00-7.00	2.00-8.00	1.13-3.63	0.00-1.50
Lymphocytes	(%)	30.0-53.0	28.5-52.5	73.0-90.9	25.0-83.0
Monocytes	(%)	4.00-12.0	2.00-16.0	0.73-3.25	2.00-10.0

Source: Burns and DeLannoy (1966); Schermer (1967); Bushby (1970); Fetters (1972); Weisbroth et al. (1974); Jones (1975); Melby and Altman (1976); Benirschke et al. (1978).

Table 14. (continued)   Hematological Values of Different Species of Rabbits

Test	Unit	Wild Jack Rabbit		NZW
		1 year	Adult	
Erythrocytes (RBC)	$(\times 10^6/mm^3)$	6.17-9.29	6.77-8.81	4.20-7.10
Hemoglobin	(g/dl)	13.4-18.6	13.6-18.3	10.8-16.1
MCV	(μ^3)	58.7-68.6	56.2-66.0	–
MCH	$(\mu\mu g)$	18.6-22.8	18.3-22.7	–
MCHC	(%)	30.4-34.6	31.1-35.9	–
Hematocrit (PCV)	(%)	41.2-57.0	41.8-53.4	36.0-44.0
Platelets	$(\times 10^3/mm^3)$	–	–	–
Leukocytes (WBC)	$(\times 10^3/mm^3)$	2.72-7.10	3.31-10.8	2.63-11.9
Neutrophils	(%)	11.8-57.4	11.4-71.8	11.9-55.1
Eosinophils	(%)	0.90-8.10	0.20-4.20	0.00-2.60
Basophils	(%)	0.00-1.00	0.00-0.90	0.00-6.40
Lymphocytes	(%)	31.0-77.4	22.9-81.3	39.8-85.4
Monocytes	(%)	2.80-9.80	1.30-7.30	0.00-5.40

Table 15. Hematological Values of Certain Species of Opossums

Test	Unit	Virginia opossums (Didelphis virginiana)	Wooly opossums (Caluosmys deribianus)	Pouchless opossums (Marmosa mitis)
Erythrocytes (RBC)	$(\times 10^6/mm^3)$	3.40-5.50	3.50-5.70	5.70-7.10
Hemoglobin	(g/dl)	12.1-16.2	10.1-16.4	12.2-19.8
MCV	(μ^3)	80.5-106.	65.4-104.	67.9-94.3
MCH	$(\mu\mu g)$	25.8-36.5	21.6-37.2	19.7-30.2
MCHC	(%)	29.0-39.5	25.7-42.3	23.9-38.8
Hematocrit (PCV)	(ml%)	35.6-47.0	30.0-48.0	44.0-58.0
Platelets	$(\times 10^3/mm^3)$	400.-1050	235.-995.	780.-1235
Leukocytes	$(\times 10^3/mm^3)$	3.00-15.5	4.50-23.8	8.90-26.9
Neutrophils	(%)	15.0-65.5	22.8-44.4	15.0-39.4
Basophils	(%)	0.00-2.00	0.00-3.80	0.00-0.75
Eosinophils	(%)	2.00-15.0	0.00-22.8	2.0-20.4
Lymphocytes	(%)	20.0-76.0	19.0-92.0	18.0-80.0
Monocytes	(%)	2.00-6.50	0.00-5.00	0.50-5.20

Source: Didisheim et al. (1959); Youatt (1961); Lewis et al. (1967); Mays and Loew (1968); Sherwood (1969); Timmons (1969); Wolf et al. (1971); Rothstein and Hunsaker (1972).

Table 16. Comparative Hematological Values of Other Small Mammalian Species

Test	Unit	Chinchilla (Chinchilla laniger)	Mink (Mustela vision)	Racoon (Procyon lotor)	Armadillo (Dasypus novemcinctus)
Erythrocytes (RBC)	$(\times 10^6/mm^3)$	5.20-10.3	6.90-11.1	6.25-9.20	4.79-6.92
Hemoglobin	(g/dl)	8.00-15.4	16.4-24.2	9.20-13.6	12.9-14.1
MCV	(μ^3)	32.1-69.2	49.0-76.1	38.4-59.6	58.0-84.0
MCH	$(\mu\mu g)$	10.4-19.8	18.0-28.4	11.2-17.8	20.0-27.4
MCHC	(%)	20.0-38.5	29.0-42.8	24.0-36.0	26.4-36.5
Hematocrit (PCV)	(ml%)	25.0-54.0	44.7-68.6	30.0-46.0	40.0-49.0
Leukocytes (WBC)	$(\times 10^3/mm^3)$	1.60-45.1	8.4-14.8	8.7-12.9	4.90-10.2
Neutrophils	(%)	1.00-78.0	54.0-82.0	37.6-56.4	52.0-68.0
Basophils	(%)	0.00-11.0	0.00-0.60	- - -	3.00-8.00
Eosinophils	(%)	0.00-9.00	0.00-1.20	- - -	3.00-16.0
Lymphocytes	(%)	19.0-98.0	21.9-32.8	40.5-61.5	11.0-26.0
Monocytes	(%)	0.00-5.00	0.00-1.00	1.50-2.60	6.00-10.0

Source: Kennedy (1935); Newberne (1953); Albritton (1961); Casella (1963); Lewis and Doyle (1964); Anderson and Benirschke (1966); Blair (1970); Strike (1970); Altman and Dittmer (1974).

(1–2%) are seen in the peripheral blood of these animals, especially in the chinchilla and mink. The presence of nucleated red cells in peripheral blood usually indicates an acute or chronic condition reflecting a temporary physiological problem or the early stages of a disease (Strike, 1970). Each of the blood values listed in Table 16 is not significantly different from species to species. Although there are physiological variations, the MCHC values of each animal species are similar—a fact indicating that despite other variance, all of the species represented have roughly the same amount of hemoglobin concentration per erythrocyte volume.

Chickens

Erythrocytes and other cellular components of the blood of chickens are nucleated like the blood of reptiles, amphibians, and fishes. The RBC count varies from 1.5 to 4.5 million, which is in good agreement with previously reported values (Schermer, 1967; Ross et al., 1978; Lucas and Jamroz, 1961; Schalm, 1965). The cells are elliptical in outline and biconvex in profile with a mean diameter of 12×7.5 μ. The nucleus is oval, usually lies in the center of the cell, and measures $3–7 \times 3–4$ μ. The color of the individual cells, as in lower vertebrates, is yellowish-green.

In the blood of chickens the type of erythrocyte known as the reticulocyte is found. Frequently such cells are present as a fairly high percentage (7–8%) of the red cells. The large number of red cells that can be found in the immature stages of the lower vertebrates are not found in the blood of birds. However polychromatic erythrocytes are normally found in birds of all ages. Erythroblasts are often present in small numbers. The number of erythrocytes in the chicken is influenced by physiological conditions, sex, age, nutrition, laying performance, moulting, and physical exercise.

Adult males have higher erythrocyte counts and slightly lower platelet counts than young chickens (Table 17). In freshly hatched chicks the erythrocyte count is lower, and after 7 days the values of adult chickens are reached.

The hemoglobin concentration varies between 7.40 and 13.1 g/dl. Males have a slightly higher hemoglobin concentration than females; Albritton (1961) indicates a mean of 11.2 g/dl.

The average leukocyte count in birds is about 20,000/mm^3, which is higher compared to counts of humans and other mammalian blood. As in lower vertebrates, they are divided into the two great groups—the nongranular or agranulocytes (lymphocytes and monocytes), and the granular or granulocytes (neutrophils, basophils, and eosinophils). The range of the lymphocyte count is considerable, varying from 40 to 80%, and the size varies from 4.1 to 12.4 μ. The lymphocyte nucleus often appears flattened on the side where a wider zone of the nuclear membrane may appear. The cytoplasm is basophilic. The small lymphocytes are always more numerous and account for over 50% of blood cells. Their appearance does not differ from that of mammalian lymphocytes. In the large lymphocytes the nucleus is usually round, but depressions and S-shaped nuclei may occur. The pyknosis is less pronounced than that in the small lymphocytes. The cytoplasm has a strong blue color, and it seems to contain vacuoles which give it a foam-like structure.

The average monocyte is larger than the large lymphocyte, but the range in the size of the two cell types does overlap. The monocytes often show delicate bluish granules, whereas lymphocytes sometimes contain a few large, darker colored granules called magenta bodies. Unlike the cytoplasm of lymphocytes, the cytoplasm of monocytes frequently has a slightly bluish color in the reticulum.

Neutrophils (pseudoeosinophils or het-

Table 17. Hematological Values of the Normal Chicken*

Test	Unit	Male (52 animals)			Female (46 animals)		
		Mean	S.D.	Range	Mean	S.D.	Range
Erythrocytes (RBC)	$(\times 10^6/mm^3)$	3.50	0.30	2.90-4.10	2.70	0.56	1.58-3.82
Hemoglobin	(g/dl)	10.3	1.40	7.50-13.1	9.80	1.20	7.40-12.2
MCV	(μ^3)	102.	1.75	100.-128.	122.	2.34	102.-129.
MCH	$(\mu\mu g)$	29.4	2.00	25.4-33.4	36.3	2.20	31.9-40.7
MCHC	(%)	28.9	1.80	25.3-32.5	29.9	2.00	25.9-33.9
Hematocrit (PCV)	(ml%)	35.6	4.82	26.0-45.2	32.8	3.95	24.9-40.7
Sedimentation rate	(mm/hr)	2.50	0.92	0.66-4.34	2.90	1.10	0.80-5.10
Platelets	$(\times 10^3/mm^3)$	31.0	8.50	14.0-48.0	38.0	11.2	15.6-60.4
Leukocytes (WBC)	$(\times 10^3/mm^3)$	20.4	5.32	9.76-31.0	18.9	4.85	9.20-28.6
Neutrophils	$(\times 10^3/mm^3)$	6.43	1.66	3.07-9.76	4.57	1.17	2.23-6.92
	(%)	31.5	6.20	19.1-43.9	24.2	4.30	15.6-32.8
Eosinophils	$(\times 10^3/mm^3)$	1.63	0.42	0.78-2.48	1.37	0.35	0.67-2.07
	(%)	8.00	0.84	6.34-9.66	7.25	0.50	6.25-8.25
Basophils	$(\times 10^3/mm^3)$	0.87	0.23	0.41-1.32	0.74	0.19	0.36-1.12
	(%)	4.25	0.55	3.15-5.36	3.90	0.70	2.50-5.30
Lymphocytes	$(\times 10^3/mm^3)$	11.4	2.98	5.45-17.3	12.1	3.09	5.91-18.4
	(%)	55.8	5.45	43.9-67.7	64.2	8.50	47.2-81.2
Monocytes	$(\times 10^3/mm^3)$	0.10	0.02	0.05-0.16	0.08	0.02	0.04-0.12
	(%)	0.50	0.22	0.06-0.94	0.42	0.18	0.06-0.78

*White Leghorn, young adults weighing 1.5-2.5 kg.

erophils) vary greatly in their staining characteristics. The specific granules are eosinophilic and are often in the form of rods. The average diameter of heterophils in the chicken is about 8.7 μ. The nucleus is segmented and has up to five lobes. The number of heterophils varies between 15 and 45% of the total leukocytes. Eosinophils are about the same size as the pseudoeosinophils (neutrophils or heterophils); however they stain more intensely and the granules are round and almost always the same size. The granules of eosinophils are purple, whereas those of pseudoeosinophils are yellow-red. Also, eosinophil granules give a strong oxidase reaction. The number of eosinophils in the blood of chickens is less than 10% of the total leukocyte count.

Basophils are also the same size as the neutrophils (about 8.2 μ diameter), but they have a large round nucleus which is completely concealed by the granulation. Basophils have a ground cytoplasm which is almost colorless. The granules have a marked affinity for basic dyes and are metachromatic. Their number in the chicken is less than 2% of total leukocyte count, but in some species of birds, such as in pheasants, there may be as many as 10%.

The thrombocytes (platelets) in chickens vary from 14,000 to 60,000/mm³ of blood, and there are slightly more in females than in males (Table 17). The typical thrombocyte is a cell slightly smaller than the erythrocyte, somewhat spindle-shaped but considerably variable in form. The cytoplasm is stained pale blue or purple and contains a few specific granules of a pink to a reddish-purple color. Clotting of the blood of birds proceeds at a rate similar to that in mammals.

The blood morphology of birds does not differ greatly from species to species. Comparative hematological values in four species of birds are presented in Table 18. Literature values on many other species of birds and the values in Table 18 show minor differences from the blood morphology of the chicken. The most reliable data, as in chickens, are those on the erythrocyte count and the hemoglobin concentration. However, greater differences are found with the leukocytes and spindle cells because of physiological variations.

Reptiles, Amphibians, and Fishes

Because of the similarity with the blood morphology of chickens, the blood morphology of reptiles, amphibians, and fishes is only briefly described here. Table 19 gives normal hematological values of representative species.

Reptiles (Snakes and Turtles). In most reptiles spleen and bone marrow are the main hemopoietic organs. The circulating blood of reptiles usually includes many primitive cells. The erythroblasts, present in all stages of development, differentiate finally into rather small erythrocytes with a width of about 5 μ and a length of about 12 μ. The thrombocytes (platelets) are still smaller, only about half the size of the red cells. The lightly basophilic cytoplasm has fine lilac granules. The eosinophils have either rod-shaped or spherical granules. The rod-shaped granulated eosinophils have more complex, polymorphous nuclei. Basophils are small cells containing fragile granules. There are frequently relatively few neutrophils in reptilian blood compared to mammalian blood. In the common box turtle, lymphocytes are numerous in the circulating blood (45–67%) (Table 19). They are similar in morphology to the mammalian lymphocytes. Further details on the blood morphology of reptiles may be found in the works of Bergman (1957), Atland and Kirkland (1962), Heady and Rogers (1962), and Andrews (1965).

Amphibians (Frogs). Avian blood most closely resembles frog blood in that the blood cells are nucleated and contain spin-

Table 18. Hematological Values of Different Species of Birds

Test	Unit	Pigeon (Columba livia)	Duck (Anas Platyrhynchos domesticus)	Turkey (Meleagris gallopavo)	Quail (Coturnix coturnix coturnix)
Erythrocytes (RBC)	$(\times 10^6/mm^3)$	2.13-4.20	1.80-3.82	1.74-3.70	4.00-5.15
Hemoglobin	(g/dl)	10.7-14.9	9.00-21.0	8.80-13.4	10.7-14.3
MCV	(μ^3)	118.-144.	115.-170.	112.-168.	60.0-100.
MCH	$(\mu\mu g)$	32.0-48.0	32.0-71.0	32.0-49.3	23.0-35.0
MCHC	(%)	20.0-30.0	20.1-52.5	23.2-35.3	28.0-38.5
Hematocrit (PCV)	(ml%)	39.3-59.4	32.6-47.5	30.4-45.6	30.0-45.1
Leukocytes (WBC)	$(\times 10^3/mm^3)$	10.0-30.0	13.4-33.2	16.0-25.5	12.5-24.6
Neutrophils	(%)	15.0-50.0	19.3-49.8	29.0-52.0	25.0-49.5
Basophils	(%)	0.00-1.00	0.00-4.50	1.00-9.00	0.00-1.50
Eosinophils	(%)	0.00-1.50	1.60-2.65	0.00-5.00	0.00-15.0
Lymphocytes	(%)	25.0-70.0	13.0-73.5	35.0-48.0	50.0-70.0
Monocytes	(%)	1.00-3.00	0.50-11.5	3.00-10.0	0.50-3.80

Source: Lucas and Jamraz (1961); Schalm (1965); Shellenberger et al. (1965); Schermer (1967); Ernst and Ringer (1968); Atwal et al. (1969); Mirnalan and Robinson (1971); Hamrick (1975).

Table 19. Hematological Values of Reptiles, Amphibians and Fishes

Test	Unit	Snake (Garter) (Eutania sirtales)	Turtle (Box) (Terapene carolina)	Frog (Rana catesbeiana)	Fish (Perch, white) (Morone americana)
Erythrocytes (RBC)	$(\times 10^6/mm^3)$	0.71-1.39	0.61-0.70	0.43-0.45	2.70-3.63
Hemoglobin	(g/dl)	5.80-11.3	5.40-6.20	7.40-8.20	6.70-9.70
MCV	(μ^3)	264.-268.	398.-384.	625.-716.	104.-121.
MCH	$(\mu\mu g)$	78.0-86.0	82.1-100.	174.-184.	25.0-27.0
MCHC	(%)	28.1-34.4	24.8-32.3	26.0-28.0	21.0-26.0
Hematocrit (PCV)	(ml%)	19.0-37.0	22.4-26.9	26.6-32.0	32.7-37.8
Leukocytes (WBC)	$(\times 10^3/mm^3)$	45.2-56.4	24.0-48.0	14.7-21.9	23.1-46.5
Neutrophils	(%)	2.7-5.4	0.00-1.00	5.60-8.40	3.00-6.00
Eosinophils	(%)	- - -	8.80-12.6	21.6-32.4	- - -
Basophils	(%)	0.00-0.60	6.40-9.60	5.60-8.40	0.10-0.50
Lymphocytes	(%)	31.4-47.6	45.1-67.2	47.0-71.0	20.0-40.0
Monocytes	(%)	5.70-8.40	7.60-11.2	1.00-5.00	7.50-23.6

Source: Bergman (1957); Altland and Kirkland (1962); Heady and Rogers (1962); Andrews (1965); Schermer (1967); Altman and Dittmer (1971, 1974).

85

dle cells. Considerable individual and physiological variation exists in the blood morphology of amphibians because of a large number of species and seasonal variations. However many publications on frog blood agree that the red cell count ranges from 0.5 to 0.7 millions/mm³.

The erythrocytes of amphibians are relatively large, and in some species they may be over 70 μ long. But usually they are between 20 to 30 μ long length and 14 to 17 μ wide. The nucleus has the same shape as the entire cells (almost globular), measures about $8 \times 5 \mu$, is pyknotic, and has a clearly marked pattern. During all seasons, but more in summer than in winter, the blood contains numerous polychromatic cells with a rounder and less compact nucleus (Schermer, 1967). There are also erythroblasts, round cells with a large, round, distinct nucleus and a narrow, gray or deep blue homogeneous cytoplasm.

The number of leukocytes in frogs' blood is in the range of 14,000–21,000/mm³. Although the red cells are produced in the spleen and in the bone marrow (particularly following metamorphosis and hibernation), the granulocytes are formed in the bone marrow, in the intestinal mucosa, and in the kidneys. The parent cells of the granulocytes appear to be medium-sized lymphoid hemoblasts or medium-sized lymphocytes.

Basophils and eosinophils are numerous. Basophils are sphero-nuclear cells whereas the eosinophils have lobed nuclei. There are fewer neutrophils than eosinophils, and they are analogues of mammalian neutrophils. Their average diameter is about 15 μ, and they are polymorphonuclear in that some nuclei are segmented but others are not segmented. Juvenile forms are relatively numerous, and typical myelocytes with round nuclei and basophilic cytoplasm also occur. The neutrophil count is subject to considerable individual variation. The lowest counts occur in hibernating frogs.

Both large and small lymphocytes are found in frogs' blood. The small lymphocytes (8–12 μ in diameter) resemble those of mammals. Their cytoplasmic zone is very narrow; it often covers only one side of the nucleus like a cap and has a slightly granular, dark blue color. The nucleus is pyknotic. The large lymphocytes have a broader zone of cytoplasm and a round, less pyknotic nucleus. They are often difficult to differentiate from monocytes. The nucleus of the large lymphocyte is usually in a central position and occupies half the cell.

There are very few monocytes in frog blood. They are the largest cell, with the diameter of about 18–20 μ, and contain large amounts of cytoplasm. The cytoplasm is basophilic and appears as a finely webbed or reticulated structure, sometimes with small vacuoles.

Fishes. The erythrocytes of the great majority of fishes are flattened, oval, nucleated cells, and in this respect they resemble those of all nonmammalian blood. The number of erythrocytes in the blood of adult fishes varies from 0.5 to 5 million/mm³. The size varies considerably in different species, but in the individuals of any one species, the numbers and size of erythrocytes are quite constant. There are a number of immature cells in the blood circulation with fine nuclei and a smaller content of hemoglobin. Occasionally mitotic figures are seen in such cells. Very rarely, rounded anucleate erythrocytes are found.

The white blood cells are numerous in fishes, ranging from 20,000 to 50,000/mm³. In some species they are as high as 100,000/mm³. Most of the leukocytes are lymphocytes and monocytes, but a few neutrophils and basophils are also present. Among the white blood cells, the lymphocytes may be considered the most primitive. The lymphocyte has a round nucleus with one or more nucleoli, and basophilic, usually scanty cytoplasm, azure granules, and a negative oxidase reaction.

Cats

The average red cell count of the cat is about 7.3 million/mm³ and ranges between 5 and 9 million (Table 20). Schalm (1965) reported the average red cell count of the cat as 7.5 million and the range 5.5–10.0 million/mm³. The excitable nature of the cat is related to a sharp rise in erythrocyte counts and platelet counts under certain stimuli such as the barking of dogs. The spleen is known to be the reservoir from which the increased number of blood cells are forced out. Hematology is not significantly affected by the breed or sex of cats. However the influence of age is pronounced. Newborn and very young kittens show considerably lower erythrocyte, leukocyte, monocyte, and hemoglobin values and a much higher reticulocyte count than adult animals. Adult values are reached at the age of 2–3 months. The total leukocyte count is elevated by muscular activity, fear, apprehension, and rage.

The erythrocytes are highly anisocytotic. The diameter of red cells is about 5–7 μ (Albritton, 1961). A common feature in the red cells of the cat is a small spheroid body, which stains blue, near the periphery of the corpuscle. These resemble Howell-Jolly bodies.

According to the literature, the number of leukocytes in the blood of cats varies tremendously, from 5,000 to 46,400 in apparently normal cats. Our data indicate that the mean leukocyte counts are 17,400/mm³ for males and 16,600 mm³ for females. In young cats values ranging from 5,400 to 10,800 (mean 7,600) were reported by Schermer (1967). Recent hematological data reported in literature are given in Table 21.

In the differential count, neutrophils constitute the majority of leukocytes (40–74%). The neutrophils have a diameter of 8–12 μ and a pyknotic nucleus which is often coiled, annular, or horseshoe-shaped with frequent indentations. About 1–8% of neu-

trophils are juvenile forms (stab cells). The granules of neutrophils are very delicate and peroxidase-positive. The lymphocytes amount to 20–45% of the total leukocyte count. The small lymphocytes are slightly larger than erythrocytes. All transitions between large and small lymphocytes are found.

Eosinophils constitute about 6% of the total leukocyte counts of the cat. The eosinophils are slightly larger than the neutrophils (10–17 μ in diameter). The nucleus often shows segmentation (two to three segments), but more usually has a horseshoe shape without the marked indentations of the neutrophils. The eosinophilic granules are nonrefractile and rod-like.

Basophils are very rare in the adult cat, but in the blood of young cats they occur more frequently. Basophils in cats are the size of eosinophils, with a large lobulated nucleus which fills most of the cells as a violet band. The cytoplasm is reddish-gray and contains a few violet granules of varying sizes or sometimes minute vacuoles. Monocytes are about 1–9% of the total leukocyte count. They are the largest blood cells (12–18 μ in diameter), with a markedly depressed nucleus. The zone of cytoplasm is broader than that in the lymphocytes and it stains blue. Its structure is often foam-like.

Dogs

Like the cat, the dog has red cells that increase in number and decrease in individual cell size when the animal followed from birth to adulthood. Adult dogs usually have red cell counts between 5 to 8 million/mm³ of blood (Table 22). The erythrocytes are relatively large, about 6.6–7.7 μ in diameter, and show anisocytosis. The erythrocytes often appear as annular forms with a weakly stained center. Slight variations in shape may be observed and rouleaux formation is commonly present. Nucleated red

Table 20. Hematological Values of the Normal Cat*

Test	Unit	Male (32 animals)			Female (30 animals)		
		Mean	S.D.	Range	Mean	S.D.	Range
Erythrocytes (RBC)	$(\times 10^6/mm^3)$	7.40	0.20	7.00-7.80	7.10	0.90	5.30-8.90
Hemoglobin	(g/dl)	10.7	0.40	9.90-11.5	10.3	0.40	9.50-11.1
MCV	(μ^3)	56.6	1.10	54.5-58.8	56.6	1.20	54.2-59.0
MCH	$(\mu\mu g)$	14.4	0.40	13.6-15.2	14.5	0.50	13.5-15.5
MCHC	(%)	25.8	0.80	24.2-27.4	26.0	0.80	24.4-27.6
Hematocrit (PCV)	(ml%)	41.4	1.20	39.0-43.8	39.6	1.20	37.2-42.0
Sedimentation rate	(mm/hr)	4.00	1.20	1.60-6.40	3.00	1.10	0.80-5.20
Platelets	$(\times 10^3/mm^3)$	235.	65.0	105.-365.	220.	72.0	76.0-364.
Leukocytes (WBC)	$(\times 10^3/mm^3)$	17.4	1.20	15.0-19.8	16.6	1.40	13.8-19.4
Neutrophils	$(\times 10^3/mm^3)$	10.1	0.68	8.75-11.5	9.48	0.80	7.88-11.1
	(%)	58.3	7.80	42.7-73.9	57.1	8.10	40.9-73.3
Eosinophils	$(\times 10^3/mm^3)$	1.20	0.08	1.04-1.37	0.98	0.09	0.81-1.14
	(%)	6.90	1.75	3.40-10.4	5.90	1.90	2.10-9.70
Basophils	$(\times 10^3/mm^3)$	0.00	0.02	0.00-0.10	0.02	0.01	0.01-0.02
	(%)	0.00	0.20	0.00-0.60	0.10	0.20	0.00-0.60
Lymphocytes	$(\times 10^3/mm^3)$	5.57	0.39	4.80-6.34	5.40	0.46	4.48-6.30
	(%)	32.0	5.80	20.4-43.6	32.5	4.60	23.3-41.7
Monocytes	$(\times 10^3/mm^3)$	0.49	0.04	0.42-0.55	0.81	0.07	0.68-0.96
	(%)	2.80	0.75	1.30-4.30	4.90	1.84	1.22-8.58

*Mongrels, conditioned, weighing 2-4 kg.

Table 21. Hematological Values of the Normal Cat

Test	Unit	Coles[1] (1980) Range	Benirschke et al.[1] (1978) Mean	Benirschke et al.[1] (1978) Range	Benjamin[1] (1978) Range	Melby & Altman[1] (1976) Mean	Melby & Altman[1] (1976) % C.V.	Mitruka & Rawnsley[2] (1977) Range
Erythrocytes (RBC)	$(\times 10^6/mm^3)$	5.00-10.0	7.50	5.00-10.0	5.50-10.0	5.10	11.0	5.30-8.90
Hemoglobin	(g/dl)	8.00-15.0	12.0	8.00-15.0	8.00-12.0	13.1	13.0	9.50-14.5
MCV	(μ^3)	39.0-55.0	45.0	39.0-55.0	40.0-55.0	82.0	8.00	54.2-59.0
MCH	$(\mu\mu g)$	-	15.5	12.5-17.5	13.0-17.0	26.0	9.00	13.5-16.5
MCHC	(%)	31.0-34.0	33.2	30.0-36.0	31.0-35.0	31.0	7.00	29.2-35.6
Hematocrit (PCV)	(%)	24.0-45.0	37.0	24.0-45.0	24.0-45.0	42.0	10.0	37.2-43.8
Sedimentation rate	(mm/hr)	7.00-23.0	-	-	-	-	-	0.80-6.40
Platelets	$(\times 10^3/mm^3)$	-	-	-	-	300	-	105 - 365
Leukocytes (WBC)	$(\times 10^3/mm^3)$	5.50-19.5	12.5	5.50-19.5	8.00-25.0	13.2	30.0	13.8-19.8
Neutrophils	$(\times 10^3/mm^3)$	-	-	-	-	7.80	35.0	7.88-11.5
	(%)	35.0-75.0	59.0	35.0-75.0	35.0-75.0	59.0	-	40.9-73.9
Eosinophils	$(\times 10^3/mm^3)$	-	-	-	-	0.60	50.0	0.81-1.37
	(%)	2.00-10.0	5.50	2.00-12.0	2.00-12.0	4.60	-	2.10-10.4
Basophils	$(\times 10^3/mm^3)$	-	-	-	-	0.00	0.00	0.00-0.10
	(%)	Rare	0.00	Rare	0.00-0.00	0.00	0.00	0.00-0.60
Lymphocytes	$(\times 10^3/mm^3)$	-	-	-	-	4.50	40.0	4.48-6.34
	(%)	20.0-55.0	32.0	20.0-55.0	20.0-55.0	34.0	-	20.4-43.6
Monocytes	$(\times 10^3/mm^3)$	-	-	-	-	0.33	-	0.40-2.96
	(%)	1.00-4.00	3.00	1.00-4.00	1.00-4.00	2.50	-	1.22-8.58

Source: [1] Derived from literature
[2] Data from 30 female and 32 male conditioned Mongrel cats weighing 2-4 kg. See text for methodology.

89

Table 22. Hematological Values of the Normal Dog*

Test	Unit	Male (18 animals)			Female (10 animals)		
		Mean	S.D.	Range	Mean	S.D.	Range
Erythrocytes (RBC)	$(\times10^6/mm^3)$	6.98	0.60	5.78-8.18	6.69	0.50	5.69-7.69
Hemoglobin	(g/dl)	17.1	0.90	15.3-18.9	16.8	1.00	14.8-18.8
MCV	(μ^3)	77.2	2.25	72.3-81.3	79.4	3.50	72.4-86.4
MCH	$(\mu\mu g)$	24.5	1.00	22.5-26.5	25.1	1.42	22.3-27.9
MCHC	(%)	31.7	1.50	28.7-34.7	31.6	1.00	29.6-33.6
Hematocrit (PCV)	(ml%)	53.9	2.70	48.5-59.3	53.2	3.10	47.0-59.4
Sedimentation rate	(mm/hr)	1.00	3.20	0.00-7.40	0.70	2.00	0.00-4.70
Platelets	$(\times10^3/mm^3)$	400.	105.	190.-610.	380.	90.0	200.-560.
Leukocytes (WBC)	$(\times10^3/mm^3)$	12.9	2.40	8.10-17.7	12.1	2.40	7.34-16.9
Neutrophils	$(\times10^3/mm^3)$	7.90	1.45	5.00-10.8	7.57	1.49	4.59-10.6
	(%)	61.2	14.7	31.8-90.6	62.6	16.5	29.6-95.6
Eosinophils	$(\times10^3/mm^3)$	0.60	0.12	0.37-0.84	0.60	0.12	0.36-0.83
	(%)	4.60	2.70	0.00-10.8	4.94	2.88	0.00-11.5
Basophils	$(\times10^3/mm^3)$	0.04	0.01	0.02-0.05	0.04	0.01	0.02-0.05
	(%)	0.30	0.10	0.10-0.50	0.30	0.10	0.10-0.50
Lymphocytes	$(\times10^3/mm^3)$	3.83	0.72	2.40-5.26	3.50	0.69	2.12-4.88
	(%)	29.7	6.20	17.3-42.1	28.9	7.41	9.06-38.7
Monocytes	$(\times10^3/mm^3)$	0.30	0.06	0.19-0.41	0.30	0.06	0.18-0.42
	(%)	2.32	0.10	2.12-2.52	2.47	0.15	2.17-2.77

*Mongrels, obtained from the pound, weighing 12-16 kg.

cells are numerous in blood smears of neonatal dogs, but they are rarely observed after 40–45 days of age in healthy animals. The dog erythrocyte has a life span of 100 to 120 days. In dogs, the observed erythrocyte sedimentation rate (ESR) is corrected by subtracting it from the anticipated rate of fall, due entirely to the red cell, plasma ratio (i.e., when the packed cell volume is equal to or greater than the plasma volume, no settling out of erythrocyte will occur in 60 minutes with blood from a normal dog). Normal neonatal dogs have low fibrinogen levels and the albumin fraction of the plasma proteins exceeds the globulin fraction. Because of this, negative-corrected ESR values are normal for young growing dogs even though packed cell volume levels are in the low normal range. The reticulocyte count has been reported as 0.2–0.9% in adult dogs by Schermer (1967), and as 0–1.4% by Schalm (1965). Hemoglobin concentration varies considerably (14.8–18.9 g/dl). Dogs fed a diet including all hemopoietic factors, particularly vitamin B-complex, show on the average a higher concentration of hemoglobin. In general, any apparent differences in blood values between the sexes have not been shown to be statistically significant.

The range of the total leukocyte count in dogs has been reported as 5,400–33,000. Our data indicate a mean leukocyte count of 12,900 in males and 12,100 in females (Table 22). Among the leukocytes the differential counts in the literature are fairly consistent, with neutrophils about 70%, lymphocytes 20%, monocytes 4%, eosinophils about 5%, and basophils less than 1%. Our values generally agree with the average differential counts reported in the literature (Table 23).

The diameter of neutrophils is about 10–12 μ. The nucleus has two to five segments but usually forms a narrow ring or horseshoe with indentations or a pearl-string formation. The cytoplasm of the neutrophils contains a fine dust-like granu-

lation that is pale pink or sometimes light blue. The chromatin is coarse and clumped with dark staining clumps alternating with lighter areas. The cytoplasm is similar to that of the band forms. The female sex chromosome appears as a "drumstick" on the nucleus of some of the neutrophils. They consist of a well defined, solid chromatin nodule about 1.5 μ in diameter, which is joined by a single five strand of chromatin to one lobe of the nucleus. As in humans, "drumsticks" were rarely found in immature cells. Dog neutrophils have no alkaline phosphatase activity (Jain, 1969). Eosinophils are somewhat larger than the neutrophils with strong red granulations. The separate granules are always round and of varying size. The pale bluish cytoplasm contains many unstained round vacuoles. The nucleus is broader and less pyknotic than that of neutrophils, but like the neutrophil nucleus it is often twisted and sometimes shows segmentation. Basophils are the size of eosinophils with a lobulated nucleus, and their cytoplasm contains some dark violet granules.

There are large and small lymphocytes, the latter being more prevalent. The small lymphocytes are little larger than the erythrocytes, and the large lymphocytes are similar in size to neutrophils. The nucleus is usually round and sometimes slightly indented; it stains well and is surrounded by a narrow zone of cytoplasm. The cytoplasm sometimes shows dark granulation.

Monocytes are the largest cells in the blood (15–19 μ diameter). They are characterized by a large, usually markedly indented, sometimes lobular and slightly pyknotic nucleus. The protoplasm is light blue or grayish-blue and is often interspersed with five vacuoles that form a foamy structure.

The dog platelet is generally a small, pale blue structure containing a few purplish granules, although variation in size, shape and color are common. Giant forms with abundant irregular cytoplasmic projections

Table 23. Hematological Values of the Normal Dog

Test	Unit	Coles[1] (1980) Mean	Coles[1] (1980) Range	Lumsden et al[2] (1979) Mean	Lumsden et al[2] (1979) S.D.	Lumsden et al[2] (1979) Range	Benirschke et al[1] (1978) Mean	Benirschke et al[1] (1978) S.D.
Erythrocytes (RBC)	($\times 10^6$/mm^3)	6.80	6.00-9.00	6.80	0.70	5.50-8.20	6.91	0.60
Hemoglobin	(g/dl)	15.0	21.0-18.0	16.0	1.70	12.6-19.4	15.9	1.20
MCV	(μ^3)	70.0	60.0-77.0	66.0	2.00	62.0-70.0	71.1	4.00
MCH	($\mu\mu$g)	–	–	23.2	0.80	22.0-25.0	23.0	0.80
MCHC	(%)	33.0	31.0-34.0	34.8	0.80	33.0-36.0	32.3	1.20
Hematocrit (PCV)	(%)	45.0	37.0-54.0	46.0	4.60	36.9-55.0	47.3	3.40
Sedimentation rate	(mm/hr)	–	5.00-25.0	–	–	–	–	–
Platelets	($\times 10^3$/mm^3)	345	164 – 500	320	120	80.0-560	–	–
Leukocytes (WBC)	($\times 10^3$/mm^3)	13.3	6.70-18.3	12.5	3.00	6.60-18.4	14.0	2.27
Neutrophils	($\times 10^3$/mm^3)	–	0.00-4.00	–	–	–	–	–
	(%)	–	60.0-75.0	–	–	–	–	–
Eosinophils	($\times 10^3$/mm^3)	–	–	0.50	0.50	0.00-1.90	–	–
	(%)	–	2.00-10.0	3.60	3.70	0.00-14.5	7.30	3.50
Basophils	($\times 10^3$/mm^3)	–	–	0.00	0.00	0.00-0.20	–	–
	(%)	–	Rare	0.04	0.20	0.00-1.00	0.12	0.40
Lymphocytes	($\times 10^3$/mm^3)	–	0.00-0.00	3.20	1.20	0.80-5.60	–	–
	(%)	–	12.0-30.0	25.5	7.70	10.4-40.6	28.6	7.70
Monocytes	($\times 10^3$/mm^3)	–	–	0.90	0.50	0.10-1.80	–	–
	(%)	–	3.00-9.00	7.60	3.40	1.00-14.3	5.20	2.10

Source: [1] Derived from literature
[2] Data from 26 female and 25 male, 6-24 months old mongrel and purebred dogs.
See text for methodology.

may be occasionally observed. The platelet normal values varies, in one day old beagle Earl *et al.* (1973) reported $302 \times 10^3/\mu l$ with a range of 178 to $465 \times 10^3/\mu l$. At weaning, the count was $371 \times 10^3/\mu l$ with a range of 275 to $570 \times 10^3/\mu l$. The average count is $400 \times 10^3/\mu l$ for 18 month old beagles and $300 \times 10^3/\mu l$ for 10 year old beagles. Earl *et al.* (1973) reported platelet counts of 210 to $612 \times 10^3/\mu l$ for adult beagles with an average of $411 \times 10^3/\mu l$. Schalm *et al.* (1975) found the normal range of platelet counts in dogs to be 200 to $500 \times 10^3/\mu l$ of blood. Our data indicate that the platelet count in male Mongrel dogs are $190 - 610 \times 10^3/\mu l$ and $200 - 560 \times 10^3/\mu l$ in female dogs. Variations in hematologic values are expected in different breeds of the dog. However, most values are within the physiological variations and no unusual differences have been reported in the blood values of different breeds of dogs. Table 24 compares hematological values of four different breeds of dogs reported in the literature.

Monkeys

The red cell count in the Rhesus monkey varies between 3.3 and 7.0 million/mm³ of blood (Table 25). The average red cell diameter is 7.5 μ. According to Albritton (1961) the diameter is 7.4 in Rhesus monkeys, and 7.1–7.5 μ in other species. Polychromatic erythrocytes are always found in the monkey's blood. The reticulocyte count, however, is low—0.3–0.7% in adult animals. The hemoglobin concentration is between 10.2 and 16.6 g/dl, which agrees with the values reported in the literature (Table 26).

The leukocyte counts reported in the literature vary from 1,300 to 40,000/mm³; our values range between 6,000 and 16,000/mm³. The neutrophil count fluctuates considerably but is always below the lymphocyte count. Neutrophils are large, round cells, 12–14 μ in diameter, with a pyknotic,

twisted, indented, and often segmented nucleus. The cytoplasm is closely packed with delicate pink granules of the rod type. Eosinophils are difficult to distinguish from neutrophils. The granules of eosinophils are twice as large as in the neutrophils and they are always round and show a much brighter yellow red color. The basic color of the cytoplasm is gray-blue and it is darker than the cytoplasm of neutrophils. The eosinophil count varies from 0 to 8% of the total leukocyte count.

Basophils are the same size as neutrophils. They have a broad horseshoe or annular nucleus which fills the large part of the cell. The cytoplasm has an irregular basic dull gray color and seems closely packed with pale gray rods. The basophil count ranges from 0 to 2% of the leukocyte count. Lymphocytes comprise the largest percentage of the leukocytes in monkey's blood, 25–91%. The small lymphocytes are larger than the erythrocytes. They have a round, sometimes slightly depressed nucleus with a lumpy pyknotic aspect and a highly basophilic cytoplasm which becomes lighter toward the nucleus. The large lymphocytes are only slightly larger than the small lymphocytes. Their nucleus and plasma seem lighter and they may contain azure granules. Monocytes resemble the monocytes of most mammals. However the monocyte count varies between 0 and 4% of the total leukocytes in monkey's blood.

Individual variation in hematologic values within the monkey species is considerable, particularly among the leukocytes. The range of variation within the species is as wide as that between the species. Tables 27 and 28 present hemotological values of a few selected species of the so-called old world primates and the new world primates. The data are summarized from the vast information available in the literature. On the basis of statistical analysis, most authors concluded that the blood morphology in various monkey species is similar and that there are no significant differences

Table 24. Hematological Values of Different Breeds of Dogs

Test	Unit	Beagle	Laborador Retrievers	Conditioned Mongrel	Hetrogeneous Population (Mongrel stock dogs)
Erythrocytes (RBC)	$(\times 10^6/mm^3)$	5.38-7.78	4.76-8.16	5.02-8.00	4.33-8.34
Hemoglobin	(g/dl)	13.8-18.6	12.3-18.7	11.6-17.6	10.0-13.6
MCV	(μ^3)	63.7-76.5	55.4-84.5	51.7-88.2	54.0-80.7
MCH	$(\mu\mu g)$	20.9-28.3	18.9-28.8	17.8-27.0	15.8-21.3
MCHC	(%)	29.9-40.3	27.2-41.4	25.5-38.7	23.5-31.9
Hematocrit (PCV)	(ml%)	41.9-50.3	35.8-54.6	33.7-57.4	34.2-51.1
Leukocytes (WBC)	$(\times 10^3/mm^3)$	12.7-15.1	9.00-17.8	9.50-21.3	10.2-21.8
Neutrophils	(%)	38.6-68.8	56.1-80.2	57.9-80.3	57.2-89.4
Eosinophils	(%)	0.00-7.30	0.00-6.60	0.0-6.70	0.00-17.6
Basophils	(%)	0.00-0.60	0.00-0.50	0.00-0.20	0.00-1.34
Lymphocytes	(%)	25.0-51.4	14.9-39.1	14.6-36.2	5.40-33.3
Monocytes	(%)	1.00-7.20	0.00-4.70	0.50-4.50	0.00-8.12

Source: Anderson (1958); Pick and Eubanks (1965); Michaelson et al.(1966); Robinson and Ziegler (1968); Anderson and Schalm (1970); Earl et al.(1973); Secord and Russell (1973).

Table 24. (continued)   Hematological Values of Different Breeds of Dogs

Test	Unit	Basenji	Greyhound	Fox-Hound type
Erythrocytes (RBC)	($\times 10^6/mm^3$)	5.30-7.70	5.20-8.13	4.14-10.7
Hemoglobin	(g/dl)	12.6-18.4	10.5-20.9	11.5-20.0
MCV	(μ^3)	59.0-86.0	50.6-92.4	36.2-125.6
MCH	($\mu\mu g$)	19.1-28.6	16.4-28.9	–
MCHC	(%)	26.1-37.8	19.1-37.3	22.7-48.0
Hematocrit (PCV)	(%)	38.0-55.0	40.0-66.0	48.0-59.0
Leukocytes (WBC)	($\times 10^3/mm^3$)	11.7-16.3	4.00-17.9	6.50-15.3
Neutrophils	(%)	50.7-66.1	43.0-81.0	57.0-74.9
Eosinophils	(%)	3.80-10.8	0.00-7.00	0.00-6.50
Basophils	(%)	0.08-0.52	0.00-1.00	0.00-1.50
Lymphocytes	(%)	20.9-36.3	17.0-50.0	14.8-37.6
Monocytes	(%)	3.10-7.30	0.00-7.00	0.00-8.20

Source: Portkay and Zinn (1968), Porter and Canaday (1971), Ewing et al. (1972)

95

Table 25. Hematological Values of the Norm

Test	Unit	Male (38 animals)			Female (30 animals)		
		Mean	S.D.	Range	Mean	S.D.	Range
Erythrocytes (RBC)	$(\times 10^6/mm^3)$	5.15	0.91	3.33-6.97	5.19	0.78	3.63-6.75
Hemoglobin	(g/dl)	13.2	1.35	10.5-15.9	13.4	1.58	10.2-16.6
MCV	(μ^3)	80.6	9.88	59.8-99.0	83.5	12.7	58.1-99.0
MCH	$(\mu\mu g)$	25.6	3.17	19.3-31.9	25.8	4.15	17.5-34.1
MCHC	(%)	31.8	1.97	27.9-35.7	32.0	3.20	25.6-38.4
Hematocrit (PCV)	(ml%)	41.5	4.20	33.1-49.9	41.8	2.70	36.4-47.2
Sedimentation rate	(mm/hr)	2.12	0.28	1.56-2.68	4.08	0.36	3.36-4.80
Platelets	$(\times 10^3/mm^3)$	144.	15.0	114.-174.	130.	53.0	24.0-236.
Leukocytes (WBC)	$(\times 10^3/mm^3)$	10.6	1.62	7.36-13.8	11.1	2.50	6.10-16.1
Neutrophils	$(\times 10^3/mm^3)$	3.98	0.60	2.77-5.19	4.41	0.99	2.42-6.39
	(%)	37.6	9.80	18.0-47.2	39.7	10.7	18.3-61.1
Eosinophils	$(\times 10^3/mm^3)$	0.19	0.03	0.13-0.25	0.21	0.04	0.12-0.30
	(%)	1.80	2.50	0.00-6.80	1.90	2.90	0.00-7.70
Basophils	$(\times 10^3/mm^3)$	0.02	0.01	0.01-0.03	0.02	0.01	0.01-0.03
	(%)	0.20	0.80	0.00-1.80	0.20	0.60	0.00-1.40
Lymphocytes	$(\times 10^3/mm^3)$	6.23	0.94	4.33-8.11	6.39	1.44	3.51-9.27
	(%)	58.8	16.2	26.4-91.2	57.6	16.2	25.2-90.0
Monocytes	$(\times 10^3/mm^3)$	0.17	0.02	0.12-0.22	0.07	0.02	0.04-0.10
	(%)	1.60	1.20	0.00-4.00	0.60	1.20	0.00-3.00

*Rhesus, young adult monkey (20-30 months age), weighing 2-5 kg.

Table 26. Hematological Values of the Normal Rhesus Monkey (Macaca Mulatta)

Test	Unit	Benirschke[1] et al (1978) Mean	S.D.	Melby and[1] Altman (1976) Mean	% C.V.	Mitruka and[2] Rawnsley (1977) Range
Erythrocytes (RBC)	(x10^6/mm^3)	5.61	0.58	5.60	11.0	3.10-8.60
Hemoglobin	(g/dl)	12.3	1.10	12.3	9.00	9.00-16.5
MCV	(μ^3)	76.0	0.64	76.0	7.00	58.4-104
MCH	(μμg)	22.0	2.40	22.0	9.00	18.0-35.4
MCHC	(%)	29.1	2.00	29.0	7.00	24.8-39.0
Hematocrit (PCV)	(%)	42.1	3.10	42.0	8.00	32.0-50.0
Platelets	(x10^3/mm^3)	418 (10^3/μl)	115	344	21	30.0-440
Leukocytes (WBC)	(x10^3/mm^3)	10.1	3.60	15.2	40	4.60-21.6
Neutrophils	(x10^3/mm^3) (%)	- 40.0	- 17.0	5.40 36.0	- 47.0	- 14.0-71.0
Eosinophils	(x10^3/mm^3) (%)	- 3.50	- 3.0	0.40 2.60	- 90.0	- 0.00-16.0
Basophils	(x10^3/mm^3) (%)	- 0.30	- 0.50	0.03 0.20	- 84.0	- 0.00-4.00
Lymphocytes	(x10^3/mm^3) (%)	- 55.0	- 16.5	9.20 61.0	- 28.0	- 23.0-95.0
Monocytes	(x10^3/mm^3) (%)	- 1.10	- 1.60	0.11 0.70	- 53.0	- 0.00-8.00

Source: [1] Derived from Literature
[2] Data from 30 female and 38 male, young adult (20-30 months age) Rhesus monkeys weighing 2-5 kg. See text for methodology.

Table 27. Hematological Values of Certain Species of Old World Primates

Test	Unit	Rhesus (Macaca mulatta)	Stump-tailed (M. arctoides)	Pig-tailed (M. nemestrina)	Cynomolgus (M. fascicularis)	Bonnet Monkey (M. radiata)
Erythrocytes (RBC)	$(\times10^6/mm^3)$	3.10-8.60	4.40-5.90	4.10-7.45	3.90-7.10	3.17-4.80
Hemoglobin	(g/dl)	9.00-16.5	11.3-14.7	11.7-15.7	11.6-14.5	10.6-14.8
MCV	(μ^3)	58.4-104.	70.0-88.0	63.3-79.7	69.1-90.0	85.0-120.
MCH	$(\mu\mu g)$	18.0-35.4	20.0-30.0	17.5-28.1	21.1-26.4	26.5-37.0
MCHC	(%)	24.8-39.0	28.0-38.0	27.9-37.4	26.4-33.0	25.8-36.1
Hematocrit (PCV)	(ml%)	32.0-50.0	36.0-43.0	38.0-46.0	38.0-50.0	34.0-48.0
Platelets	$(\times10^3/mm^3)$	30.0-440.	317.-588.	110.-190.	90.0-140.	---
Leukocytes (WBC)	$(\times10^3/mm^3)$	4.60-21.6	5.20-14.9	5.30-13.1	8.15-21.3	12.0-13.7
Neutrophils	(%)	14.0-71.0	17.0-74.0	23.0-58.0	13.8-81.0	44.0-76.0
Eosinophils	(%)	0.00-16.0	2.00-17.0	0.00-8.00	0.00-11.0	1.70-2.50
Basophils	(%)	0.00-4.00	0.00-1.00	0.00-1.00	0.00-1.00	---
Lymphocytes	(%)	23.0-95.0	22.0-79.0	37.0-69.0	35.0-83.0	36.0-73.0
Monocytes	(%)	0.00-8.00	0.00-2.00	0.00-4.00	0.00-1.00	0.00-2.00

Table 27 (con't.). Hematological Values of Certain Species of Old World Primates

Test	Unit	Chacma baboon (Papio ursimus)	African green (Cercopithecus aethiops)	Chimpanzee (Pan troglodytes)	Silvered leaf (Prebytis cristatus)	Slender loris (Loris tardigradus)
Erythrocytes (RBC)	$(\times 10^6/mm^3)$	4.49-6.34	4.34-4.89	3.40-6.00	4.00-7.60	2.45-4.52
Hemoglobin	(g/dl)	7.60-14.0	10.8-13.2	8.50-15.1	9.70-12.9	7.70-13.9
MCV	(μ^3)	53.4-83.0	70.0-96.0	70.0-108.	69.7-89.0	61.4-88.6
MCH	$(\mu\mu g)$	14.0-25.8	23.4-28.6	18.0-32.0	16.7-22.2	22.0-39.7
MCHC	(%)	20.8-38.3	27.0-34.0	19.0-35.0	21.1-28.0	29.6-53.5
Hematocrit (PCV)	(ml%)	30.0-43.0	35.0-44.0	36.0-51.0	40.4-51.6	21.5-31.0
Platelets	$(\times 10^3/mm^3)$	219.-453.	120.-270.	145.-280.	--	--
Leukocytes	$(\times 10^3/mm^3)$	4.75-17.1	9.70-17.0	12.5-16.6	5.50-15.9	27.7-31.1
Neutrophils	(%)	27.0-73.0	20.0-61.0	39.0-71.0	22.0-81.0	11.0-33.0
Eosinophils	(%)	1.00-4.00	0.00-11.0	0.00-5.70	0.00-8.00	2.30-4.30
Basophils	(%)	0.00-0.50	0.00-1.00	0.00-0.50	0.00-3.50	--
Lymphocytes	(%)	26.0-59.0	32.0-75.0	30.4-45.6	15.0-63.0	53.0-81.0
Monocytes	(%)	0.00-5.80	0.00-2.00	0.00-3.00	0.00-6.00	0.00-2.00

99

Source: Krise and Wald (1959); Krise (1964); Anderson (1966); King and Gargus (1967); Melville et al. (1967); Robinson and Ziegler (1968); Stanley and Cramer (1968); Oser et al. (1970); Rollins et al. (1970); Vondruska (1970); Rahaman et al. (1975); Spiller et al. (1975); Melby and Altman (1976); Benirschke et al. (1978); Hanser (1970); Hambleton et al. (1979).

Table 27. (continued)   Hematological Values of Certain Species of Old World Primates

Test	Unit	BABOON			
		(P. cynocephalus)	(P. anubis)	(P. hamadryas)	(P. doguera) (anubis)
Erythrocytes (RBC)	$(\times 10^6/mm^3)$	4.42-5.03	-	3.87-6.07	3.45-5.45
Hemoglobin	(g/dl)	12.5-13.9	7.50-18.2	11.4-14.7	8.60-15.0
MCV	(μ^3)	87.8-97.0	-	69.0-88.0	71.6-88.4
MCH	$(\mu\mu g)$	-	-	-	23.9-29.1
MCHC	(%)	27.9-33.9	25.9-43.3	-	28.8-35.2
Hematocrit (PVC)	(%)	41.0-44.0	24.5-49.0	35.0-49.0	18.4-53.2
Platelets	$(\times 10^3/mm^3)$	-	-	-	-
Leukocytes	$(\times 10^3/mm^3)$	4.10-12.3	3.90-20.4	4.90-15.4	9.50-18.7
Neutrophils	(%)	24.0-71.0	8.00-93.0	21.0-68.0	26.5-94.5
Eosinophils	(%)	1.00-5.00	0.00-25.0	0.00-16.0	1.00-10.0
Basophils	(%)	0.00-1.00	0.00-5.00	0.00-3.00	0.40-0.70
Lymphocytes	(%)	25.0-73.0	2.00-86.0	21.0-74.0	20.5-51.5
Monocytes	(%)	1.00-4.00	0.00-13.0	0.00-9.00	0.00-1.50

Table 27. (continued)   Hematological Values of Certain Species of Old World Primates

Test	Unit	NORMAL ADULT MACACA		
		(M. Cyclopis)	(M. Speciosa)	(M. Irus)(Fascicularis)
Erythrocytes (RBC)	$(x10^6/mm^3)$	3.80-6.44	3.10-6.30	0.00-13.1
Hemoglobin	(g/dl)	12.1-15.7	9.30-14.9	9.30-13.7
MCV	(μ^3)	71.0-88.0	70.6-89.0	61.1-71.5
MCH	$(\mu\mu g)$	23.0-30.0	22.9-28.9	17.0-19.4
MCHC	(%)	31.0-36.0	29.2-35.2	25.8-31.4
Hematocrit (PCV)	(%)	33.5-44.3	32.5-42.3	34.6-45.0
Platelets	$(x10^3/mm^3)$	330 - 622	-	359 - 427
Leukocytes	$(x10^3/mm^3)$	6.20-14.4	7.60-31.6	4.30-21.1
Neutrophils	(%)	32.2-67.0	22.5-56.5	25.0-79.0
Eosinophils	(%)	0.00-3.60	1.50-18.5	0.00-6.00
Basophils	(%)	0.00-1.65	0.00-1.00	0.00-1.50
Lymphocytes	(%)	30.3-64.6	32.5-62.5	16.0-68.0
Monocytes	(%)	0.00-1.67	1.00-4.00	1.00-5.00

Table 27. (continued)   Hematological Values of Certain Species of Old World Primates

Test	Unit	BABOON		
		(Papio sp.)	(P.c. cynocephalus)	(P.c. anubis)
Erythrocytes (RBC)	$(\times 10^6/mm^3)$	4.17-5.36	4.99-6.95	4.40-6.37
Hemoglobin	(g/dl)	10.5-14.7	11.0-16.6	11.1-14.7
MCV	(μ^3)	73.7-90.7	66.2-82.6	67.6-88.0
MCH	$(\mu\mu g)$	21.7-30.6	20.6-26.6	20.9-26.9
MCHC	(%)	26.3-38.9	27.6-34.4	28.2-33.4
Hematocrit (PCV)	(%)	34.1-45.1	36.3-52.3	34.6-49.0
Platelets	$(\times 10^3/mm^3)$	-	-	-
Leukocytes	$(\times 10^3/mm^3)$	5.20-11.2	4.00-13.4	5.84-14.0
Neutrophils	(%)	-	-	-
Eosinophils	(%)	-	-	-
Basophils	(%)	-	-	-
Lymphocytes	(%)	-	-	-
Monocytes	(%)	-	-	-

Table 28. Hematological Values of certain Species of New World Primates

Test	Unit	Squirrel monkey (Saimiri sciureus)	Prosimian Bush baby (Galago cras- sicaudatus)	Panamanian Marmoset (Saguinus geoffroyi)	White face monkey (Cebus capucinus)
Erythrocytes (RBC)	$(\times 10^6/mm^3)$	5.90-9.40	5.60-8.40	5.47-8.68	4.15-6.68
Hemoglobin	(g/dl)	9.30-16.1	8.20-14.1	10.6-18.7	12.4-19.8
MCV	(μ^3)	39.0-60.0	40.0-67.1	48.0-87.6	70.0-117.
MCH	$(\mu\mu g)$	12.2-21.0	11.0-20.1	15.0-26.5	21.6-37.3
MCHC	(%)	24.8-44.2	21.9-37.6	22.3-40.9	24.2-39.1
Hematocrit (PCV)	(ml%)	30.5-43.6	28.0-47.0	34.0-62.0	40.0-63.0
Platelets	$(\times 10^3/mm^3)$	60.0-2.30	---	231.-650.	110.-600.
Leukocytes (WBC)	$(\times 10^3/mm^3)$	4.60-11.7	3.30-26.7	7.30-24.6	6.30-34.3
Neutrophils	(%)	22.5-80.0	11.0-78.0	30.0-90.0	40.0-70.0
Eosinophils	(%)	0.00-10.0	0.00-9.00	0.00-2.00	0.00-5.00
Basophils	(%)	0.00-1.00	0.00-1.00	0.00-6.00	0.00-1.00
Lymphocytes	(%)	20.0-68.0	18.0-88.0	9.00-66.0	25.0-55.0
Monocytes	(%)	1.00-6.00	0.00-2.00	0.00-11.0	0.00-4.00

Table 28 (con't.). Hematological Values of Certain Species of New World Primates

Test	Unit	Black spider monkey (Ateles fusciceps)	Red spider monkey (Ateles geoffroyi)	Black howler (Alouatta villosa)	Night monkey (Aotus trivirgatus)
Erythrocytes (RBC)	$(x10^6/mm^3)$	1.97-4.22	3.68-4.98	2.89-4.65	3.50-7.74
Hemoglobin	(g/dl)	5.00-12.5	13.0-15.1	8.60-13.8	11.9-16.0
MCV	(μ^3)	54.8-154.	98.0-117.	76.9-117.	57.8-91.8
MCH	$(\mu\mu g)$	16.1-40.3	28.8-34.9	22.7-36.9	21.7-29.5
MCHC	(%)	15.4-38.5	27.0-39.8	23.7-37.8	28.6-39.4
Hematocrit (PCV)	(ml%)	17.0-48.0	44.0-50.0	29.0-44.0	32.5-51.6
Platelets	$(x10^3/mm^3)$	---	---	100.-800.	204.-734.
Leukocytes	$(x10^3/mm^3)$	4.30-28.5	7.30-13.5	7.20-22.9	3.20-28.5
Neutrophils	(%)	46.0-82.0	26.5-4.20	51.0-64.0	13.0-91.0
Eosinophils	(%)	0.00-4.00	0.00-3.00	0.00-2.00	0.00-37.0
Basophils	(%)	0.00-2.00	0.00-1.00	0.00-1.00	0.00-1.00
Lymphocytes	(%)	13.0-54.0	50.0-76.1	32.0-49.0	5.00-80.0
Monocytes	(%)	0.00-3.00	0.00-6.00	0.00-3.00	---

Source: Garcia and Hunt (1966); Anderson et al. (1967); Burns et al. (1967); New (1968); Manning et al. (1969); Porter (1969, 1970, 1971); Vogin and Oser (1971); Wellde et al. (1971); Middleton and Rosal (1972); Melby and Altman (1976); Benirschke et al. (1978); Whitney et al. (1973); McClure et al. (1973).

Table 28. (continued)   Hematological Values of Certain Species of New World Primates

Test	Unit	Callithricidae (marmosets)			
		(Saguinus mystax)	(Saguinus nigricollis)	(Saguinus oedipus)	(Senegalensis)
Erythrocytes (RBC)	$(\times 10^6/mm^3)$	4.70-7.70	-	5.47-8.68	5.96-13.4
Hemoglobin	(g/dl)	8.70-19.6	-	-	13.2-20.4
MCV	(μ^3)	-	-	-	-
MCH	$(\mu\mu g)$	18.3-29.9	-	-	19.0-27.5
MCHC	(%)	26.1-35.1	-	-	30.4-40.0
Hematocrit (PCV)	(%)	35.3-60.7	-	35.0-62.0	40.0-58.0
Platelets	$(\times 10^3/mm^3)$	-	-	331 - 650	431 - 684
Leukocytes (WBC)	$(\times 10^3/mm^3)$	7.70-20.8	6.80-20.8	7.30-24.6	3.63-16.5
Neutrophils	(%)	-	4.00-72.5	30.0-90.0	6.00-46.0
Eosinophils	(%)	-	0.00-11.5	0.00-2.00	1.00-7.00
Basophils	(%)	-	0.00-7.50	0.00-6.00	-
Lymphocytes	(%)	-	20.5-91.0	9.00-66.0	49.0-81.0
Monocytes	(%)	-	0.00-11.0	0.00-11.0	-

Table 28. (continued)   Hematological Values of Certain Species of New World Primates   (Lemur)

Test	Unit	Variegatus	Mongoz	Catta	Macaco
Erythrocytes (RBC)	$(\times10^6/mm^3)$	8.10-11.1	11.7-13.4	6.20-9.80	8.30-13.0
Hemoglobin	(g/dl)	17.2-19.0	16.2-19.8	15.6-20.2	16.1-21.1
MCV	(μ^3)	-	-	-	-
MCH	$(\mu\mu g)$	-	-	-	-
MCHC	(%)	-	-	-	-
Hematocrit (PCV)	(%)	53.0-61.0	49.6-59.4	48.0-53.0	50.0-59.0
Platelets	$(\times10^3/mm^3)$	-	-	-	-
Leukocytes	$(\times10^3/mm^3)$	9.00-19.0	9.90-11.4	6.20-16.9	6.00-28.0
Neutrophils	(%)	14.0-60.0	12.6-15.4	14.0-40.0	14.0-26.0
Eosinophils	(%)	0.00-3.00	0.00-6.00	0.00-4.00	1.00-5.00
Basophils	(%)	-	-	-	-
Lymphocytes	(%)	36.0-78.0	63.0-93.0	33.0-97.0	65.0-79.0
Monocytes	(%)	2.00-4.00	1.00-11.0	1.00-9.00	2.00-4.00

Table 28. (continued)   Hematological Values of Certain Species of New World Primates   (Lemur)

Test	Unit	(F. fulvus)	(F. rufus)	(F. collaris)
Erythrocytes (RBC)	$(\times 10^6/mm^3)$	7.10-11.9	6.30-11.4	5.90-10.1
Hemoglobin	(g/dl)	13.0-20.3	13.6-18.8	12.8-17.5
MCV	(μ^3)	-	-	-
MCH	$(\mu\mu g)$	-	-	-
MCHC	(%)	-	-	-
Hematocrit (PCV)	(%)	39.0-51.0	42.0-53.0	48.0-50.0
Platelets	$(\times 10^3/mm^3)$	-	-	-
Leukocytes	$(\times 10^3/mm^3)$	5.00-17.0	7.10-22.5	10.4-25.6
Neutrophils	(%)	11.0-31.0	11.0-29.0	17.0-49.0
Eosinophils	(%)	1.00-11.0	1.00-12.0	1.00-15.0
Basophils	(%)	-	-	-
Lymphocytes	(%)	45.0-83.0	47.0-91.0	35.0-79.0
Monocytes	(%)	2.00-4.00	2.00-4.00	0.00-4.00

in cell counts, with the exception of platelets. Large variations noted within the species may be attributable to differences in nutrition, maintenance, climate, seasonal influences, animal behavior, and blood sampling methods. The hematology is not influenced by nutritional variations or by the age of the nonhuman primates. However studies of sex differences reported by many investigators show that, in general, the erythrocyte counts in male animals are slightly higher than those in females; leukocytes showed the reverse pattern and the hemoglobin concentration was about the same. The percentage of leukocyte distribution showed irregular differences which were not statistically significant.

Pigs

The number of erythrocytes in the pig varies with age and husbandry during the suckling and growing period (Table 29). At birth the range is 4–7 million with a mean of 5.6 million. The count is reduced by 30% by the tenth day of life as a result of rapid growth and lack of a significant body reserve of iron. If the pigs are in contact with a natural source of iron, as in soil, the erythrocyte count increases, and by 2 months of age the pig reaches adult levels of 5–9 million with means of 6.7 million (Table 30). The red cells of the newborn are about 10 μ^3 larger than that of mature pigs (Schalm, 1965). The diameter of the red cell in the mature pig ranges from 4 to 8 μ with a mean of about 6 μ. Rouleau formation in the erythrocytes is characteristic, and sharp crenation is common. Polychromatic, nucleated erythrocytes and erythrocytes with Howell-Jolly bodies are found in the blood of piglets and growing swine up to 3 months of age. When the erythrocyte count approaches the adult level, the nucleated red cells disappear and the reticulocyte percentage falls to less than 1.0%. Swine erythrocytes are very susceptible to

hemolysis by mechanical means, such as the method of drawing blood or the handling of blood after it is drawn. Reticulocytes are a prominent feature in blood of miniature swine of all strain from day 1 until after weaning has occurred. They decreased gradually to about 3 years of age when they stabilize at about 0.2 to 2 percent.

Fluctuation in erythrocyte count, hematocrit (PCV), and hemoglobin concentration are considerable during the growing period of the pig. Hemoglobin values have been reported to change from birth (8–13 g/dl); to 10 days (4–9 g/dl); to 20 days (9–11 g/dl); to 1–2 months (11–12 g/dl); and to maturity (10–15 g/dl). Our data agree with the values reported in the literature.

The total leukocyte counts reported in the literature have ranged from 5,600 to 19,500 during the first 3 weeks of life. At 2 months total counts are from 16,000 to 25,000. Our values in the adult pigs range from 10,300 to 20,700 in males and from 7,800 to 20,600 in females. The differential leukocyte counts are characterized by a very high percentage of neutrophils at birth (70%). By the end of the first week of life the neutrophils and lymphocytes are about equal (45%), but the former continue to drop (35%) and the latter continue to rise (55%) when the pig reaches adult age. In the neutrophils the nuclear chromatin stains intensely. The nucleus is often coiled and the nuclear outline is very irregular, but filaments commonly are not present. The cytoplasm is a very light blue and is diffusely filled with dust-like granules that stain pink. Band neutrophils can constitute up to 1–2% of the differential count. The nucleus is U or S-shaped with a smooth nuclear membrane. Otherwise the cell is similar to the mature neutrophils. Metamyelocyte neutrophils may appear in the peripheral blood during the first day postpartum. In the eosinophil the granules are round to ovoid, stain a dull orange, and completely fill the cytoplasm with a few

Table 29. Hematological Values in Miniature Swine of Various Age Groups

Test	Unit	Pitman-Moore (8-15 weeks)	Hormel 3-12 mos.	Pitman-Moore 19 mos.	Pitman-Moore Adult	Hormel 5 mos.
Erythrocytes (RBC)	$(\times 10^6/mm^3)$	7.62-10.3	6.40-9.60	5.80-9.20	5.50-8.80	5.08-9.36
Hemoglobin	(g/dl)	12.0-14.6	12.5-18.1	8.90-20.5	14.0-15.0	10.7-14.3
MCV	(μ^3)	-	51.8-63.2	54.9-67.1	57.6-70.4	42.3-60.7
MCH	$(\mu\mu g)$	-	17.2-21.0	17.1-20.9	19.8-24.2	13.6-20.0
MCHC	(%)	-	27.1-33.1	28.8-35.2	34.0-35.0	31.1-34.3
Hematocrit (PCV)	(%)	37.0-43.5	38.2-53.8	28.0-64.0	41.0-43.0	26.3-48.3
Platelets	$(\times 10^3/mm^3)$	-	-	-	-	-
Leukocytes (WBC)	$(\times 10^3/mm^3)$	9.50-28.9	13.2-22.4	7.80-13.4	10.8-13.2	10.1-24.5
Neutrophils	(%)	12.0-41.0	11.0-31.0	-	36.0-44.0	-
Eosinophils	(%)	1.00-6.00	0.00-6.00	-	-	-
Basophils	(%)	0.00-1.00	0.00-0.40	-	-	-
Lymphocytes	(%)	53.0-78.0	54.0-94.0	-	45.0-55.0	-
Monocytes	(%)	0.00-4.00	0.00-4.00	-	-	-

Sources: Osborne and Meredith(1971), Loeb et al.(1978), Benirschke et al.(1978)

Table 30. Hematological Values of the Normal Pig*

Test	Unit	Male (12 animals)			Female (10 animals)		
		Mean	S.D.	Range	Mean	S.D.	Range
Erythrocytes (RBC)	$(\times10^6/mm^3)$	7.09	0.80	5.49-8.69	6.90	0.50	5.90-7.90
Hemoglobin	(g/dl)	13.0	0.25	12.5-13.5	11.8	0.21	11.4-12.2
MCV	(μ^3)	60.0	1.00	58.0-62.0	57.0	1.21	54.6-59.4
MCH	(μμg)	18.3	0.25	17.8-18.8	17.1	0.30	16.5-17.7
MCHC	(%)	30.1	2.00	26.1-34.1	30.3	1.25	27.8-32.8
Hematocrit (PCV)	(ml%)	43.2	0.50	42.4-44.2	38.9	0.72	37.5-40.3
Sedimentation rate	(mm/hr)	4.50	1.10	2.30-6.70	5.30	1.40	2.50-8.10
Platelets	$(\times10^3/mm^3)$	300.	34.0	232.-368.	215.	26.8	161.-269.
Leukocytes (WBC)	$(\times10^3/mm^3)$	15.5	2.95	10.3-20.7	14.2	3.20	7.80-20.6
Neutrophils	$(\times10^3/mm^3)$	6.04	1.01	4.02-8.07	4.12	0.93	2.26-5.97
	(%)	39.0	4.35	30.3-47.7	29.0	3.65	21.7-36.3
Eosinophils	$(\times10^3/mm^3)$	0.70	0.12	0.46-0.93	0.71	0.16	0.39-1.03
	(%)	4.50	1.20	2.10-6.90	5.00	1.40	2.20-7.80
Basophils	$(\times10^3/mm^3)$	0.19	0.04	0.12-0.25	0.28	0.06	0.16-0.41
	(%)	1.20	0.30	0.60-1.80	2.00	0.35	1.30-2.70
Lymphocytes	$(\times10^3/mm^3)$	8.08	1.35	5.37-10.8	8.35	1.88	4.59-12.1
	(%)	52.1	5.40	41.3-62.9	58.8	4.87	49.1-68.5
Monocytes	$(\times10^3/mm^3)$	0.51	0.08	0.34-0.68	0.74	0.17	0.40-1.07
	(%)	3.30	0.42	2.46-4.14	5.20	0.32	4.56-5.84

*Mini-pig, young adult weighing 70-100 kg.

granules also over the nucleus. In the basophils the nucleus takes a lavender stain and the chromatin is smooth. The granules may stain similarly to the nucleus or they make take a much darker stain. They are generally limited to the cytoplasm with some in the nucleus. The granule has a coccoid to dumbbell shape.

The nucleus occupies most of the cell area of the small lymphocyte. It is oval and the chromatin is smooth but with enough condensation to exhibit a distinct pattern. The large lymphocyte's nucleus stains somewhat lighter and shows some clumping of the chromatin. In the monocyte the nucleus is irregular with folding and convolutions of its surface. The chromatin is stringy with uniformly distributed condensations. The cytoplasm is fairly voluminous and may completely surround the nucleus. It has a blue-gray, ground-glass, somewhat mottled appearance because of variation in the intensity of staining. In the eosinophil the nucleus is commonly nonsegmented and either band or oval in shape. The granules stain a dull orange, are rounded in shape, and pack the cytoplasm; they seldom cover much of the nucleus. Basophils in the pig are characterized by large, deeply basophilic granules, coccoid in shape. The granules are usually confined to the cytoplasm although a few may be scattered over the nucleus. The nucleus stains pale purple with smooth chromatin. The platelets of minipig are small, oval and anucleate structures, with faintly blue or almost unstained cytoplasm containing a cluster of fine purple granules. They are frequently observed in irregular clumps of varying size.

Goats

The main characteristics of the goat's hematology are high erythrocyte counts (15–20 million/mm³) and low platelet counts (35,000–60,000/mm³) (Table 31). The erythrocyte is only 4.25 μ in diameter. No re-

ticulocytes are found in the peripheral blood of healthy goats. Rouleau formation is also absent and no sedimentation takes place in the first hour (Andrews, 1965). The total leukocyte count of the goat is several thousand above that of the sheep and cow. Their mean neutrophil count is greater on a percentage basis than that in sheep or cows, whereas the mean lymphocyte percentage is slightly lower (Schalm, 1965). Our data basically agree with those reported in the literature (Table 32). Other than the characteristics mentioned above, the blood morphology of the goat is similar to that of the sheep, and it will be discussed briefly in the following section (Lewis, 1976).

Sheep

The erythrocyte counts of sheep range from 9.5 to 11.1 million in males and from 10.3 to 12.9 million in females (Table 33). Differences in breed are not known to affect the counts. However age differences have been found to affect erythrocyte counts by many investigators. The mean erythrocyte counts in lambs up to 1 year of age generally have been reported as being between 11 and 13 million/mm³ of blood. Albritton (1961), however, gave the following data: at 60 days of pregnancy 1.83 million; at 90 days 5.1 million; at 120 days 8 million; and at 135 days 9 million. The erythrocytes of the sheep are very small (4.4 μ in diameter), round (annular form), and show pronounced anisocytosis. The mean hemoglobin concentration exceeds that of adults and by the second week rises to about 18% above the value at birth. The fall to the adult value does not occur until the animal is about 1 year old. At birth the hemoglobin still consists of 50% of fetal hemoglobin.

In sheep the physiological range of variation of the leukocytes is very wide (4,200–19,800/mm³). Our observations show that the leukocyte count in most animals is between 5,000 and 10,000. The platelet count

Table 31. Hematological Values of the Normal Goat*

Test	Unit	Male (12 animals)			Female (8 animals)		
		Mean	S.D.	Range	Mean	S.D.	Range
Erythrocytes (RBC)	$(\times10^6/mm^3)$	18.0	1.15	15.7-20.3	17.3	1.20	14.9-19.7
Hemoglobin	(g/dl)	10.5	0.65	9.20-12.8	12.6	0.81	11.0-14.2
MCV	(μ^3)	19.3	0.40	18.5-20.1	16.5	0.38	15.7-17.3
MCH	$(\mu\mu g)$	5.83	0.15	5.33-6.13	7.28	0.20	6.88-7.68
MCHC	(%)	31.8	1.15	29.5-34.1	42.0	1.20	39.6-44.4
Hematocrit (PCV)	(ml%)	33.0	1.90	29.2-36.8	30.0	2.20	25.6-34.4
Sedimentation rate	(mm/hr)	2.20	0.50	1.20-3.20	2.80	0.45	1.90-3.70
Platelets	$(\times10^3/mm^3)$	46.5	5.50	34.0-60.0	50.0	5.00	40.0-60.0
Leukocytes (WBC)	$(\times10^3/mm^3)$	9.50	2.25	5.00-14.0	7.42	1.86	3.70-11.1
Neutrophils	$(\times10^3/mm^3)$	3.81	0.90	2.00-5.61	2.71	0.68	1.35-4.05
	(%)	40.1	5.25	29.6-50.6	36.5	4.10	28.3-44.7
Eosinophils	$(\times10^3/mm^3)$	0.22	0.05	0.12-0.32	0.26	0.07	0.13-0.39
	(%)	2.30	0.44	1.42-3.18	3.50	0.52	2.46-4.54
Basophils	$(\times10^3/mm^3)$	0.00	0.01	0.00-0.02	0.00	0.01	0.00-0.02
	(%)	0.00	0.10	0.00-0.20	0.00	0.10	0.00-0.20
Lymphocytes	$(\times10^3/mm^3)$	5.37	1.28	2.82-7.91	4.30	1.08	2.15-6.44
	(%)	56.5	5.12	46.3-66.7	58.0	6.20	45.6-70.4
Monocytes	$(\times10^3/mm^3)$	0.10	0.02	0.05-0.14	0.15	0.04	0.07-0.22
	(%)	1.00	0.20	0.60-1.40	2.00	0.22	1.56-2.44

*Domestic, young adult weighing 20-50 kg.

Table 32. Hematological Values of the Normal Goat

Test	Unit	Coles[1] (1980) Mean	Coles[1] (1980) Range	Benjamin[1] (1978) Mean	Benjamin[1] (1978) Range	Melby and Altman[1] (1976) Mean	Melby and Altman[1] (1976) %C.V.	Mitruka and Rawnsley[2] (1977) Range
Erythrocyte (RBC)	(x10^6/mm^3)	13.0	8.00-17.5	15.0	12.0-20.0	12.7	20.0	14.9 — 20.3
Hemoglobin	(g/dl)	11.0	8.00-14.0	11.0	8.00-14.0	11.1	16.0	9.20 — 14.2
MCV	(μ^3)	19.0	16.0-25.0	20.0	18.0-24.0	23.0	-	15.7 — 20.1
MCH	(μμg)	-	-	-	5.00-7.40	9.00	-	5.33 — 7.68
MCHC	(%)	32.0	2.80-34.0	32.0	30.0-35.0	36.0	-	29.5 — 44.4
Hematocrit (PCV)	(%)	28.0	20.0-38.0	35.0	24.0-48.0	29.0	16.0	25.6 — 36.8
Sedimentation rate	(mm/hr)	-	-	-	-	-	-	1.20 — 3.70
Platelets	(x10^3/mm^3)	-	-	-	-	-	-	34.0 — 60.0
Leukocytes (WBC)	(x10^3/mm^3)	12.0	6.00-16.0	12.0	6.00-16.0	8.10	31.0	3.70 — 14.0
Neutrophils	(x10^3/mm^3)	-	0.00-2.00	0.50	0.00-2.00	4.00	22.0	1.35 — 5.61
	(%)	-	30.0-48.0	36.5	30.0-48.0	49.0	22.0	28.3 — 50.6
Eosinophils	(x10^3/mm^3)	-	-	-	-	0.10	-	0.12 — 0.39
	(%)	-	3.00-8.00	5.00	3.00-8.00	2.00	-	1.42 — 4.54
Basophils	(x10^3/mm^3)	-	-	-	-	Rare	-	0.00 — 0.02
	(%)	-	0.00-2.00	0.50	0.00-2.00	Rare	-	0.00 — 0.20
Lymphocytes	(x10^3/mm^3)	-	-	-	-	3.70	43.0	2.15 — 7.91
	(%)	-	50.0-70.0	55.0	50.0-70.0	42.0	25.0	45.6 — 70.4
Monocytes	(x10^3/mm^3)	-	-	-	-	0.20	-	0.05 — 0.22
	(%)	-	1.00-4.00	2.50	1.00-4.00	3.00	80.0	0.60 — 2.44

Source: [1]Derived from literature
[2]Data from 8 female and 12 male goats (domestic, young adult) weighing 20-50 kg.
See text for methodology.

Table 33. Hematological Values of the Normal Sheep*

Test	Unit	Male (32 animals)			Female (26 animals)		
		Mean	S.D.	Range	Mean	S.D.	Range
Erythrocytes (RBC)	$(\times10^6/mm^3)$	10.3	0.42	9.50-11.1	11.6	0.67	10.3-12.9
Hemoglobin	(g/dl)	10.9	0.45	10.0-11.8	11.0	0.62	10.7-12.2
MCV	(μ^3)	31.0	0.50	30.0-32.0	30.0	0.80	28.4-31.6
MCH	$(\mu\mu g)$	10.6	0.20	10.2-11.0	9.50	0.15	9.20-9.80
MCHC	(%)	34.4	1.22	32.0-36.8	32.2	1.08	30.0-34.4
Hematocrit (PCV)	(ml%)	31.7	0.92	29.9-33.6	34.1	1.30	31.5-36.7
Sedimentation rate	(mm/hr)	1.00	0.25	0.50-1.50	1.20	0.30	0.60-1.80
Platelets	$(\times10^3/mm^3)$	350.	62.0	226.-474.	320.	71.0	178.-462.
Leukocytes (WBC)	$(\times10^3/mm^3)$	7.80	1.25	5.30-10.3	7.30	1.20	4.90-9.70
Neutrophils	$(\times10^3/mm^3)$	2.80	0.46	1.89-3.68	2.04	0.34	1.37-2.72
	(%)	35.7	6.25	33.2-48.2	28.0	6.76	14.5-41.5
Eosinophils	$(\times10^3/mm^3)$	0.20	0.04	0.13-0.28	0.58	0.10	0.39-0.78
	(%)	2.50	0.80	0.90-4.10	8.00	2.20	3.60-12.4
Basophils	$(\times10^3/mm^3)$	0.03	0.01	0.02-0.04	0.00	0.01	0.00-0.02
	(%)	0.40	0.50	0.00-1.40	0.00	0.10	0.00-0.20
Lymphocytes	$(\times10^3/mm^3)$	4.44	0.40	3.02-5.86	4.53	0.75	3.04-6.01
	(%)	56.9	5.00	46.9-66.9	62.0	6.55	48.9-75.1
Monocytes	$(\times10^3/mm^3)$	0.35	0.06	0.24-0.46	0.14	0.03	0.09-0.19
	(%)	4.50	1.75	1.00-8.00	2.00	0.64	0.72-3.28

*Dorset-Delane, adult sheep weighing 40-60 kg.

114

Table 34. Hematological Values of the Normal Sheep

Test	Unit	Coles[1] (1980) Mean	Coles[1] (1980) Range	Benjamin[1] (1978) Mean	Benjamin[1] (1978) Range	Melby and Altman[1] (1976) Mean	Melby and Altman[1] (1976) %C.V.	Mitruka & Rawnsley[1] (1977) Range
Erythrocytes (RBC)	$(x10^6/mm^3)$	12.0	8.00-15.0	12.0	8.00-16.0	12.0	–	9.50-13.9
Hemoglobin	(g/dl)	12.0	8.00-16.0	12.0	8.00-16.0	12.0	9.00	10.0-14.2
MCV	$(µ^3)$	33.0	23.0-48.0	32.0	23.0-48.0	32.0	–	28.4-36.0
MCH	$(µµg)$	–	–	–	9.00-13.0	10.0	–	9.20-11.0
MCHC	$(\%)$	32.0	29.0-35.0	32.0	29.0-35.0	34.0	–	30.0-36.8
Hematocrit (PCV)	$(\%)$	38.0	24.0-49.0	38.0	24.0-50.0	38.0	8.0	29.9-36.7
Sedimentation rate	(mm/hr)	–	–	–	–	–	–	0.50-1.80
Platelets	$(x10^3/mm^3)$	441	284-659	–	–	–	–	278-494
Leukocytes (WBC)	$(x10^3/mm^3)$	7.40	6.30-8.60	9.00	4.00-12.0	7.40	30.0	4.90-10.3
Neutrophils	$(x10^3/mm^3)$	–	0.00-2.00	0.50	0.00-2.00	2.00	–	1.37-3.68
	$(\%)$	–	10.0-50.0	30.0	10.0-50.0	27.0	30.0	14.5-48.2
Eosinophils	$(x10^3/mm^3)$	–	–	–	–	0.40	–	0.13-0.78
	$(\%)$	–	1.00-8.00	4.50	1.00-10.0	6.00	–	0.90-12.4
Basophils	$(x10\ mm^3)$	–	–	–	–	0.10	–	0.00-0.02
	$(\%)$	–	0.00-3.00	0.50	0.00-3.00	1.00	–	0.00-1.40
Lymphocytes	$(x10^3/mm^3)$	–	–	–	–	4.60	–	3.02-6.01
	$(\%)$	–	40.0-75.0	62.0	40.0-75.0	63.0	14.0	46.9-75.1
Monocytes	$(x10^3/mm^3)$	–	–	–	–	0.20	–	0.09-0.46
	$(\%)$	–	1.00-5.00	2.50	1.00-6.00	3.00	–	0.72-8.00

Source: [1] Derived from literature

[2] Data from 26 female and 32 male, Dorset-Delane, adult sheep weighing 40-60 kg. See text for methodology.

115

is given as 180,000 to 475,000. Schalm (1965) reported 10–14 million RBCs and leukocytes ranging from 4,000 to 12,000/mm³. The differential average is lymphocytes 62% (range 40–75%); neutrophils 30% (range 10–50%); eosinophils 4.5% (range 1–10%); monocytes 2.5% (range 1–6%); and basophils 0.5% (range 0–3%). Our data fall within the normal range of differential counts reported for the blood of sheep (Table 34).

As in the pig, the neutrophil count is highest at birth, exceeding 50% of the total. In a few weeks it drops to approximately the adult level. Grunsell (1955) reported that the total white count decreases and that percentages of neutrophils and eosinophils increase with increasing age in adult sheep.

The neutrophils have a diameter of 9–12 μ. The nucleus is exceedingly polymorphous and frequently multilobed; the chromatin is plaqued, filaments are short, and in the female a small nuclear sex bud may be observed in some cells. The cytoplasm stains faintly and usually slightly pinkish. The granulation is dust-like and diffuse. In the band neutrophils, the nucleus is fat and U-shaped with occasional elongation to the S form. Chromatin plaques are not as condensed as in the mature cell. The cytoplasm is slightly bluer and less granular.

The eosinophils have a diameter of 10–16 μ and are somewhat larger than neutrophils, from which they are distinguished by a more gross and more intensive red granulation. The granules are uniform in size, ovoid, and refractile, and they stain orange-red. The cytoplasm is densely packed with granules and a few are scattered over the nucleus. A small amount of pale blue cytoplasm is visible, especially near the margin of the cell.

The basophils have a variable number of dark granules with a reddish halo contrasting well with the lighter blue-staining nucleus. The cell membrane may be irregular because it is pushed into the spaces between the erythrocytes. Basophils are as large as the eosinophils and are recognized by their violet granulation.

The lymphocytes have a diameter of 12.1–19.8 μ (large lymphocytes, about 2–4% of the total WBC count), and 6.6–11 μ (small lymphocytes). The small lymphocytes resemble those of other species. However a special feature is the presence of large blue granules in 5–6% of lymphocytic cells. The large lymphocytes have a round or oval nucleus which is lighter and less pyknotic than that of the small lymphocyte; the zone of cytoplasm is also lighter and broader.

The monocytes are large round cells with a diameter of about 11 to 17.6 μ. The nucleus shows a chromatin pattern of lacy strands which stain relatively intensely. The nuclear shape is usually ameboid, often three-lobed, and the lobes always have wide dimensions. The cytoplasm is rather coarse and blue-gray with ground-glass texture. The staining of the cytoplasm is more intense than that of the large lymphocyte.

Cattle

The average red cell count of the male bovine species is about 8.1 million/mm³, and the cow has about 7.1 million (Table 35). The mean values of erythrocytes and differential values for leukocytes are influenced by the age of the animal. Breed differences have not been reported (Mammerick et al., 1978). There is general agreement that the erythrocyte counts are highest in calves and undergo a gradual decrease until the adult level is reached (between 18 and 36 months of age). The volume percentage of erythrocytes (PCV) and hemoglobin are relatively high at birth and are reduced as soon as the calf ingests colostrum as a result of dilution of plasma. During the nursing period, however, when iron intake is low and the calf is growing,

Table 35. Hematological Values of the Normal Cattle*

Test	Unit	Male (6 animals)			Female (10 animals)		
		Mean	S.D.	Range	Mean	S.D.	Range
Erythrocytes (RBC)	$(\times 10^6/mm^3)$	8.10	1.15	5.80-10.4	7.10	1.10	4.90-9.30
Hemoglobin	(g/dl)	11.5	1.45	8.60-14.4	12.9	2.28	8.20-17.3
MCV	(μ^3)	50.0	1.75	46.5-53.5	54.4	1.50	51.4-57.4
MCH	$(\mu\mu g)$	14.2	1.00	12.2-16.2	18.2	1.25	15.7-20.7
MCHC	(%)	28.8	1.60	25.6-32.0	33.4	1.45	30.5-36.3
Hematocrit (PCV)	(ml%)	40.0	3.50	33.0-47.0	38.6	5.75	37.1-50.1
Sedimentation rate	(mm/hr)	1.45	0.33	0.79-2.11	1.20	0.24	0.72-1.68
Platelets	$(\times 10^3/mm^3)$	350.	87.5	175.-525.	465.	77.5	310.-620.
Leukocytes (WBC)	$(\times 10^3/mm^3)$	9.25	1.40	6.45-12.0	8.55	1.12	6.31-10.8
Neutrophils	$(\times 10^3/mm^3)$	2.82	0.43	1.97-3.66	2.30	0.30	1.70-2.90
	(%)	30.5	5.01	20.5-40.5	26.9	4.80	17.3-35.3
Eosinophils	$(\times 10^3/mm^3)$	0.88	0.14	0.61-1.14	0.60	0.08	0.44-0.76
	(%)	9.50	2.45	3.60-15.4	7.00	2.23	2.54-11.5
Basophils	$(\times 10^3/mm^3)$	0.05	0.01	0.03-0.06	0.00	0.01	0.00-0.02
	(%)	0.50	0.21	0.08-0.92	0.00	0.15	0.00-0.30
Lymphocytes	$(\times 10^3/mm^3)$	4.75	0.72	3.32-6.17	4.97	0.65	3.67-6.27
	(%)	51.4	4.60	42.2-60.6	58.1	5.20	47.7-68.5
Monocytes	$(\times 10^3/mm^3)$	0.75	0.12	0.52-0.97	0.72	0.09	0.55-0.86
	(%)	8.10	2.75	2.60-13.6	8.00	2.50	3.00-13.0

*Holstein, adult cattle, weighing 500-600 kg.

the PCV and Hb may decline steadily over a period of weeks and even months, reaching levels as low as 28–30% for PCV and 8–9 g/dl for Hb. According to Schalm (1965), a range of 5–10 million erythrocytes per mm³ of blood will encompass nearly all of the normal cattle above 1 year of age. The mean values of PCV lie between 37 and 40%, and the mean Hb value for cattle is between 11.5 and 13.0 g/dl. Some authors have reported significant differences between the sexes in hemoglobin values. The diameter of the erythrocytes in normal cattle ranges between 3.6 and 8.0 μ. Reticulocytes are not found beyond the second day of life in the peripheral blood of healthy cattle. Normal bovine erythrocytes do not participate in Rouleau formation, and therefore little or no sedimentation should be present during the period of 1 hour.

Slight anisocytosis is normal and an occasional erythrocyte appears to be twice the size of the smallest cells. The thrombocytes (platelets) appear as rosettes of prominent reddish-purple granules enclosed in a delicate membrane. The individual thrombocyte is usually small, but giant forms up to the size of a red cell may be seen. They often occur in small clusters or loose groups, or individual thrombocytes may be superimposed on erythrocytes. They may occur in large numbers and this correlates well with the size of the buffy coat of the Wintrobe hematocrit.

The mean total leukocyte count of adult cattle is between 7,000 and 8,000/mm³. In aged cows the mean total count may be lower. The lymphocytes are not as numerous in newborn and very young animals as in older ones. At birth, according to Holmann (1955, 1956), 33% of the leukocytes are lymphocytes, at 4 months they have risen to 72%, and then they undergo a slow decline and reach the 50% adult level at about 2 years of age. In the adult cattle Schalm (1965) reported the following leukocyte and differential count ranges: total leukocytes 4,000–12,000/mm³; lymphocytes 45–75%; neutrophils 15–45%; eosinophils 2–20%; monocytes 2–7%; and basophils 0–2%. Our data reported are generally in good agreement with the values given in literature (Table 36).

In mature neutrophils the nucleus is well plaqued and the nuclear membrane is often smooth but may be irregular. Short filaments between nuclear lobes are seen with relative frequency. The cytoplasm is pale pink with a dust-like granulation. Toxic granulation is seen in severe infections and is characterized by the occurrence of a few small black granules widely scattered in the cytoplasm. In banded neutrophils the nucleus is short, fat, and slightly curved. The chromatin clumps are more dispersed than in mature cells and the cytoplasm is paler and shows less granulation.

In metamyelocyte neutrophils the nucleus is quite thick and sometimes angular or bulbous, but it rarely has a kidney bean shape. The chromatin is always plaqued. The cytoplasm is pale gray and the granulation indistinct.

The eosinophils have small, round, regular jewel-like refractile granules. The granules usually fill the cytoplasmic space with occasional slight bulging of the cell margin. The nucleus may be partly covered by the granules.

In the basophils the granules are tightly packed and may be scattered over the nucleus so as to partially obscure it. The granules are purplish and the nuclei are less intensely stained.

The large lymphocytes may be confused with monocytes. They have a much lighter-staining nucleus and cytoplasm. The nucleus has a smooth chromatin pattern and the cytoplasm may occasionally contain small vacuoles. The small lymphocyte, on the other hand, has a granular yet pyknotic compact round nucleus. The cytoplasm is sparse and stains deep blue. Azurophilic granules are seen relatively frequently, but they are not present in every bovine blood film. These granules

Table 36. Hematological Values of the Normal Cow

Test	Unit	Coles[1] (1980) Mean	Range	Lumsden et al[2] (1980) Mean	S.D.	Benjamin[1] (1978) Mean	Range	Mitruka & Rawnsley[3] (1977) Range
Erythrocytes (RBC)	(x10^6/mm^3)	7.00	5.00-8.00	6.40	0.70	7.00	5.00-10.0	4.90-10.4
Hemoglobin	(g/dl)	11.00	8.00-14.0	10.8	1.20	11.0	8.00-14.0	8.20-17.3
MCV	(µ^3)	52.0	40.0-60.0	47.4	4.80	50.0	40.0-60.0	46.5-57.4
MCH	(µµg)	–	–	17.2	1.50	–	14.4-18.6	12.2-20.7
MCHC	(%)	31.0	26.0-34.0	36.1	2.20	30.0	26.0-34.0	25.6-36.3
Hematocrit (PCV)	(%)	34.0	26.0-42.0	30.0	3.20	35.0	24.0-48.0	33.0-50.1
Sedimentation rate	(mm/hr)	–	–	–	–	–	–	0.72-2.11
Platelets	(x10^3/mm^3)	684	542-975	430	110	–	–	175-620
Leukocytes (WBC)	(x10^3/mm^3)	8.90	4.00-12.0	7.40	1.80	8.00	4.00-12.0	6.31-12.0
Neutrophils	(x10^3/mm^3)	–	0.00-1.00	2.80	1.10	–	–	1.70-3.66
	(%)	–	15.0-45.0	38.0	12.0	28.0	15.0-45.0	17.3-40.5
Eosinophils	(x10^3/mm^3)	–	–	0.60	0.50	–	–	0.44-1.14
	(%)	–	2.00-15.0	7.90	6.70	9.00	2.00-20.0	2.54-15.4
Basophils	(x10^3/mm^3)	–	–	0.03	0.20	–	–	0.00-0.06
	(%)	–	0.00-2.00	0.05	0.20	0.50	0.00-2.00	0.00-0.92
Lymphocytes	(x10^3/mm^3)	–	–	3.40	1.20	–	–	3.32-6.27
	(%)	–	48.0-75.0	46.9	10.6	58.0	45.0-75.0	42.2-68.5
Monocytes	(x10^3/mm^3)	–	–	0.30	0.30	–	–	0.52-0.97
	(%)	–	2.00-7.00	4.10	2.50	58.0	2.00-7.00	2.60-13.6

Source: [1]Derived from Literature

[2]Data from 172 female Holstein animals, 4 age groups.   See text for methodology.
[3]Data from 10 Holstein cows weighing 500-600 kg.   See text for methodology.

119

vary considerably in size and shape; some are rod-like and others are very large and resemble blobs of nuclear material.

The nucleus of the monocyte varies in shape from round to convoluted. The chromatin is usually diffuse and appears stringy or granular. It often appears foamy because of vacuoles. True azurophilic granules are not commonly found in the cytoplasm.

Horses

The difference in the red cell count of horses because of sex is similar to that in humans. The RBC count for stallions is around 9.5 million/mm³ and for mares about 6.5 million (Table 37). The WBC count for stallions is 8,000 and for mares 6,900. The differential count is reported as: neutrophils 37–86%; eosinophils 0.6–7.5%; basophils 0–1.3%; lymphocytes 11–45.7%; and monocytes 1.6–12.8% (Table 37). These values essentially agree with the values cited in the literature (Tables 38 and 39).

A distinction should be made, however, in describing the quantitative aspects of the blood of the horse. The data are different when the "hot-blooded" horses, those of essentially Arabian ancestry, and the "cold-blooded" ones are compared. The thoroughbreds, including mustangs and cow ponies of the West, are hot-blooded, as are the American saddle horse, the standard-bred trotter, and the Morgan horse. Schalm (1965) reported the average erythrocyte count for hot-blooded horses as 9.75 million, with a range of 7 to 13 million; for cold-blooded horses the count was 7.5 million with a range of 5.5–9.5 million. The hemoglobin values are correspondingly greater in hot-blooded horses. The leukocyte count for the hot-blooded horse is 10,000/mm³ and for the cold-blooded, 8,500. The erythrocytes are smaller and more numerous in the hot-blooded. In addi-

tion to the 2–3 million difference in RBC count per mm³, the hemoglobin is 2–4 g higher per deciliter of blood in hot-blooded horses, and the mean PCV is 42 for hot-blooded and 35 for cold-blooded.

In thoroughbreds the erythrocyte number increases up to the third month of life, reaching as high as 15 million. The MCV decreases during this time from 33 to 30.4. After the third month there is a constant decline to the mean adult level of about 10 million, while at the same time the MCV increases. Temperament and pulse rate influence the red cell count in that excitable states and an increased pulse rate lead to a greater concentration of erythrocytes in the peripheral blood stream. Erythrocytes appear to enter the blood of the horse in a mature state. Even after severe hemorrhage, there are only a very few reticulocytes.

Rouleau formation is the most prominent feature of the erythrocyte in equine blood, and this phenomenon is associated with a rapid sedimentation rate. Cells in rouleau stain uniformly and deeply. A Howell-Jolly-like body may be found in healthy animals in a small number of erythrocytes. The body varies in size and staining, but it is generally black and in an eccentric position. The platelets in horse's blood vary in shape from the more common oval form to elongated sperm-like structures. Giant forms equal to the red cell size may be observed.

The nuclear chromatin is very heavily plaqued in the mature neutrophils of equine blood, and this causes the nuclear membrane to be jagged and tends to give it the impression of multilobulation. The cytoplasm takes a very pale stain and is filled with very fine, evenly dispersed pinkish granules. Because of the heavy chromatin plaque formation at the nuclear margins, the nuclear outline of the equine band neutrophils is more irregular than those of other species. In comparison with the mature neutrophils, the nucleus is thicker, less

Table 37. Hematological Values of the Normal Horse*

Test	Unit	Male (15 animals)			Female (6 animals)		
		Mean	S.D.	Range	Mean	S.D.	Range
Erythrocytes (RBC)	$(\times 10^6/mm^3)$	9.40	0.55	8.30-10.5	6.50	0.21	6.08-8.92
Hemoglobin	(g/dl)	11.1	1.50	8.10-14.1	10.7	1.15	8.40-13.0
MCV	(μ^3)	35.5	4.80	26.6-45.1	48.6	3.40	42.0-54.4
MCH	$(\mu\mu g)$	11.8	0.45	12.9-19.7	16.5	0.55	15.4-17.6
MCHC	(%)	33.2	2.15	28.9-37.5	33.9	2.20	29.5-38.3
Hematocrit (PCV)	(ml%)	33.4	3.50	26.4-40.4	31.6	2.42	26.8-36.4
Sedimentation rate	(mm/hr)	7.20	3.40	0.40-14.0	9.50	4.50	0.00-18.5
Platelets	$(\times 10^3/mm^3)$	250.	50.0	150.-350.	220.	65.0	90.0-350.
Leukocytes (WBC)	$(\times 10^3/mm^3)$	8.00	1.50	5.00-11.0	6.90	1.41	4.08-9.72
Neutrophils	$(\times 10^3/mm^3)$	4.95	0.98	3.10-6.81	3.92	0.80	2.32-5.52
	(%)	61.9	12.2	37.5-86.3	56.8	10.4	36.0-77.6
Eosinophils	$(\times 10^3/mm^3)$	0.32	0.06	0.20-0.44	0.26	0.06	0.15-0.36
	(%)	4.00	1.72	0.62-7.44	3.70	1.15	1.40-6.00
Basophils	$(\times 10^3/mm^3)$	0.05	0.01	0.03-0.07	0.03	0.01	0.03-0.05
	(%)	0.62	0.31	0.00-1.24	0.50	0.20	0.10-0.90
Lymphocytes	$(\times 10^3/mm^3)$	2.20	0.41	1.38-3.02	2.10	0.42	1.24-2.95
	(%)	27.5	8.20	11.1-43.9	30.4	7.65	15.1-45.7
Monocytes	$(\times 10^3/mm^3)$	0.48	0.09	0.30-0.66	0.59	0.12	0.35-0.83
	(%)	6.00	2.21	1.58-10.4	8.50	2.15	4.20-12.8

*Light horses of mixed breed weighing 350-450 kg.

Table 38. Hematological Values of Thoroughbred Horses

Test	Unit	Coles (1980)[1] Mean	Coles (1980)[1] Range	Lumsden et al (1980)[2] Mares Mean	Mares S.D.	Lumsden et al (1980)[2] Foals Mean	Foals S.D.	Benjamin (1978)[1] Thoroughbred Foals 1-26 mths. Mean	Range
Erythrocytes (RBC)	(x10^6/mm^3)	9.5	7-13	9.0	1.3	11.90	0.80	11.77	8.52-15.2
Hemoglobin	(g/dl)	15.0	11-18	14.8	2.0	14.10	0.50	12.2	9.9-16
MCV	(µ^3)	-	-	45.3	2.1	33.60	1.40	42	37-50
MCH	(µµg)	-	-	16.6	0.8	12.0	0.40	13	
MCHC	(%)	-	-	36.1	1.2	34.6	0.90	33	31-35
Hematocrit (PCV)	(%)	41.5	32-52	40.80	6.0	39.90	1.60	36.5	23-49
Sedimentation rate	(mm/hr)	15	-	-	-	-	-		
Platelets	(x10^3/mm^3)	-	-	239	80	284	22		
Leukocytes (WBC)	(x10^3/mm^3)	10	7-14	8.1	1.5	11.80	1.30	13.1	4.8-13.4
Neutrophils	(x10^3/mm^3)	-	0-2	4.13	1.06	7.33	1.25		
	(%)	-	30-65	51.80	8.90	57.80	4.20	53.3	26.82
Eosinophils	(x10^3/mm^3)	-	1-10	0.20	0.20	0.10	0.20		
	(%)	-	-	2.70	2.60	1.10	1.40	2.1	0-10
Basophils	(x10^3/mm^3)	-	-	0.02	0.03	0	0		
	(%)	-	0-3	0.20	0.40	0	0	0.5	0-1.0
Lymphocytes	(x10^3/mm^3)	-	-	3.40	0.80	4.90	0.90		
	(%)	-	15-50	41.80	7.80	39.40	6.70	42.2	15-73
Monocytes	(x10^3/mm^3)	-	-	0.20	0.10	0.30	0.30		
	(%)	-	2-10	2.90	1.70	2.40	2.80	2.4	0-13

Source: [1]Derived from literature
[2]Data from 60 Thoroughbred mares, early to late pregnancy and foals between 2 to 8 weeks of age. See text for methodology.

Table 38. (con't)   Hematological Values of Thoroughbred Horses

		Sato et al[3] (1978)		Melby and Altman[1] (1976)	
		Mean	S.D.	Mean	%C.V.
Erythrocytes (RBC)	(x10^6/mm^3)	9.15	0.61	10.1	13.0
Hemoglobin	(g/dl)	12.8	0.37	15.0	9.00
MCV	(μ^3)	42.0	2.10	44.0	12.0
MCH	(μμg)	13.9	0.80	15.0	14.0
MCHC	(%)	33.9	0.60	34.0	8.00
Hematocrit (PCV)	(%)	38.0	1.00	44.0	11.0
Sedimentation rate	(mm/hr)	–	–	–	–
Platelets	(x10^3/mm^3)	180	–	–	–
Leukocytes (WBC)	(x10^3/mm^3)	9.36	0.20	9.20	23.0
Neutrophils	(x10^3/mm^3)	4.12	0.81	4.70	–
	(%)	43.0	7.90	51.0	15.0
Eosinophils	(x10^3/mm^3)	0.16	0.06	0.37	–
	(%)	1.00	1.00	4.00	85.0
Basophils	(x10^3/mm^3)	0.00	0.00	0.04	–
	(%)	0.00	0.00	0.40	85.0
Lymphocytes	(x10^3/mm^3)	5.10	0.69	3.90	–
	(%)	53.3	7.50	47.0	20.0
Monocytes	(x10^3/mm^3)	0.19	0.09	0.28	–
	(%)	2.00	1.00	3.00	66.0

Source: [1] Derived from Literature.

[3] Data from 5 thoroughbred foals during the first six months of life. Blood samples were analyzed with manual methods.

Table 39. Hematological Values of Normal Standardbred Horses

Test	Unit	Cole(1980)[1] Mean	Cole(1980)[1] Range	Lumsden et al[2] (1980) Mean	Lumsden et al[2] (1980) S.D.	Benjamin[1] (1978) Mean	Benjamin[1] (1978) Range	Melby&Altman[1] (1976) Mean	Melby&Altman[1] (1976) %C.V.	Mitruka & Rawnsley[3] (1976) Range
Erythracytes (RBC)	$(\times10^6/mm^3)$	7.5	5.5-9.5	8.8	1.0	-	6.5-9.4	8.4	-	6.08-10.5
Hemoglobin	(g/dl)	11.5	8.0-14	14.6	1.7	-	9.0-14	13.0	8	8.10-14.1
MCV	(μ^3)	46.0	34-58	45	2.4	44	39-52	-	-	26.60-54.4
MCH	$(\mu\mu g)$	-	-	16.6	0.9	16.9	15.2-18.6	-	-	12.9-19.7
MCHC	$(\%)$	35.0	31-37	37.2	1.2	33	31-35	35	1	28.9-38.3
Hematocrit (PCV)	$(\%)$	35	24-44	39	4.3	-	30-44	36	7	26.4-40.4
Platelets	$(\times10^3/mm^3)$	335	249-461	139	44	-	-	235	50	90-350
Leukocytes (WBC)	$(\times10^3/mm^3)$	9.26	6-12	7.5	1.3	-	5-11	7.3	-	4.08-11.0
Neutrophils	$(\times10^3/mm^3)$	-	0-2	-	-	-	-	4.2	-	2.32-6.81
	$(\%)$	-	35-75	-	-	56	50-65	54	16	36-86.3
Eosinophils	$(\times10^3/mm^3)$	-	-	0.2	0.2	-	-	0.33	-	0.15-0.44
	$(\%)$	-	1-10	2.8	2.2	4	1-5	5	57	0.63-7.44
Basophils	$(\times10^3/mm^3)$	-	-	0.02	0.04	-	-	0.04	-	0.03-0.07
	$(\%)$	-	0-3	0.30	0.6	0.5	0-1	0.60	130	0-124
Lymphocytes	$(\times10^3/mm^3)$	-	-	3.10	0.7	-	-	2.3	-	1.24-3.02
	$(\%)$	-	15-50	41.8	7.1	30	20-40	29	22	11.1-45.7
Monocytes	$(\times10^3/mm^3)$	-	-	0.2	0.1	-	-	0.33	-	0.3-0.83
	$(\%)$	-	2-6	2.6	1.6	8	2-12	5.0	50	1.58-12.80

Source: [1]Derived from Literature, male and female adult animals, standard hematological methodology.

[2]Data from 50 animals in light to heavy training. Values were obtained with Coulter Model-S. See text for methodology.

[3]Data from 6 female and 15 male light horses of mixed breed weighing 350-450 kg. See text for methodology.

Table 40. Comparative Hematological Values of Small Experimental Animals*

Test	Unit	Mouse	Rat	Hamster	Guinea pig	Rabbit	Chicken	Cat	Dog
Erythrocytes (RBC)	$(\times 10^6/mm^3)$	6.70-12.5	5.00-12.0	3.00-10.0	3.00-7.00	4.00-8.60	1.25-4.50	4.50-9.00	4.30-8.77
Hemoglobin	(g/dl)	10.2-16.6	11.1-18.0	10.0-20.2	11.2-18.1	9.30-19.3	7.00-18.6	9.00-12.7	10.0-19.0
MCV	(μ^3)	31.0-62.0	44.5-69.0	54.5-78.5	61.0-98.0	57.0-90.0	100.-139.	49.0-59.0	50.0-88.0
MCH	$(\mu\mu g)$	9.20-20.8	12.0-24.5	15.4-26.8	22.5-28.5	16.0-31.0	25.0-48.0	13.0-17.0	15.0-29.0
MCHC	(%)	22.0-35.5	21.6-42.0	26.5-37.4	28.0-39.0	22.0-38.7	20.0-34.0	24.0-30.0	23.0-42.0
Hematocrit (PCV)	(ml%)	32.0-54.0	36.0-52.0	36.0-59.0	37.0-51.0	30.0-53.0	23.0-55.0	34.0-46.0	32.0-60.0
Sedimentation rate	(mm/hr)	0.00-1.00	0.50-2.00	0.20-1.00	0.00-2.95	0.90-3.70	0.50-6.50	0.50-7.30	0.00-8.00
Platelets	$(\times 10^3/mm^3)$	150.-500.	140.-600.	300.-680.	225.-800.	120.-800.	13.0-70.0	70.0-400.	161.-650.
Leukocytes (WBC)	$(\times 10^3/mm^3)$	5.40-16.0	3.00-15.0	2.55-11.6	5.00-18.0	2.00-15.0	9.00-32.0	11.0-20.0	7.00-22.0
Neutrophils	$(\times 10^3/mm^3)$	1.89-3.62	1.10-4.00	1.00-4.06	1.50-8.00	1.00-8.00	2.00-10.0	6.50-12.5	4.00-11.5
	(%)	8.00-42.9	4.00-50.0	3.00-42.0	20.0-60.0	10.0-85.0	15.1-50.0	40.0-82.0	22.0-96.0
Eosinophils	$(\times 10^3/mm^3)$	0.00-0.51	0.00-0.08	0.00-0.18	0.00-1.00	0.00-0.80	0.00-5.25	0.80-1.50	0.00-1.50
	(%)	0.00-2.90	0.00-4.00	0.00-4.36	0.00-8.00	0.00-8.00	0.00-16.0	2.00-11.0	0.00-18.0
Basophils	$(\times 10^3/mm^3)$	0.00-0.15	0.00-0.04	0.00-0.43	0.00-0.30	0.00-0.75	0.30-2.60	0.00-0.20	0.00-0.10
	(%)	0.00-0.85	0.00-2.00	0.00-6.00	0.00-2.00	0.00-8.00	0.00-8.00	0.00-1.00	0.00-1.00
Lymphocytes	$(\times 10^3/mm^3)$	8.00-18.0	4.00-10.0	3.00-7.86	2.00-11.0	2.00-8.60	4.80-19.3	2.00-9.00	1.00-7.50
	(%)	55.0-95.0	40.0-95.0	50.1-95.7	37.0-81.0	25.0-95.0	29.0-84.0	15.0-48.0	5.00-52.0
Monocytes	$(\times 10^3/mm^3)$	0.00-1.50	0.00-0.10	0.00-0.42	0.05-1.00	0.10-1.80	0.03-1.20	0.35-1.40	0.00-1.20
	(%)	0.00-8.00	0.00-8.00	0.00-4.90	1.00-9.00	0.50-16.0	0.05-7.00	1.00-9.00	0.00-8.00

*Mixed sex, breed and age. Values are pooled from our data and from those reported in literature.

Table 41. Comparative Hematological Values of Large Experimental Animals* and Humans**

Test	Unit	Monkey	Pig	Goat	Sheep	Cattle	Horse	Man
Erythrocytes (RBC)	$(\times 10^6/mm^3)$	3.00-9.00	5.00-10.0	14.0-22.0	8.00-14.0	4.00-11.0	6.00-11.0	4.20-6.20
Hemoglobin	(g/dl)	09.0-17.3	10.0-15.0	8.80-15.4	8.50-13.2	7.70-18.5	8.00-15.0	12.0-17.0
MCV	(μ^3)	50.0-105.	50.0-62.0	15.0-22.0	28.0-34.9	44.0-59.0	26.0-58.0	82.0-92.0
MCH	$(\mu\mu g)$	17.1-35.7	16.0-19.0	5.00-8.00	8.20-12.3	12.0-22.0	10.0-20.0	27.0-31.0
MCHC	(%)	24.0-40.0	25.0-36.0	28.0-45.0	30.0-38.0	25.0-38.0	28.0-40.0	32.0-36.0
Hematocrit (PCV)	(ml%)	30.0-50.0	36.0-47.0	23.0-38.6	29.0-38.5	33.0-55.0	26.0-42.0	37.0-52.0
Sedimentation rate	(mm/hr)	1.00-5.50	2.00-9.00	0.30-3.80	0.50-2.50	0.70-2.80	0.00-20.0	0.00-20.0
Platelets	$(\times 10^3/mm^3)$	20.0-600.	115.-425.	20.0-70.0	170.-530.	150.-650.	150.-400.	145.-375.
Leukocytes	$(\times 10^3/mm^3)$	4.00-22.0	6.00-25.0	3.00-14.0	4.00-12.0	6.00-12.0	5.00-11.9	4.50-11.0
Neutrophils	$(\times 10^3/mm^3)$	2.00-8.10	2.00-8.85	1.00-6.35	1.00-4.50	1.50-3.90	2.00-6.97	4.10-6.50
	(%)	14.0-71.0	17.0-49.0	26.1-55.3	10.0-50.0	16.0-42.0	36.0-89.0	46.9-64.9
Eosinophils	$(\times 10^3/mm^3)$	0.10-1.20	0.18-1.32	0.10-1.10	0.08-0.80	0.30-2.30	0.04-0.50	0.06-0.40
	(%)	0.00-16.0	1.00-9.00	1.00-6.00	0.60-17.0	3.00-18.0	0.50-8.80	0.90-5.10
Basophils	$(\times 10^3/mm^3)$	0.00-0.06	0.00-0.47	0.00-0.06	0.00-0.15	0.00-0.09	0.00-0.11	0.00-0.09
	(%)	0.00-4.00	0.00-3.00	0.00-0.80	0.00-2.00	0.00-1.00	0.00-2.00	0.04-1.10
Lymphocytes	$(\times 10^3/mm^3)$	3.40-9.90	4.00-13.8	2.00-9.20	3.00-6.50	3.10-6.90	1.10-3.44	2.10-3.32
	(%)	20.0-95.0	40.0-70.0	40.0-72.0	40.0-78.0	40.0-70.0	10.0-48.0	28.0-42.0
Monocytes	$(\times 10^3/mm^3)$	0.00-0.90	0.30-2.03	0.05-0.40	0.08-0.60	0.04-1.00	0.30-1.00	0.32-0.68
	(%)	0.00-8.00	2.00-8.00	0.50-4.30	0.50-8.00	2.00-15.0	1.00-15.0	4.20-8.20

*Mixed sex, breed and age. Values are pooled from our data and from those reported in literature.

**Adult males and females.

Table 42. Representative Hematological Reference Values in Humans

Age	Erythrocytes 10¹²/l	Hemoglobin g/dl	Hematocrit %	MCV fl	MCH pg	MCHC %	Platelet 10³/µl
Newborn	-	13.5-20.5	41.7-62.9	-	-	-	140 - 290
1 day	3.74-6.54	16.8-21.2	53.6-68.4	110 - 128	36.7	29.2-33.5	-
4 weeks	2.80-4.40	9.50-15.9	31.2-40.8	92.9- 109	35.3	31.7-38.1	175 - 380
6 months	3.89-55.3	9.70-13.5	29.4-41.8	67.0-89.0	21.7-30.5	29.3-37.7	-
12 months	3.74-5.78	7.70-14.5	25.9-43.1	57.0-89.0	16.2-30.8	27.7-36.9	220 - 460
2 years	3.93-5.63	8.90-14.5	29.7-41.9	61.0-89.0	17.6-31.8	28.4-36.8	-
10 years	3.85-5.75	10.7-15.7	32.2-44.8	68.0-94.0	22.1-33.1	31.4-37.2	-
Adult male	4.70-6.10	14.0-18.0	42.0-52.0	81.0-99.0	21.0-31.0	32.0-36.0	150 - 400
female	4.20-5.40	12.0-16.0	37.0-47.0	81.0-99.0	21.0-31.0	32.0-36.0	150 - 400
male	4.47-6.05	13.8-18.4	41.0-54.0	79.0- 100	26.7-33.9	31.0-36.0	-
female	3.96-5.36	11.6-15.9	35.0-47.0	76.0-99.0	25.2-33.4	31.0-36.0	-
Elderly: (70-90 years)							
male	3.27-5.33	10.1-19.6	30.0-55.0	96.5	33.7	34.9	-
female	3.33-5.38	10.4-18.3	31.0-51.0	95.2	32.9	34.6	-

Source: Altman and Dittmer (1961), Altman and Katz (1977), Henry et al (1979), Giomo R. et al (1980), Sonnenwirth and Jarett (1980).

* Values given are 95% ranges

127

Table 43. Representative Hematological Reference Values in Humans

Age	Total White Count $10^3/\mu l$	Neutrophils %	Lymphocytes %	Monocytes %	Eosinophils %	Basophils %
Newborn	9.0-30.0	61	31	5.4	2.1	0.5
1 day	9.5-34.0	61	31	5.4	2.1	0.5
4 weeks	5.0-19.5	35	55	6.6	2.9	0.5
6 months	6.0-17.5	32	60	5.0	2.6	0.4
12 months	6.0-17.5	31	61	4.8	2.7	0.5
10 years	4.5-13.5	54	39	4.2	2.4	0.4
Adults	3.0-11.0	60	33	4.0	2.7	0.3

Sources: Altman and Dittmer (1961), Altman and Katz (1977), Henry et al. (1979), Giomo R et al. (1980),Sonnenwirth and Jarett (1980)

Table 44. Reference Values - Westergren Erythrocyte
Sedimentation Rate (ESR)

Male under 50 years	< 15 mm/hr
Male over 50 years	< 20 mm/hr
Female under 50 years	< 20 mm/hr
Female over 50 years	< 30 mm/hr

Source: Henry et al.(1979)

knobby, and not coiled in the band neutrophils. The cytoplasm is similar in both types of neutrophils. The band neutrophils may be as many as 1 to 10% in bacterial infections, but metamyelocyte neutrophils are not found in the peripheral blood of horses. Neutrophilia with a marked shift to the left appears to be the common response to bacterial infection in the equine.

In the eosinophils the granules are very large but may vary in size, and they are usually tightly packed. Because of the large size of the granules, the cell wall conforms to the granule contour and gives the cell a raspberry-like appearance. The granules stain bright orange, and a pale blue cytoplasm is sometimes visible between the granules and along the edge of the cell.

The basophils have granules of irregular size and shape, and they stain darkly. The granules may be tightly packed or scattered irregularly in the cytoplasm with many open spaces; some may be over the nucleus and other granules may cause the cell wall to bulge.

The lymphocytes have a heavily staining nucleus and limited cytoplasm. Azurophilic granules, when present, are few in number, small in size, irregular in shape, and scattered over the cell cytoplasm without being evenly dispersed.

In the monocytes the nucleus is in the form of a broad kidney bean located to one side of the cell. Folding of the nucleus is not common. The chromatin is lacy or stringy, the cytoplasm is blue-gray and granular, and pinpoint pinkish azurophilic granules are scattered throughout.

The comparative hematologic values of 14 species of normal experimental animals and humans are summarized in Tables 40 and 41. The data are extracted from the voluminous literature and compared with our own observations. Extreme low or high values and questionable data have been excluded from these tables. No effort was made to analyze the values statistically, and the data are presented in the minimum and maximum range values either directly from the reported work or calculated by the mean value ± 2 standard deviations. More detailed hematological values in different age groups of humans are presented in Tables 42–44.

Methods in Clinical Biochemistry

Most of the methods used to determine the normal values of clinical biochemical and cellular constituents of blood from experimental animals are standard methods described in the textbooks of human clinical pathology or veterinary medicine. In our studies more than one technique was frequently used to check the reliability of the methods. Newer methods, especially microtechniques and automated methods, were also used in addition to the routine manual methods conventionally used in the hematology and clinical chemistry laboratory. The purpose of this section is not to describe the methods in detail, but to discuss briefly the various techniques and to list the specific methods used in our studies. References are given for the alternative methods in hematological and chemical determinations. The methods currently used to establish the normal values in laboratory animals are briefly described, and different available methods are listed to emphasize that variability in the results is often caused by variability in the materials and methods employed.

Experimental Animals

Experimental animals used for clinical biochemical determinations were selected, clinically normal and laboratory conditioned as described in Section III. Blood samples were drawn from the animals for hematology and chemistries at the same time whenever possible.

Manual Methods of Clinical Biochemical Determinations

Methods in clinical biochemical determinations of blood (serum or plasma) have been extensively discussed in excellent textbooks and laboratory manuals (Henry *et al.*, 1974, 1979; Tietz, 1976; Miller and Weller, 1971; Hoffman, 1970; Hepler, 1968; Bauer *et al.*, 1975; Frankel *et al.*, 1970; Wolf *et al.*, 1972; Wolf *et al.*, 1973; Lamela, 1971; Goodale and Widmann, 1979; Werner, 1976; Kaplan and Szabo, 1979; Coles, 1980; Cornelius *et al.*, 1971; Medway, 1969, and others). There are many methods for the determination of a chemical component in blood, particularly enzymes. For example, at least six methods for determining serum phosphatase activity are available (Table 1). Therefore it is unnecessary to repeat the information on the methodology of clinical chemical determinations and interpretations in this section. Table 2 summarizes the test methods used

131

Table 1. Comparison of Methods for Assay of Serum Alkaline Phosphatase Activity

	Bodansky	Reinhart	King	Brock	Babson	McComb
Type	Manual	Manual	Manual	Manual	Manual	Manual
Buffer	Barbitol	Barbitol	Carbonate	Glycine	2-methyl, 2-amino-propanol-1	2-methyl, 2-amino-propanol-1
pH of Reaction	8.5-8.7	9.4-9.8	9.5-9.7	9.8-10.1	10.1	10.1
Assay Unit	1 mg P produced in 60 min	1 mg P produced in 60 min	1 mg phenol produced in 15 min	1 m mol p-nitro-phenol formed in 60 min	1 μ mol phenol-phthalein produced in 1 min	1 μ mol. p-nitrophenol split in 1 min
Substrate	β-glycero-phosphate	β-glycero-phosphate	Phenyl-phosphate	p-nitro-phenyl phosphate	Phenol-phthalein monophos-phate	p-nitrophenyl phosphate
Normal Range	1.5-4.0 unit/dl	2.2-6.5 unit/dl	3.5-13. unit/dl	0.7-2.7 unit/l	9.0-35. mIU/ml	6.0-110 mIU/ml
Normal Range in International units (μ mol/min/ml)	8-22 mIU/ml	15-35 mIU/ml	25-92 mIU/ml	13-38 mIU/ml	9-35 mIU/ml	6-100 mIU/ml

Source: Wilkinson (1962); King (1965); Bowers and McComb (1966); Tietz (1970).

Table 2. Methods Used in the Determination of Biochemical Constituents in Sera of Experimental Animals

Test	Method	Reference
Amylase	Spectrophotometric determination of reducing substances from starch derivatives	Somogyi (1960); Henry and Chiamori (1960)
Bicarbonate	Titrimetric determination (micromethod)	Segal (1955)
Bilirubin (Total)	Diazo reaction (Malloy and Evelyn)	Malloy and Evelyn (1937); With (1968)
Bromosulfophthalein Retention (BSP)	Spectrophotometric determination - BSP dissociation by alkaline buffer containing P-toluene sulfonic acid	Seligson et al (1957)
Calcium	Atomic absorption spectrophotometric	Trudeau and Freier (1967)
Chloride	Chloridometric	Cotlove (1961)
Cholesterol	Spectrophotometric, Liebermann-Buchard reagent (acetic anhydride - H_2SO_4 - acetic acid)	Abell et al (1958)
Creatinephosphokinase (CPK)	Spectrophotometric measurement of linked reaction (creatine + ATP to creatine phosphate)	Rosalki (1967)
Creatinine	Spectrophotometric determination of alkaline creatinine picrate reaction (Jaffe reaction)	Taussky (1954); Owen et al (1954)
Glucose	O-Toluidine method	Dubowski (1962); Hyvarinen and Nikkila (1962)
Lactic dehydrogenase (LDH)	Spectrophotometric measurement of linked reaction (conversion of lactate to pyruvate with simultaneous reduction of NAD to NADH)	Amador et al (1963)

133

Table 2 (con't). Methods Used in the Determination of Biochemical Constituents in Sera of Experimental Aminals

Test	Method	Reference
Magnesium	Atomic absorption spectrophotometric	Perkin Elmer Corp. Handbook (1971)
Phosphatase, acid	Spectrophotometric determination of hydrolysis of sodium β-glycerophosphate (pH 5.0) at 37 C	Jones and Reinhart (1942); Babson et al (1959)
Phosphatase, alkaline	Spectrophotometric determination of hydrolysis of phenylphosphate (pH 10.0) at 37 C	Amador et al (1963)
Phosphorus, inorganic	Spectrophotometric determination of Molybdi-phosphate reduction by Fe^{++}	Goldenberg and Fernandez (1966)
Potassium	Flame photometric (Li background)	Vallee and Thiers (1965); Instru-ment Lab., Inc. Manual (1973)
Protein, Total	Spectrophotometric determination of colored com-plex with Cu^{++} in alkaline solution (Biuret re-action)	Henry et al (1957)
Protein Electro-phoretic fractions Albumin α₁ α₂ β γ	Cellulose acetate; Tris-Barbital buffer pH 8.6; 110 volts for 20 minutes; Ponceau S staining; Densitometer scanning	Kohn (1957, 1958)
Sodium	Flame photometric (Li background)	Valle and Thiers (1965)
Transaminase, alanine (SGPT)	Spectrophotometric determination of linked re-action (rate of formation of pyruvate from alanine-linked with LDH-NADH system)	Henley and Pollard (1955); Henry et al (1960)

134

Table 2 (con't.) Methods Used in the Determination of Biochemical Constituents in Sera of Experimental Animals

Test	Method	Reference
Transaminase, aspartate (SGOT)	Spectrophotometric determination of linked reaction (rate of formation of oxaloacetic acid from aspartate-linked with MDH-NADH system)	Ladue et al (1954); Henry et al (1960)
Urea nitrogen	Spectrophotometric measurement of diacetyl monoxime reaction	Wybenga et al (1971)
Uric acid	Spectrophotometric measurement of reaction with phosphotungstic acid	Pileggi et al (1972)

135

in this work to determine the chemical constituents of blood or serum from experimental animals. The table also lists other works in which the methods have been discussed in detail. A general discussion of the chemical values determined in normal animal species is presented elsewhere in this book.

Lumsden et al. (1979, 1980) reported biochemical reference values of certain animal species which we have included in this book for comparison with other data. Their methods of determinations were: Serum calcium and magnesium by atomic absorption spectrophotometer (Trudeau and Freier, 1967); Phosphorus content by the reduction of phosphomolybdate (Goldenburg and Fernandez, 1966); Sodium and potassium in heparinized plasma were measured by flame emission (Instrument Laboratory, Inc. Fort Lee, New Jersey); Chloride in heparinized plasma by Coulometric dilution (Buchler-Cotlove Chloridometer, Buchler Instruments, Inc., Fort Lee, New Jersey); Plasma glucose was determined by O–toluidine method of Feteris (1965); Blood Urea Nitrogen (BUN) by diacetylmonoxime thiosemicarbazide method (Crocker, 1967); Serum Creatinine by a modified Jaffe reaction (Owen et al., 1954); Total Cholesterol by the method of Wybenga et al. (1970); Direct, Indirect and Total Bilirubin by the method of Jendrassik as modified by Nosslin (Jendrassik Bilirubin Set American Monitor Corp., Indianapolis, Indiana (Michaelson, 1961); Serum Iron by Ferrocheck II system (Hyland Division, Travenol Lab., Inc., Costa Mesa, California); Total Protein in K–EDTA plasma using a hand refractometer (American Optical Corp., Buffalo, New York) and in Serum by the manual Biuret Method (Weichselbaum, 1946); the Serum Proteins were separated on agarose gel by electrophoresis at pH 8.6, stained with amido black 10B and quantitated by transmission densitometry (Clifford Electrophoresis Densitometer, Clifford Instruments, Natick, Massachusetts); Serum GOT (Ast–aspartate aminotransferase and GPT (Alt–Alanine aminotransferase) were measured by a modified method of Karmen et al. (1955) and Wroblewski and Ladue (1956) at 30°C in an LKB Model 8600 Reaction Rate Analyzer (LKP Produkter AB, Brommal, Sweden); Serum Lactic Dehydrogenase (LDH) was measured with an LDH–L Test Combination kit (Boehringer Manheim, Ville St., Laurent, Quebec) at 30°C as described by Gay et al. (1968); Serum Alkaline Phosphatase according to Babson et al. (1966); Amylase by the method of Caraway (1959); Lipase as described by Tietz and Fiereck (1966); CPK by using a CPK activated UV Kit at 30°C according to Rosalki (1967); and Gamma-glutamyl transferase was measured by the method described by Szasz (1969).

Automated Chemical Analyzers

Three distinctively different operating principles were utilized in the development of most of the automated analyzers available to the clinical laboratory. These three types include continuous flow, discrete sampling and centrifugal fast analyzers. A brief description of some of the instruments in these categories will be given in the following section. The reader is referred to Raphel (1976) and Ferris (1980) for further details.

1. **Continuous Flow Analyzers.** Technicon (Technicon Instrument Corporation, Tarrytown, New York) manufactures instrument which include Auto Analyzers, SMA (6/60), SMA (12/60), SMA II and SMAC utilizing continuous flow principle. In these systems, in succession, the sample is drawn into the instrument and step by step goes through dialysis, if needed, through addition of reagents, into incubation for completion of the reaction and

through the colorimeter flow cell, from which signals are recorded. Samples are separated by air bubbles and as they flow through the colorimeter flow cell, the reacted sample is read at a steady state plateau.

Auto Analyzer II: Since the introduction of the Technicon Auto Analyzer continuous flow analytical instrument to the clinical laboratory, many modifications have taken place in order to improve the capability and versatility of the system. The second generation Technicon Auto Analyzer II continuous flow analytical instrument provides a number of advantages over its predecessor, the Auto Analyzer system. These include steady-state analysis with improved precision, linear results requiring single-point calibration, simultaneous blank measurement and correction, and improved chemical procedures. However, the dedicated analytical cartridge limits the methods that can be run on a single channel machine to those for which cartridges are available. The cost of such cartridges, together with the operator involvement in method changeover are not suitable economically for the laboratories which have small numbers of samples for analysis over a wide range of methods. The modification of multitest cartridge in Auto Analyzer II allows any one of 15 possible analyses to be carried out by push-button selection of a given pathway within a permanently assembled analytical cartridge. The Auto Analyzer II is thus capable of running three different tests and a blank at desirable rates at 60–80 samples/hour.

Sequential Multiple Analyzers (SMA). The SMA are a series of instruments manufactured by Technicon and designed to process many tests at the same time. The SMA instrument has a numerical qualification to indicate the number of tests that may be made simultaneously at a specified rate per hour. The SMA 12/60 which does 12

tests (calcium, phosphorus, creatinine, creatine phosphokinase (CPK), uric acid, cholesterol, total protein, albumin, total bilirubin, alkaline phosphatase, lactic dehydrogenase and glutamic oxaloacetic transaminase–GOT) at the rate of 60 samples per hour and the SMA 6/60 which does six tests (glucose, blood urea nitrogen, sodium, potassium, chloride and carbon dioxide) at the rate of 60 samples per hour.

SMA II. The SMA II is a computerized sequential multiple analyzer. The chemical assays used in the cartridges are identical to those with SMA 12/60 and SMA 6/60 systems. However, the analysis rate is increased from 60 to 90 samples per hour. Difference between SMA II and SMA 12/60 would be due to electromechanical, hydraulic or software changes from the peak monitor and signal processor. Schatz (1976) reported an excellent correlation between SMA II and SMA 12/60 performance by analyzing human serum at a specified concentration level for selected methods.

SMAC. The Technicon SMAC high speed, computer controlled biochemical analyzer is the third generation of continuous flow instruments for clinical laboratory. Snyder and Leon (1976). SMAC operating at 150 samples per hour has reduced the sample consumption to 250 to 325 μl for a 20 test profile depending on the choice of tests. Selectivity on the SMAC system is due to the low reagent and sample consumption for example, the average amount of sample per test is 12 μl and the average volume of reagent used per test for the same chemistries is 250 μl/test at 150 samples per hour. To insure minimal contamination of one specimen by another in the SMAC Analyzer, the probe is moved up and down rapidly to introduce four air segments in the leading part of the sample as it enters the analytical system. This reduces contamination from the specimen analyzed previously and the water used in the rinse cycle, and enables a much higher rate of analysis. In the SMAC analyzer the pump consists of small replaceable modules dedicated to one analytical cartridge. Air segmentation of the sample diluent is performed to minimize sample interaction by the introduction of 90 bubbles per minute at a uniform rate through a sample valve mechanism. There are several unique features incorporated in the SMAC system. Unlike all other Technicon chemistry analyzer systems, the air bubbles in the reagent stream of SMAC system are not removed before it enters the cuvet. Elimination of noise is accomplished electronically. The sudden large changes in the voltage caused by an air bubble is differentiated from the more gradual increase, associated with the development of a peak and is filtered out by computer monitoring of the absorbance. The passage of frequent air bubbles through the cuvet enhances its cleaning and reduces sample interaction. The cuvets in the SMAC system are mounted in such a way that a single photomultiplier tube is used. It has the capability of interrogating each cuvet four times per second. Interrogation is determined by a scanning disc, so that a signal is recorded from each cuvet in turn. Fiber optics are used to transmit light to and from the cuvets. Interference filters and a reference channel are used as in other Auto Analyzer colorimeters. Other unique features of the SMAC system is the use of flow through ion specific electrodes for sodium and potassium determinations. A process control computer is an integral part of the SMAC system. It governs the phasing of the analytical process and also regulates the display of results. The SMAC system provides a printed output of a numerical result as well as a graphical display relating the measured value to the normal range and abnormal results are identified by a symbol to differentiate them from those falling within the normal range.

Technicon methodologies used for each test procedure in Autoanalyzer, SMA and SMAC systems are summarized in Table 3. SMA and Auto Analyzers have been used for testing specimens from experimental animals, medical research and veterinary pathology. Coles (1980), Benirschke (1978), Benjamin (1978), and Melby and Altman (1977) have reported blood chemistry data of various animal species which were analyzed with Technicon instruments. In recent years, normal reference values have been established in large private commercial laboratories analyzing large number of specimens from various animal species and humans with SMAC instrument.

2. **Discrete Sampling Analyzer.** In this type of instrument, exact duplication of manual procedures are handled automatically. An aliquot of each sample for any specific determination progresses through the system in separate sample cups until completion of analysis. This type of instrument includes Hycel Mark X and XVII, Beckman DSA560 and DSA564, DuPont ACA, Harleco Clinicard American Optical Corporation, Robot Chemist American Monitor Programachem 1040, and Perkin Elmer C–4 automatic analyzer. Other instruments offered by foreign manufacturer include the Vickers M300, Joyce Loehl Mecolab, and AGA Autochemist.

Automatic Clinical Analyzer (ACA). The DuPont ACA permits many different tests on the same specimens while retaining the capability of performing the same test on different specimens. The overall output is approximately 100 tests per hour. The simplicity of operation of the system and short time of analysis enables emergency requests to be processed swiftly and accurately even by relatively unskilled staff. The instrument is quite suitable to handle bulk of the work in a small hospital or private laboratory, although the cost per test is high. The ACA I and ACA II did not have the capability of performing sodium and

Table 3. Summary of Methodologies Used With Technicon's Automated Chemistry Analyzers

| Test (Unit) | Methods and References | |
	Autoanalyzers & SMAs	SMAC
Albumin (g/dl)	Dye Binding BCG(AA I and AA II), SMA II, SMA 12/60 and all other SMAs (Doumas et al.,1971; Rodkey 1965)	same
Bilirubin, Total (mg/dl)	Diazo-other coupling (Jendrassik-Grof) without blank (AA I, AA II, SMA II, 12/60, all SMA) and same with blank (Jendrassik and Grof, 1938; Gambino, 1965)	Diazotized Sulfonic acid (Gambino and Schreiber, 1964)
Bilirubin, Direct (mg/dl)	-	Diazotized Sulfonic acid (Gambino and Schreiber, 1964)
Calcium (mg/dl)	O-Cresolphthalein Complexone (AA I, II, SMA II, 6/60, 12/60) (Kessler and Wolfman, 1964)	Similar to SMA Method (Gitelman 1967)
Carbon Dioxide (mEq/l)	Colorimetric (Cresol Red) (SMA II, 12/60, 6/60) (Skeggs and Hochstrasser, 1960)	Phenolphthalein (Hochstrasser, 1964.
Chloride (mEq/l)	a) Ion Selective Electrode (Technicon STAT/ION; STAT/LYTE) b) Mercuric Nitrate/Thiocyanate (Technicon AA I, AA II, SMA 6/60, 12/60, SMA II (Skeggs & Hochstrasser, 1964; Zall et al., 1956.	Mercuric-thiocyanate, Ferric Ion (Skeggs and Hochstrasser 1964)
Cholesterol (mg/dl)	a) Without extraction/ Leiberman-Burchard (Technicon SMA 12/60, all other SMAs)	same (Huang et al., 1961; Wefler[r] and Raferty, 1976)

139

Table 3. (con't) Technicon Methodologies

Cholesterol (mg/dl)	b) Enzymatic (Technicon AA II, AA I, SMA II, SMA 12/60) Leon and Stasin, 1976; Klose et al, 1975)	
Creatinine (mg/dl)	a) Alkaline Picrate with Lloyds reagent (all Technicon instruments)	same
	b) Alkaline Picrate with Lloyds reagent (all Technicon instruments) (Chasson et al, 1961; Jaffe, 1886)	
Glucose (mg/dl)	a) Glucose Oxidase (MBTH/DMA) (SMA 12/60)	a) Oxidase/ Peroxidase (Gochman and Schmitz, 1972)
	b) G-Oxidase (PAP) (AA I, AA II, SMA 6/60)	
	c) Ferricyanide (AA I, AA II)	b) Hexokinase (Tietz, 1976)
	d) Hexokinase (SMA 6/60, 12/60, SMA II)	
	e) Copper Neocuproine (SMA instruments) (Trinder, 1969; Tietz, 1976)	
Inorganic Phosphorus (mg/dl)	a) Phosphomolybdate (with any reduction method) (AA I, AA II, SMA 12/60) (Goodwin 1970)	Phosphomolybdate UV (Amador and Urban, 1972)
Iron, Total	a) With prior protein removal - Bathophenanthroline or related compound (AA I and II)	Ferrachrome/ Ferrozine (Giovaniello et al,1968.
	b) With prior protein removal - Tripyridyltriazine TPZ (AA I, AA II)	
	c) Without prior protein removal using Ferrachrome/Ferrozine (AA I, AA II)	

Table 3. (con't) Technicon Methodologies

Protein, Total	Biuret (all Technicon instruments) (Skeggs & Hochstrasser, 1964; Leon et al, 1976)	same (Weichsel-baun, 1946)
Triglycerides (mg/dl)	a) With blank for free glycerol/enzymatic UV/ (SMA II, all SMAs) b) Without blank for free glycerol/ enzy-matic/UV (all SMAs) (Leon et al, 1976; Bucolo & David, 1973)	same as (a) (Bucolo and David, 1973)
Urea nitrogen (BUN) (mg/dl)	Diacetyl monoxime (all Technicon instrument) (Marsh et al, 1965)	same
Uric Acid (mg/dl)	Phosphotungstate (all Technicon instrument) (Musser & Ortigoza, 1966).	same
Sodium and Potassium (mEq/l)	a) Flame Photometer with separate Auto Dilutor (all Tech-nicon SMA) b) Flame with integ-rated Dilutor (SMA 6/60, SMA II, SMA 12/60, all Technicon SMAs) (Berry et al, 1946)	Ion Selective Electrode (Rao, et al, 1972)
Alkaline Phosphatase (IU/l)	P-Nitrophenolphosphate (Morgenstern, et al, 1965; Bessey et al, 1946) (SMA 12/60 and SMA II)	same
Creatine Phosphokinase (CPK) (37°C)(IU/l)		Creatine-Diacetyl-Orcinol (Rosalki, 1967)

Table 3. (con't) Technicon Methodologies

Glutamic-Oxalo acetic Trans- aminase (GOT) 37°C (IU/1) and Glutamic- Pyruvic Trans- aminase (GPT) 37°C (IU/1)	Coupled Reaction NADH/MDH and NADH/LDH (SMA 12/60, SMA II) Kessler et al, 1971; Henry et al, 1960)	same
Lactic Dehydro- genase (LDH) 37°C (IU/1)	Lactate Oxidation (Kessler et al, 1970; Wacker et al, 1956)	same

potassium however, recently the ACA III was introduced which has the capability of doing 60 tests including sodium and potassium. ACA employs a heat-sealed transparent plastic envelop containing reagents in separate compartments specific for each test to be performed. The envelop or test pack becomes the reaction chamber and, subsequently, also the cuvet for the photometric measurement. To perform a test, the serum specimen is placed in a cup to which an identification card is attached. The cup is then inserted into the input tray of the ACA. A pack specific for each of the tests requested on the same specimen is placed immediately behind the specimen cup. All packs are marked with human-readable identification and also a machine readable binary code that programs the analyzer. As the machine code is read, the appropriate volume of sample (20–600 microliters) is aspirated and delivered with a specific diluent (purified water, AMP–HCl buffer, phosphate buffer, tri–HCl buffer or glycine buffer) into the respective reagent pack. The pack is then transported on a continuous chain through two processing stations at which the diluted sample is mixed with the reagents first by breaking the reagent compartments, and then by vibrating the pack. Before the pack reaches the photometer, the machine code is read again and the correct filter is moved into place. Once in the photometer, the pack is compressed so that a cuvet with a 1.0 cm pathlength is formed, containing a portion of the reaction mixture. After completion of the photometric measurement, the packs are automatically discarded. Results in appropriate concentration units, are printed on paper tape together with the reproduction of the identification card which was used to identify the specimen cup. Various codes may also be printed on the report form to indicate possible malfunction of the system. All reactions occur in air bath in which the temperature is maintained at 37°C. Specificity of analysis is increased by the use of adsorption columns in the analytical packs to remove interfering substances which otherwise might invalidate results. The instrument employs two photometric readings for every test. For fixed endpoint reactions, the instrument employs for every test two photometric readings at two different wavelengths to minimize the influence of turbidity on the results. For enzyme measurements, the change in absorbance during 17.1 seconds is used to calculate results in International Units. The instrument must be calibrated and control samples must be run everyday to insure precision and accuracy of test results. The methodology used with each test performed with the ACA instrument is outlined below:

Acid Phosphatase (ACP): Hydrolysis of thymolphthalein monophosphate (Roy et al., 1971).

Albumin (ALB): BCG dye binding (Doumas et al., 1975).

Alkaline Phosphatase (ALKP): PNPP hydrolysis at pH 10.15 (Bowers and McComb, 1966).

Amylase (AMYL): Hydrolysis of maltopentose followed by hexokinase (Schmidt, 1965).

Urea nitrogen (BUN): Urease glutamate dehydrogenase (340 nm absorbance) (Talke and Schubert, 1965).

Calcium (CAL): Reaction with O–Cresolphthalein complexone (Sarkar and Chauhan, 1967).

Carbon Dioxide (CARB): Phosphoenolpyruvate carboxylase, malate dehydrogenase coupled reaction (Menson et al., 1974).

Conjugated (Direct) Bilirubin (DBIL): Direct Van den Bergh diazo reaction; p-Nitrobenzene diazonium tetrafluoroborate is used to couple PNB (Van den Bergh and Snapper, 1913) (Doumas et al., 1973).

Cholesterol (CHOL): Enzymatic (Stadtman); Cholesterol-esterase/oxidase/horse radish peroxidase (Stadtman, 1957).

Chloride (CL): Modified Hamilton colorimetric measurement. Reaction with mercuric thiocyanate and ferric ion (Hamilton, 1966) (Cotlove, 1961).

Creatine Phosphokinase (CPK): UV enzymatic determination (Oliver, 1955; Rosalki, 1967).

Creatinine (CREA): Jaffe's reaction (alkaline picrate) (Larsen, 1972).

Glucose (GLU): Hexokinase/G-6-PDH method (Slein and Bergmeyer, 1965).

GOT: Coupled Aspartate/MDH/NADH transaminase reaction method (Karmen modified by Henry) (Henry *et al.*, 1960) (Bergmeyer, 1974).

GPT: Coupled Alanine/LDH/NADH transaminase reaction (Wroblewski and La Due modified by Bergmeyer and Bernt) (Bergmeyer and Bernt, 1965) (Bergmeyer, 1974).

HBDH: Alpha-Ketobutyrate to alpha-OH butyrate (Rosalki and Wilkinson, 1960).

PCHE: Butyrthiocholine (BTC) is hydrolysed to butyric acid and thiocholine (Gal and Roth, 1957).

Total Bilirubin (TBIL): Diazo reaction of Van den Bergh (Jacobsen and Wennberg, 1974).

Triglyceride (TGL): Enzymatic (Lipase/NAD/GLDH) (Rautela *et al.*, 1974).

Total Protein (TP): Modified Biuret reaction (Henry, 1974).

Uric Acid (UA): Uricase method (Kalckar, 1947).

Gamma Glutamyltransferase (GT): Transfer of glutamyl group from gammaglutamyl-p-nitroanilide hydrochloride to glycylglycine thereby releasing p-nitroaniline which absorbs at 405 nm (Szasz, 1969).

Phosphorus, inorganic (PHOS): Fiske Subbarow Phosphomolybdate reaction (Goodwin, 1970).

Lipase (LIP): Hydrolysis of triolein (Shihabi and Bishop, 1971).

Ammonia (AMON): Enzymatic (coupled amination of alphaketoglutarate/NADPH) (Van Anken and Schiphorst, 1974).

Iron, total (IRN): Release of iron from transferrin at pH 4.2 in presence of NH_2OH (Megraw *et al.*, 1973; Bouda, 1968).

Lactic acid (LA): Oxidation of lactate in presence of hydrazine (Marbach and Weil, 1967).

Lactic dehydrogenase (LDH): Oxidation of lactate with NAD (McComb and Bowers, 1968).

Magnesium (MG): Methylthymol blue complexometric procedure (Couverty *et al.*, 1971).

ACA introduced in 1980 a number of new tests, in addition to those described above. The new tests include HDL Cholesterol, Creatinine kinase MB isoenzyme, antileptic drugs (premidone, phenobarbital, phenytoin, carbamazepine now available and ethosuximide soon). In research and development are TDM tests, coagulation factor tests, thyroid function tests, and tests for immunoglobulins, hemoglobin and glycosated hemoglobin.

Coulter Chemistry System

The Coulter Chemistry Analyzer permits from one to 22 different tests simultaneously, on up to 60 samples per hour. Sample volumes range from 15 to 40 μl. Reagent volume is between 1 to 6 ml. Each test procedure is performed in its own set of modules so that there are separate sampling, reagent addition, and mixing stations. Each test procedure also has its own dedicated colorimeter with an appropriate interference filter. A flame photometer with internal standardization is used for sodium and potassium analysis. Specimen identification is maintained throughout the instrument by mechanical interlocking of specimen and reaction containers. Analyses are performed without protein separation, using the methods outlined below. The reagents are available for use in manual back-up procedures if required. Specimens for emergency analysis may be inserted during a run. Calibration of each test procedure is performed automatically and in most cases with aqueous solutions of pure chemicals. Serum blank corrections are performed automatically. Results are matched with the specimen identification previously entered into the system and are printed in concentration units on paper tape. Coulter Chemistry II (Coulter Electronics, Hialeah, Florida) an automated, computer controlled discrete analyzer uses the following test methodology (Tasker, 1978):

Albumin (ALB): Dye-binding technique using bromocresol green.

Alkaline Phosphatase (ALKP): Colorimetric using sodium thymolphthalein monophosphate as substrate.

Calcium: Complexing with O–cresolphthalein in the presence of 8–quinolinol to bind magnesium.

Cholesterol (CHOL): A modified Lieberman-Burchard reaction.

Chloride (CL): Displacement of thiocyanate ions from mercuric thiocyanate by chloride ion forming ferric thiocyanate complex.

Creatine Phosphokinase (CPK): Colorimetric procedure degradation of creatine phosphatase and coupled reaction with indicator measured in the visible spectrum.

Creatinine (CREA): Jaffe reaction.

Glucose (GLU): Orthotoluidine reaction.

Lactic Dehydrogenase (LDH): Colorimetric reaction using lactate as substrate and measuring NADH coupled reaction with a pink formasan.

Inorganic Phosphorus (PHOS): Fiske-Subbarow method.

Total Bilirubin (TBIL): Diazotization in the presence of caffeine-sodium benzoate and quantitation of the blue alkaline.

Total Protein (TP): Biuret reaction.

Urea nitrogen: Diacetyl monoxime reaction.

Sodium and Potassium: Flame Photometry (integral part of the instrument).

GOT: Chromogenic reaction with diazonium salt.

3. **Centrifugal Fast Analyzer.** This re- cently developed principle makes use of centrifugal force to transfer and mix samples and reagents. A single light source and detector is used to record data from many curves. The major manufacturers supplying these systems include Electro-Nucleonics, Inc.—GEMSAEC instrument; Union Carbide Corporation—CentrifiChem; and American Instrument Co., Inc.—Rotochem. All these instruments offer microsampling (20 μl) and automatic loading of samples and reagents onto the rotor.

Other Automated Chemistry Analyzers

Kodak Ektachem 400 is the first system using Kodak's innovative dry film reagent technology that is potentially useful in a large number of clinical laboratories. At this writing, tests available in this system include glucose, BUN, total bilirubin, amylase, uric acid, total protein, albumin, sodium, potassium, chloride, and carbon dioxide; and triglycerides, creatinine and cholesterol are in advanced development stage; and enzyme capabilities will be added in the future. Because of the stability of the reagents, frequent calibrations are not required in this system. This instrument may be very useful for stat analysis. Although human engineering is well conceived, slide manipulation requires a lot of mechanisms.

Ektachem Electrolyte Analyzer, also introduced recently by the Eastman Kodak Company, performs electrolyte analysis precisely and accurately without wet reagents, open flames, or glass electrodes. The key to this new method of measuring serum electrolytes is the Ektachem slide. It is a single use analysis slide that is essentially a small disposable battery-line cell in a plastic mount. Each slide contains two thin film, ion selective electrodes joined by a paper bridge. A patient sample of 10 μl and a like amount of reference fluid are metered to the slide and interact with the reagent layers to create a pair of electrochemical half cells. A liquid junction is formed by capillary flow of the two fluids through a small bridge. An electrometer within the analyzer measures the electrical voltage generated and gives a precise indication of the electrolyte concentration in the sample. After three minutes of analysis, the electrometer measures the potential difference between the two ion-selective electrodes and transmits the information to a microprocessor. The electrolyte concentration is then calculated and reported on a digital display and printed out for a permanent record. The electrolyte analyzer is capable of running tests for sodium, potassium, chloride, and carbon dioxide in any of four channels that can be used interchangeably.

In addition to these automated chemistry analyzers available to clinical laboratories for performing routine as well as special tests, a wide range of semiautomated and automated instruments serving a combination of routine and stat needs have been introduced in recent years. These instruments include Beckman ASTRA, Beckman Model 42 Clinical UV/Visible Spectrophotometers; Abbott–ABA (ABA 100 and ABA 200) Clinical Chemistry Bichromatic Analyzers; Instrument Laboratory Clinicard 368 Analyzers and 504/508 Analyzers; Diagnostic Aids Company CompuChem Analyzer; Biokimetix Corporation Ministat–S; Ames Company's Seralyzer; American Monitor Company's KDA; Ortho/Damon AccuChem Microanalyzer; Gilford 3400 Automatic Enzyme Analyzer; LKB–8600 Reaction Rate Analyzer; Greiner Scientific (a Swiss Company) G-300 Automated Chemistry Analyzer; AutoMed Systems Clinispin Chemistry Analyzer.

Table 4. Prefixes for SI Units

Factor	Name	Symbol
10^{12}	tira-	T
10^9	giga-	G
10^6	mega-	M
10^3	kilo-	K
10^2	hecto-	h
10^1	deca-	da
10^{-18}	atto-	a
10^{-15}	femto-	f
10^{-12}	pico-	p
10^{-9}	nano-	n
10^{-6}	micro	u
10^{-3}	milli	m
10^{-2}	centi	c
10^{-1}	deci	d

Source: Dybkaer (1970).

Table 5. Names and Symbols for Basic SI Units

Physical Quantity	Symbol	Name of SI Unit	Symbol
Length	l	meter	m
Mass	m	kilogram	kg
Time	t	second	S
Electric current	I	ampere	A
Temperature (Thermodynamic)	T	kelvin	K
Luminous intensity	I	candela	cd
Amount of substance	n	mole	mol
Amount of enzyme	—	enzyme unit	U

Source: Dybkaer (1970).

Table 6. Special Names and Symbols for Derived SI Units

Quantity	Name of SI Unit	Symbol	Expression in Terms of SI Base Units or Derived Units
Frequency	hertz	HZ	1 HZ = 1/S
Force	newton	N	1 N = 1 kg·m/S^2
Work, energy, quantity of heat	Joule	J	1 J = 1 N·m
Power	watt	W	1 W = 1 J/S
Quantity of electricity	coulomb	C	1 C = 1 A·S
Electric potential, potential differences, tension electromatic force	volt	V	1 V = 1 W/A
Electric capacitance	farad	F	1 F = 1 A·S/V
Electric resistance	ohm	Ω	1 = 1 V/A
Flux of magnetic induction, magnetic flux	weber	Wb	1Wb = 1 V·S
Magnetic flux density, magnetic induction	tesla	T	1 T = 1 Wb/m^2
Induction	henry	H	1 H = 1 V·S/A
Pressure	pascal	Pa	1Pa = 1 N/m^2 = 1 kg/m·S^2

Source: Baron et al (1974).

Table 7. Conversion of Laboratory Values from Traditional Units into SI Units

Constituent	Traditional Units	Multiplication Factor	SI Units
Amylase	Units/l	1.0	arb. unit
Bilirubin (direct) conjugated total	mg/100 ml mg/100 ml mg/100 ml	43.06 17.10 17.10	μ mol/l μ mol/l μ mol/l
Calcium	mg/100 ml	0.2495	m mol/l
Carbon dioxide	m Eq/l	1.0	m mol/l
Chloride	m Eq/l	1.0	m mol/l
Creatine phosphokinase (CPK)	m U/ml	0.01667	μ mol S^{-1}/l
Creatinine	mg/100 ml	88.40	μ mol/l
Glucose	mg/100 ml	0.05551	m mol/l
Lactic dehydrogenase	m U/ml	0.01667	μ mol S^{-1}/l
Cholesterol	mg/100 ml	0.02586	m mol/l
Magnesium	m Eq/l	0.50	m mol/l
P_{CO_2}	mm Hg	0.1333	kPa
pH		1.0	1

Table 7 (con't.). Conversion of Laboratory Values from Traditional Units into SI Units

Constituent	Traditional Units	Multiplication Factor	SI Units
Po$_2$	mm Hg	0.133	kPa
Phosphatase, acid	Sigma	278.4	n mol S^{-1}/l
Phosphatase, alkaline	Bodansky	0.08967	n mol S^{-1}/l
Phosphorus, inorganic	mg/100 ml	0.3229	m mol/l
Protein, total	g/100 ml	10	g/l
Protein, electrophorises			
Albumin	% total	0.01	1
Globulin, α_1	% total	0.01	1
α_2	% total	0.01	1
β	% total	0.01	1
γ	% total	0.01	1
Potassium	m Eq/l	1.0	m mol/l
Sodium	m Eq/l	1.0	m mol/l
Transaminase (SGOT) (Aminotransferase)	Karmen	0.008051	μ mol S^{-1}/l
Urea nitrogen	mg/100 ml	0.3569	m mol/l
Uric acid	mg/100 ml	0.65948	m mol/l

Clinical Biochemical Reference Values

In most instances, disease states in the experimental animals are accompanied by biochemical alterations. Therefore use of the correct normal range of biochemical constituents in the blood of these experimental animals, established by quality controlled determinations, is required to avoid pitfalls in interpreting laboratory measurements. Other factors that may affect the results of laboratory tests include sensitivity and diagnostic specificity of methods, preparation and collection of blood specimens from the experimental subject, nutrition and environmental conditions, and other physiopathologic conditions discussed previously in this book. Although the principal physiological and biochemical mechanisms in animals and humans are similar, there are nevertheless considerable variations in the size, concentration, and activity of individual systems in different species. The clinician should therefore have a good working knowledge of the reference values of the normal experimental animals with respect to the species of the animal and to variations caused by differences in age and sex. A medical investigator must be familiar with the techniques and the precision attainable by them before he can adequately interpret the results. It must be emphasized that the values for various chemical components in the blood vary from species to species and sometimes according to the age of the animal.

This section provides some of the basic data necessary for a better utilization of the animal as a test subject for physiological, biochemical, and biological investigations. Biochemical constituents in the blood of 14 commonly used normal experimental animal species are compared with normal values in humans. Particular attention has been given to the age, sex, and breed or species of the animal in compiling blood chemistry data. Normal ranges of blood constituents in experimental animals reported in the literature are also presented to demonstrate the problems caused by the wide variability of these values.

Mice

Clinical chemical and electrophoretic protein values of sera from ICR strains of normal albino mice weighing 20–25 g are presented in Tables 1–3. These are mean values derived from test sera from 145 male and 128 female mice. The values reported in the literature are also included in the tables. In comparison to humans, the sera from normal mice generally show lower values of

Table 1. Chemical Components in Serum of the Normal Albino Mouse

Components	Values in				Literature values*
	Male		Female		
	Mean	S.D.	Mean	S.D.	Range
Bilirubin (mg/dl)	0.75	0.05	0.70	0.04	0.10-0.90
Cholesterol (mg/dl)	63.3	11.8	65.5	12.1	26.0-82.4
Creatinine (mg/dl)	0.84	0.19	0.67	0.17	0.30-1.00
Glucose (mg/dl)	92.2	10.5	85.0	9.50	62.8-176.
Urea nitrogen (mg/dl)	20.8	5.86	17.9	4.50	13.9-28.3
Uric acid (mg/dl)	4.12	1.10	3.90	0.95	1.2-5.00
Sodium (mEq/l)	138.	2.90	134.	2.60	128.-145.
Potassium (mEq/l)	5.25	0.13	5.40	0.15	4.85-5.85
Chloride (mEq/l)	108.	0.60	107.	0.55	105.-110.
Bicarbonate (mEq/l)	26.2	2.10	24.8	2.30	20.0-31.5
Phosphorus (mg/dl)	5.60	1.61	6.55	1.30	2.30-9.20
Calcium (mg/dl)	5.60	0.40	7.40	0.50	3.2-8.50
Magnesium (mg/dl)	3.11	0.37	1.38	0.28	0.80-3.90

*Source: Burns and Lannoy (1966); Burns et al. (1971); Altman and Dittmer (1974); Hill (1971); Bonilla (1972); Finch and Foster (1973); Moreland (1974); Nomura et al. (1975).

Table 2.   Enzyme Activities in Serum of the Normal Albino Mouse

| Enzyme | Values in | | | | Literature values* |
| | Male | | Female | | |
	Mean	S.D.	Mean	S.D.	Range
Amylase (Somogyi units/dl)	160.	22.0	140.	17.0	95.0-204.
Alkaline Phosphatase (I.U./1)	21.6	4.56	16.8	5.37	10.5-27.6
Acid Phosphatase (I.U./1)	14.7	3.50	11.8	2.70	4.5-21.7
Alanine Transaminase (SGPT) (ALT) (I.U./1)	13.5	5.30	12.6	4.40	2.10-23.8
Aspartate Transaminase (SGOT) (AST) (I.U./1)	36.2	6.13	35.4	6.10	23.2-48.4
Creative Phosphokinase (CPK) (CK) (I.U.L.)	370	14.5	250	15.2	50-680
Lactic Dehydrogenase (LDH) (I.U./1)	149.	19.0	119.	21.0	750.-185.

*Source: Altman and Dittmer (1974); Bonilla (1972); Nomura et al. (1975)

Table 3. Serum Proteins of the Normal Albino Mouse

Components	Values in				Literature values*
	Male		Female		
	Mean	S.D.	Mean	S.D.	Range
Total protein (g/dl)	6.25	0.75	6.10	0.68	4.00-8.62
Albumin (g/dl)	3.14	0.36	2.92	0.32	2.52-4.84
(%)	50.2	6.50	47.9	5.90	35.0-62.7
α_1 - Globulin (g/dl)	0.53	0.12	0.48	0.10	0.22-0.78
(%)	8.50	1.60	7.80	1.50	4.30-11.8
α_2- Globulin (g/dl)	0.91	0.10	1.02	0.10	0.65-1.30
(%)	14.6	3.10	16.8	2.54	8.20-23.0
β-Globulin (g/dl)	1.04	0.19	0.98	0.25	0.40-1.58
(%)	16.7	4.80	16.0	4.20	6.50-26.6
γ-Globulin (g/dl)	0.61	0.11	0.70	0.09	0.38-0.90
(%)	9.80	1.90	11.5	1.45	5.80-15.5
Albumin/Globulin	0.98	0.22	0.92	0.23	0.56-1.30

*Source: Burns and Lannoy (1966); Burns et al.(1971); Dimopoullos
(1972); Finch and Foster (1973); Moreland (1974); Nomura
et al. (1975).

cholesterol, creatinine, calcium, creatine phosphokinase (CPK) (CK) lactic dehydrogenase (LDH), γ-globulin and A/G ratio. Levels of urea nitrogen, inorganic phosphorus, amylase, acid phosphatase, serum glutamic oxaloacetic transaminase (SGOT) (AST) and α_2-globulin are higher in the sera of normal mice than in normal human sera. Williams *et al.* (1966) reported that clinical biochemical observations in ICR mice and strains of BALB$_c$ and Cinnamon mice were consistent with findings in larger animal species. They also reported that AST, chloride, and glucose values differ in different strains of mice. Some variations due to strain differences are expected, perhaps because genetic factors play some role in physiological modifications. However variations due to age, sex, or nutrition do not have a significant effect on biochemical constituents in the sera (Finch and Foster, 1973; Bonilla, 1972; Burns *et al.*, 1971; Moreland, 1974). Our data are in agreement with the reference values reported recently by other workers (Table 4).

Rats

Tables 5–7 list the normal values of a Wistar strain of albino rats weighing 180–250 g. The mean values were derived from test results with 160 male and 140 female animals. In general, the normal rat sera had lower levels of bilirubin, cholesterol, creatinine, uric acid, CK, LDH, and A/G ratio compared with levels in human sera. However the values of blood urea nitrogen (BUN), potassium, inorganic phosphorus, calcium, magnesium, amylase, alkaline phosphatase, acid phosphatase, AST, α_1- and β-globulin were higher in normal rat sera when compared to those values in human sera. Comparative biochemical reference values in the rat are presented in Table 8(a).

The large number of breeds, stocks, substocks, and strains within a species complicates the establishment of normal values in a given laboratory species such as the rat. Physiological variations between different stocks of rats makes it necessary to stress the importance of accurately identifying and properly defining the rat stocks used. In these studies the biochemical constituents of the sera of four different strains of rat were compared (Table 8(b)). Most of the values of these strains were within the normal range compared to the Wistar strain, except for uric acid which was lower in the sera of these animals. Compared to human sera, protein and magnesium values were low in germ-free (axenic) and other strains, and inorganic phosphate and BUN were high in inbred and germ-free strains. Burns *et al.* (1971) reported similar values in germ-free and SPF rats: LDH was higher in these strains of rats and AST was lower when compared to the values in human sera.

Street and Highman (1971) studied the normal blood parameters of Mastomys (African white-tailed) rats and Osborne Mendel rats in fed and fasted animals of both sexes. They reported that the Mastomys of both sexes had higher ALT and BUN, and lower LDH, alkaline phosphatase, and glucose than other female rats. In Mastomys fasting raised AST, ALT, alkaline phosphatase, and aldolase in males, but caused no change in females other than lower glucose levels. In Osborne-Mendel rats, fasting lowered alkaline phosphatase, urea nitrogen, and glucose in both sexes, and ALT was lower in males. Fed male Mastomys had higher urea nitrogen and glucose values, and fasted males had higher urea nitrogen and glucose values; fasted males had higher AST, ALT, and aldolase values than corresponding females.

Kozma *et al.* (1969) reported normal values in Long Evans (BLU, LE) rats for separated sexes in five age groups—2, 4, 12, and 24 months. They reported that total proteins were not significantly different with respect to age or sex. The percentage

Table 4. Biochemical Reference Values of the Normal Mouse

Components	Caisey and King (1980) [3]	Harrison et al (1978) [2]		Melby and Altman (1974) [1]		Mitruka and Rawnsley (1977) [4]	
	Mean	Mean	S.D.	Mean	S.D.	Mean	S.D.
Bilirubin (mg/dl)	–	0.25	0.16	–	–	0.74	0.05
Cholesterol (mg/dl)	–	–	–	–	–	64.4	12.0
Creatinine (mg/dl)	–	0.50	0.17	–	–	0.76	0.18
Glucose (mg/dl)	–	206	32.0	62.8	176	88.6	10.0
Urea nitrogen (mg/dl)	–	22.1	5.30	13.9	23.3	19.4	5.18
Uric acid (mg/dl)	–	1.76	0.50	–	–	4.01	1.03
Sodium (mEq/l)	150	150	13.0	–	–	136	2.75
Potassium (mEq/l)	5.4	6.41	1.08	–	–	5.33	0.14
Chloride (mEq/l)	114	–	–	–	–	107.5	0.58
Bicarbonate (mEq/l)	–	–	–	–	–	25.5	2.20
Phosphorus (mg/dl)	11.5	10.6	1.40	5.6	–	6.08	1.46
Calcium (mg/dl)	9.9	8.75	0.99	4.20	0.60	6.50	0.45
Magnesium (mg/dl)	–	–	–	–	–	2.25	0.33
Amylase (Somogyi units/dl)	–	–	–	–	–	150	19.5
Alkaline Phosphatase (I.U./l)	439	62.7	10.3	–	–	19.2	4.97
Acid Phosphatase (I.U./l)	–	–	–	–	–	13.3	3.10

Table (cont.) Biochemical reference values of the normal mouse

Components	Caisey and King (1980)[3] Mean	Harrison et al (1978)[2] Mean	S.D.	Melby and Altman (1974)[1] Mean	S.D.	Mitruka Rawnsley (1977)[4] Mean	S.D.
Alanine Transaminase (SGPT) (I.U./l) (ALT)	19.0	40.0	20.6	—	—	13.1	4.85
Aspartate Transaminase (SGOT) (I.U./l) (AST)	37.0	141	67.4	—	—	35.8	6.12
Creatine Phosphokinase (CPK) I.U./l (CK)	155	320	287	—	—	310	14.9
Lactic Dehydrogenase (LDH) (I.U./l)	366	172	103	—	—	134	20.0
Total Protein (g/dl)	5.30	5.10	0.42	6.40	1.11	6.18	0.72
Albumin (g/dl)	2.00	3.13	0.26	—	—	3.03	0.34
(%)	—	—	—	3.68	0.58	49.1	6.20
Alpha - 1 Globulin (g/dl)	1.50	—	—	—	—	0.51	0.11
(%)	—	1.96	0.32	—	—	8.15	1.55
Alpha -2 Globulin (g/dl)	—	—	—	—	—	0.97	0.10
(%)	—	—	—	—	—	15.7	2.82
Beta Globulin (g/dl)	1.30	—	—	0.60	—	1.01	0.22
(%)	—	—	—	—	—	16.4	4.50
Gamma Globulin (g/dl)	0.50	—	—	—	—	0.66	0.10
(%)	—	—	—	—	—	10.7	1.68
Albumin/Globulin	0.62	—	—	—	—	0.95	0.23

Source: [1] Derived from Literature
[2] Value from male B_6D_2F mice. Plasma samples were analyzed with a Basic Microanalyzer (Ortho Instruments, Westwood, Mass.)
[3] Data from 10 CRH mice 4-5 weeks old. See text for methodology.
[4] Data from 145 male and 128 female ICR, albino mice, weighing 20-25 g.

159

Table 5. Chemical Components in Serum of the Normal Albino Rat

| Components | Values in | | | | Literature values* |
| | Male | | Female | | |
	Mean	S.D.	Mean	S.D.	Range
Bilirubin (mg/dl)	0.35	0.02	0.24	0.07	0.00-0.55
Cholesterol (mg/dl)	48.3	10.2	44.7	9.62	10.0-54.0
Creatinine (mg/dl)	0.46	0.13	0.49	0.12	0.20-0.80
Glucose (mg/dl)	78.0	14.0	71.0	16.0	50.0-135.
Urea nitrogen (mg/dl)	15.5	4.44	13.8	4.15	5.0-29.0
Uric acid (mg/dl)	1.99	0.25	1.79	0.24	1.20-7.5
Sodium (mEq/l)	141.	2.65	140.	2.50	143.-156.
Potassium (mEq/l)	5.82	0.11	6.1	0.12	5.40-7.00
Chloride (mEq/l)	102.	0.85	101.	0.90	100.-110.
Bicarbonate (mEq/l)	24.0	3.80	20.8	3.60	12.6-32.0
Phosphorus (mg/dl)	7.56	1.51	8.26	1.14	3.11-11.0
Calcium (mg/dl)	12.2	0.75	10.6	0.89	7.2-13.9
Magnesium (mg/dl)	3.12	0.41	2.60	0.21	1.6-4.44

*Source: Spector (1961); Burns and Lannoy (1966); Blue Spruce Farm Tech. Bulletin (1970); Burns et al. (1971); Altman and Dittmer (1974); Hill (1971); Laird (1971)' Laird (1972); Foster (1973); Vondruska and Greco (1973); Moreland (1974); Nomura et al. (1975); Becker et al. (1979).

Table 6. Enzyme Activities in Serum of the Normal Albino Rat

Enzyme	Values in				Literature values*
	Male		Female		
	Mean	S.D.	Mean	S.D.	Range
Amylase (Somogyi units/dl)	245.	32.0	196.	34.0	128.-313.
Alkaline Phosphatase (I.U./1)	81.4	14.8	93.9	17.3	56.8-128.
Acid Phosphatase (I.U./1)	39.0	4.30	37.5	3.70	28.9-47.6
Alanine Transaminase (SGPT) (ALT) (I.U./1)	25.2	2.05	22.5	2.50	17.5-30.2
Aspartate Transaminase (SGOT) (AST) (I.U./1)	62.5	8.40	64.0	6.50	45.7-80.8
Creatine Phosphokinase (CPK) (CK) (I.U./1)	56.0	13.0	68.0	24.0	80-116
Lactic Dehydrogenase (LDH) (I.U./1)	92.5	13.9	90.0	14.5	61.0-121.

*Source: Blue Spruce Farms Communication (1970); Burns et al. (1971); Laird (1972); Union Carbide Biochemical Data (1973); Vondruska and Greco (1973); Nomura et al. (1975).

Table 7. Serum Proteins of the Normal Albino Rat

Components	Values in				Literature values*
	Male		Female		
	Mean	S.D.	Mean	S.D.	Range
Total protein (g/dl)	7.61	0.50	7.52	0.32	4.70-8.15
Albumin (g/dl)	3.73	0.53	3.62	0.52	2.70-5.10
(%)	49.0	7.10	48.1	7.40	33.3-63.8
α_1 - Globulin (g/dl)	1.03	0.22	0.89	0.25	0.39-1.60
(%)	13.5	2.20	11.9	3.80	4.30-21.1
α_2 - Globulin (g/dl)	0.71	0.14	0.70	0.32	0.20-2.10
(%)	9.3	1.80	8.60	2.70	3.20-14.7
β-Globulin (g/dl)	1.07	0.35	1.31	0.26	0.35-2.00
(%)	14.1	4.70	17.4	3.60	5.70-26.8
γ-Globulin (g/dl)	1.05	0.21	1.00	0.21	0.62-1.60
(%)	13.8	2.70	14.0	2.80	10.0-19.8
Albumin/Globulin	0.96	0.24	0.93	0.25	0.72-1.21

*Source: Spector (1961); Burns and Lannoy (1966); Blue Spruce
         Farm Communication (1970); Altman and Dittmer (1974);
         Burns et al. (1971); Hill (1971); Dimopoullos (1972);
         Laird (1972); Moreland (1974); Nomura et al. (1975).

Table 8.a. Biochemical Reference Values of the Normal Rat

Components	Caisey and King (1980)[1] Mean	Benirschke et al (1978)[2] Mean	S.D.	Melby and Altman (1974)[2] Mean	S.D.	Mitruka and Rawnsley (1977)[3] Mean	S.D.
Bilirubin (mg/dl)	-	0.35	0.02	0.15	0.15	0.30	0.05
Cholesterol (mg/dl)	47.2	49.7	9.62	49.5	17.9	46.5	9.91
Creatinine (mg/dl)	-	-	-	0.49	0.06	0.48	0.13
Glucose (mg/dl)	-	-	-	187.5	45.7	75.0	15.0
Urea nitrogen (mg/dl)	-	24.8	4.65	21.5	9.50	14.7	4.30
Uric acid (mg/dl)	-	1.79	0.04	2.40	1.17	1.89	0.25
Sodium (mEq/l)	141	140.4	5.47	140.4	5.50	140.5	2.58
Potassium (mEq/l)	4.50	5.73	0.88	5.73	0.88	5.86	0.12
Chloride (mEq/l)	103	-	-	108	1.98	101.5	0.88
Bicarbonate (mEq/l)	-	-	-	-	-	22.4	3.70
Phosphorus (mg/dl)	6.60	8.02	1.23	8.26	1.35	7.91	1.33
Calcium (mg/dl)	10.1	9.70	0.48	5.29	0.44	11.4	0.82
Magnesium (mg/dl)	-	-	-	2.16	0.08	2.86	0.31
Amylase (Somogyi units/dl)	-	-	-	-	-	220.5	33.0
Alkaline Phosphatase (I.U./l)	713 (U/l)	-	-	97.9	35.2	87.7	16.1
Acid Phosphatase (I.U./l)	-	-	-	-	-	38.3	4.00
Alanine Transaminase (SGPT) (I.U./l) (ALT)	36.0 (U/l)	26.6	5.07	31.9	10.8	23.9	2.28

Table 8.a. (con't)   Biochemical Reference Values of the Normal Rat

Components	Caisey and King (1980)[1] Mean	Benirschke et al (1978)[2] Mean	Benirschke et al (1978)[2] S.D.	Melby and Altman (1974)[2] Mean	Melby and Altman (1974)[2] S.D.	Mitruka and Rawnsley (1977)[3] Mean	Mitruka and Rawnsley (1977)[3] S.D.
Aspartate Transaminase (SGOT) (AST) (I.U./l)	83.0 (U/l)	-	-	54.3	18.2	63.3	7.45
Creatine Phosphokinase (CPK) (CK) (I.U./l)	111 (U/l)	-	-	-	-	62.0	18.5
Lactic Dehydrogenase (LDH) (I.U./l)	374 (U/l)	-	-	-	-	91.3	14.2
Total Protein (g/dl)	6.20	6.30	-	5.41	0.75	7.57	0.41
Albumin (g/dl)	2.30	4.11	0.28	3.40	0.40	3.68	0.53
(%)	-	-		-		48.6	7.25
Alpha-1 Globulin (g/dl)	1.60	0.81	-	2.47	0.62	0.96	0.24
(%)	-	-		-		12.7	3.00
Alpha-2 Globulin (g/dl)						1.06	0.23
(%)		12.8	1.96			8.95	2.25
Beta Globulin (g/dl)	1.40	0.90				1.19	0.31
(%)	-	-				15.8	4.15
Gamma Globulin (g/dl)	0.90	0.10				1.12	0.21
(%)	-	-				13.9	2.75
Albumin/Globulin	0.59	-	-	-	-	0.95	0.25

Source: [1] Data from 10 Charles River, CD, 8-10 weeks old rats. See text for methodology.
[2] Data from Literature.
[3] Data from 140 female and 160 male Wistar, Albino rats weighing 180-250 g.
See text for methodology.

Table 8 (b). Biochemical Values in Sera of Certain Strains of Rats

Component	Fisher Inbred (Strain 344/cr)	Osborne-Mendel	Long Evans BLU:(LE)	Germ Free (Axenic)
Alkaline Phosphatase	12.9-34.2 (K.A. units)	2.90-5.20 (Sigma units)	1.80-2.35 (Sigma units)	19.4-32.4 (K.A. units)
Bilibrubin, Total (mg/dl)	0.15-0.36	0.10-0.52	0.00-0.55	0.24-0.64
Calcium (mg/dl)	9.60-12.2	4.45-6.90	5.40-7.20	9.50-10.3
Cholesterol (mg/dl)	41.0-78.0	28.0-76.0	35.0-75.0	37.0-64.0
Chloride (mEq/l)	104.-115.	99.0-112.	105.-117.	89.8-103.
Glucose (mg/dl)	148.-195.	121.-157.	128.-167.	116.-188.
Magnesium (mg/dl)	1.60-4.35	1.50-4.10	2.10-4.20	1.80-2.30
Phosphate, Inorganic (mg/dl)	6.72-10.3	3.15-7.43	5.80-8.20	6.04-10.0
Potassium (mEq/l)	4.10-7.70	5.50-6.30	5.40-6.40	3.50-7.80
Protein, Total (g/dl)	4.76-6.60	5.50-6.90	5.80-7.25	4.72-5.84
Protein, Albumin (%)	41.7-59.4	46.9-54.6	47.5-68.9	41.1-51.9
Globulin (%)	38.5-54.1	45.4-53.2	31.8-59.0	49.9-59.0
Sodium (mEq/l)	125.-148.	135.-155.	143.-156.	123.-146.
Transaminase, Asparate (SGOT) (AST)	----	50.0-75.0 (S.F. units)	88.1-147. (S.F. units)	----
Transaminase, Alanine (SGPT) (ALT)	----	20.0-25.0 (S.F. units)	10.6-17.7 (S.F. units)	----
Urea nitrogen (mg/dl)	16.8-42.0	15.0-21.0	16.6-21.0	18.4-28.4
Uric acid (mg/dl)	1.18-3.56	1.80-3.00	0.90-3.40	1.22-2.44

Source: Albritton (1952); Hamada et al. (1960); Farris and Griffith (1963); Altman and Dittmer (1964); Burns et al. (1966, 1971); Carles et al. (1969); Kozma et al. (1969); Strett and Highman (1971).

165

of albumin was significantly higher in males of all ages than in females. The general relationships of the other serum proteins were similar to those reported by others. It has been reported that the percentage of γ-globulin is generally lower (4–7%) in pathogen-free CFN and CFE rats than in conventionally maintained rats (Kozma *et al.*, 1969; Schermer, 1967; Hudgins *et al.*, 1956). The BUN and glucose were also uniform with respect to age and sex. However the older rats maintained a consistently higher blood glucose level than 2-month-old rats. The serum creatinine was higher in young male rats 2–4 months old compared to young females of the same age group, but by 8 months of age this difference was abolished. Other values including AST and ALT were not significantly different with respect to age and sex.

Blood parameters are also affected by sampling time and by the condition of the animal when samples are taken. The effect of stress (restraint) and the site of blood collection may alter the physiological values in sera of rats. Upton and Morgan (1975) reported that anesthesia with ether, pentobarbital sodium, or fentanyl and properidol had no effect on acid base balance, plasma proteins, calcium, and magnesium in rats. However the use of manual restraint increased blood acidity, plasma proteins, calcium, and magnesium but decreased blood glucose. Increased calcium, magnesium, and proteins in restraint animals may be caused by plasma water loss (Dawson *et al.*, 1971). Besch and Chou (1971) have reported similar physiological responses to blood collection methods in rats.

Hamsters

Tables 9–11 list the clinical biochemical and electrophoretic protein values of 84 male and 80 female Syrian golden hamsters weighing 75–100 g. Compared to the corresponding values of humans, the hamsters had lower bilirubin, cholesterol, alkaline phosphatase, creatine phosphokinase, lactic dehydrogenase, and A/G ratio, and higher blood urea nitrogen, bicarbonate, phosphorus, amylase, acid phosphatase, AST, and α_2-globulin. Our data agreed closely with the normal range of values of hamsters reported in the literature (Table 12). Emminger *et al.* (1975) reported no significant differences in blood parameters between male and female hamsters aged 13 to 900 days.

Guinea Pigs

Laboratory-conditioned albino guinea pigs of the Hartley strains weighing 500–800 g were used to determine normal biochemical constituents in these studies. Tables 13–15 present the mean and standard deviations (S.D.) of chemical constituents, enzymes, and electrophoretically separated proteins from the sera of 110 male and 95 female guinea pigs. The values were within the normal physiological range reported in the literature (Table 16). Bilirubin, cholesterol, uric acid, creatine phosphokinase, lactic dehydrogenase, total protein, albumin, α_1-globulin, and A/G ratio were lower in guinea pigs when compared to those values of normal human sera. However blood urea nitrogen, inorganic phosphorus, amylase, alkaline phosphatase, acid phosphatase, AST, ALT, β-globulin, and α-globulin were higher in guinea pigs than in humans.

Rabbits

Tables 17–19 list the chemical components, enzymes, and electrophoretically separated proteins from the sera of normal New Zealand white rabbits weighing 3–5 kg. The values were recorded from 120 male and 80 female laboratory-conditioned

Table 9. Chemical Components in Serum of the Normal Golden Hamster (Syrian)

Components	Values in				Literature values*
	Male		Female		
	Mean	S.D.	Mean	S.D.	Range
Bilirubin (mg/dl)	0.42	0.12	0.36	0.11	0.20-0.74
Cholesterol (mg/dl)	54.8	11.9	51.5	11.0	10.0-80.0
Creatinine (mg/dl)	1.05	0.28	0.98	0.30	0.35-1.65
Glucose (mg/dl)	73.4	12.6	65.0	10.5	32.6-118.
Urea nitrogen (mg/dl)	23.4	6.74	20.8	5.64	12.5-26.0
Uric acid (mg/dl)	4.58	0.45	4.36	0.50	1.80-5.30
Sodium (mEq/1)	128.	1.90	134.	2.30	106.-146.
Potassium (mEq/1)	4.66	0.47	5.30	0.50	4.00-5.90
Chloride (mEq/1)	96.7	1.19	93.8	1.20	85.7-112.
Bicarbonate (mEq/1)	37.3	2.20	39.1	2.30	32.7-44.1
Phosphorus (mg/dl)	5.29	0.96	6.04	1.10	3.4-8.24
Calcium (mg/dl)	9.52	0.98	10.4	0.92	7.40-12.0
Magnesium (mg/dl)	2.54	0.22	2.20	0.14	1.90-3.50

*Source: Robinson (1965); Burns and Lannoy (1966); Young (1971); Lossnitzer and Bajusz (1974); Moreland (1974).

Table 10. Enzyme Activities in Serum of the Normal Golden Hamster (Syrian)

Enzyme	Values in				Literature values*
	Male		Female		
	Mean	S.D.	Mean	S.D.	Range
Amylase (Somogyi units/dl)	175.	21.0	196.	27.0	120.-250.
Alkaline Phosphatase (I.U./1)	17.5	6.10	15.4	4.20	3.20-30.5
Acid Phosphatase (I.U./1)	7.45	1.40	6.90	1.56	3.90-10.4
Alanine Transaminase (SGPT) (I.U./1) (ALT)	26.9	4.56	20.6	3.95	11.6-35.9
Aspartate Transaminase (SGOT) (I.U./1) (AST)	124.	22.0	77.6	14.5	37.6-168.
Creatine Phosphokinase (CPK) (I.U./1) (CK)	101	40.	85.0	32.0	50-190
Lactic Dehydrogenase (LDH) (I.U./1)	115.	20.0	110.	27.0	56.0-170.

*Source: Altman and Dittmer (1974); Young (1971); Byrd (1973).

Table 11. Serum Proteins of the Normal Golden Hamster (Syrian)

Components	Values in				Literature values*
	Male		Female		
	Mean	S.D.	Mean	S.D.	Range
Total protein (g/dl)	6.94	0.32	7.25	0.48	4.30-7.70
Albumin (g/dl)	3.23	0.41	3.50	0.34	2.63-4.10
(%)	46.5	5.90	48.3	4.70	34.4-60.5
α_1 - Globulin (g/dl)	0.64	0.12	0.55	0.10	0.30-0.95
(%)	9.30	1.90	7.60	1.50	3.50-13.5
α_2 - Globulin (g/dl)	1.85	0.38	1.70	0.34	0.90-2.70
(%)	26.7	6.90	23.5	5.50	10.1-40.5
β-Globulin (g/dl)	0.56	0.24	0.83	0.21	0.10-1.35
(%)	8.10	3.60	11.4	2.65	1.00-15.4
γ-Globulin (g/dl)	0.71	0.23	0.67	0.20	0.15-1.28
(%)	10.2	3.10	9.20	2.80	3.00-17.1
Albumin/Globulin	0.87	0.29	0.93	0.32	0.58-1.24

*Source: Robinson (1965); Burns and Lannoy (1966); Planas and Balasch (1970); Altman and Dittmer (1974); Young (1971); Byrd (1973).

Table 12. Biochemical Reference Values of the Normal Hamster

Components	Benirschke et al [1] (1978)		Melby and Altman [1] (1974)		Mitruka and Rawnsley [2] (1977)	
	Mean	S.D.	Mean	S.D.	Mean	S.D.
Bilirubin (mg/dl)	-	-	0.40	0.02	0.39	0.12
Cholesterol (mg/dl)	-	-	-	-	53.2	11.5
Creatinine (mg/dl)	0.95	0.04	0.95	0.04	1.02	0.29
Glucose (mg/dl)	70.0	-	73.4	-	69.2	11.6
Urea nitrogen (mg/dl)	-	-	15.3	1.10	22.1	6.19
Uric acid (mg/dl)	4.50	0.14	4.55	0.14	4.47	0.48
Sodium (mEq/l)	-	-	128.6	-	131	2.10
Potassium (mEq/l)	-	-	4.10	0.05	4.98	0.49
Chloride (mEq/l)	-	-	96.7	-	95.3	1.20
Bicarbonate (mEq/l)	27.5	-	-	-	38.2	2.25
Phosphorus (mg/dl)	-	-	-	-	5.67	1.03
Calcium (mg/dl)	-	-	9.6	-	9.96	0.95
Magnesium (mg/dl)	3.4	-	3.4	0.09	2.37	0.18
Amylase (Somogyi units/dl)	-	-	-	-	185.5	24.0
Alkaline Phosphatase (I.U./l)	-	-	12.4	0.51	16.5	5.15
Acid Phosphatase (I.U./l)	-	-	-	-	7.18	1.48

Table 12. (con't)   Biochemical Reference Values of the Normal Hamster

Components	Benirschke et al [1] (1978)		Melby and Altman [1] (1974)		Mitruka and Rawnsley [2] (1977)	
	Mean	S.D.	Mean	S.D.	Mean	S.D.
Alanine Transaminase (SGPT) (I.U./l) (ALT)	—	—	—	—	23.8	4.26
Aspartate Transaminase (SGOT) (I.U./l) (AST)	—	—	—	—	100.8	18.3
Creatine Phosphokinase (CPK) (I.U./l) (CK)	—	—	—	—	93.0	36.0
Lactic Dehydrogenase (LDH) (I.U./l)	—	—	—	—	112.5	23.5
Total Protein (g/dl)	—	—	7.30	—	7.10	0.40
Albumin (g/dl)	—	—	2.63	—	3.37	0.38
(%)	—	—	—	—	47.3	5.30
Alpha-1 Globulin (g/dl)	—	—	—	—	0.60	0.11
(%)	—	—	—	—	8.45	1.70
Alpha-2 Globulin (g/dl)	—	—	4.67	—	1.78	0.36
(%)	—	—	—	—	24.6	6.20
Beta Globulin (g/dl)	—	—	—	—	0.70	0.23
(%)	—	—	—	—	9.75	3.13
Gamma Globulin (g/dl)	—	—	—	—	0.69	0.22
(%)	—	—	—	—	9.70	2.95
Albumin/Globulin	—	—	—	—	0.90	0.31

Source: [1] Derived from Literature.
[2] Data from 80 female and 84 male Syrian, golden hamsters weighing 75-100g.
See text for methodology.

Table 13. Chemical Components in Serum of the Normal Guinea Pig

Components	Values in				Literature values*
	Male		Female		
	Mean	S.D.	Mean	S.D.	Range
Bilirubin (mg/dl)	0.30	0.08	0.32	0.07	0.00-0.90
Cholesterol (mg/dl)	32.0	10.5	26.8	11.1	16.0-43.0
Creatinine (mg/dl)	1.38	0.39	1.40	0.35	0.62-2.18
Glucose (mg/dl)	95.3	11.9	89.0	9.60	82.0-107.
Urea nitrogen (mg/dl)	25.2	6.37	21.5	5.84	9.00-31.5
Uric acid (mg/dl)	3.45	0.40	3.38	0.41	1.30-5.6
Sodium (mEq/l)	122.	0.98	125.	0.96	120.-146.
Potassium (mEq/l)	4.87	0.84	5.06	0.93	3.80-7.95
Chloride (mEq/l)	92.3	1.04	96.5	1.19	90.0-115.
Bicarbonate (mEq/l)	22.0	4.00	20.9	3.80	12.8-30.0
Phosphorus (mg/dl)	5.33	1.15	5.30	1.10	3.00-7.63
Calcium (mg/dl)	9.60	0.63	10.7	0.58	8.30-12.0
Magnesium (mg/dl)	2.35	0.25	2.46	0.27	1.80-3.00

*Source: Scarborough (1932); Albritton (1958); Altman and Dittmer (1964, 1971); Burns and Lannoy (1966); Moreland (1974).

Table 14. Enzyme Activities in Serum of the Normal Guinea Pig

Enzyme	Values in				Literature values*
	Male		Female		
	Mean	S.D.	Mean	S.D.	Range
Amylase (Somogyi units/dl)	295.	31.0	269.	28.0	237.-357.
Alkaline Phosphatase (I.U./1)	74.2	6.92	65.8	5.46	54.8-108.
Acid Phosphatase (I.U./1)	32.2	2.59	28.7	3.20	22.3-38.6
Alanine Transaminase (SGPT) (I.U./1) (ALT)	44.6	6.75	38.8	7.15	24.8-58.6
Aspartate Transaminase (SGOT) (I.U./1) (AST)	48.2	9.50	45.5	7.00	26.5-67.5
Creatine Phosphokinase (CPK) (I.U./1) (CK)	95.0	15.0	110	20.0	50-160
Lactic Dehydrogenase (LDH) (I.U./1)	46.9	9.50	52.1	11.2	24.9-74.5

*Source: Albritton (1952); Altman and Dittmer (1964, 1971);
         Nomura et al. (1975).

Table 15. Serum Proteins of the Normal Guinea Pig

Components	Values in				Literature values*
	Male		Female		
	Mean	S.D.	Mean	S.D.	Range
Total protein (g/dl)	5.60	0.28	4.80	0.34	5.00-6.80
Albumin (g/dl)	2.73	0.30	2.42	0.27	2.10-3.90
(%)	48.8	5.50	50.5	5.40	37.8-61.5
α_1 - Globulin (g/dl)	0.11	0.04	0.10	0.02	0.05-0.20
(%)	1.90	0.38	2.20	0.19	1.20-3.00
α_2 - Globulin (g/dl)	0.33	0.08	0.23	0.06	0.16-0.40
(%)	5.90	1.25	4.80	1.42	2.00-8.70
β-Globulin (g/dl)	1.14	0.20	0.82	0.17	0.40-1.54
(%)	20.4	4.10	17.1	3.60	8.90-28.6
γ-Globulin (g/dl)	1.29	0.26	1.22	0.15	0.67-2.10
(%)	23.1	4.60	25.4	3.25	12.1-35.0
Albumin/Globulin	0.95	0.16	1.02	0.18	0.72-1.34

*Source: Spector (1961); Burns and Lannoy (1966); Altman and Dittmer (1974); Dimopoullos (1972); Moreland (1974).

Table 16. Biochemical Reference Values of the Normal Guinea Pig

Components	Caisey & King (1980)[1] Mean	Coles (1980)[2] Range	Benirschke et al (1978)[2] Mean	Benjamin (1978)[2] Range	Melby & Altman (1974)[2] Range	Mitruka & Rawnsley (1977)[3] Mean	S.D.
Bilirubin (mg/dl)	-	-	-	0.00-0.60	-	0.31	0.08
Cholesterol (mg/dl)	22.8	64.0-104	43.9	-	21.0-43.0	16.1	10.8
Creatinine (mg/dl)	-	1.00-2.70	-	-	-	1.39	0.37
Glucose (mg/dl)	-	85.0-150	-	-	82.0-107	92.2	10.8
Urea nitrogen (mg/dl)	-	8.00-24.0	-	-	8.00-28.0	23.4	6.10
Uric acid (mg/dl)	-	0.50-1.95	-	-	1.30-5.60	3.42	0.41
Sodium (mEq/l)	136	135 - 150	145	140 - 160	146 - 152	134	0.97
Potassium (mEq/l)	5.50	4.40-6.70	7.40	4.90-7.10	6.80-8.90	4.97	0.89
Chloride (mEq/l)	105	94.0-106	105	100 - 105	98.0-115	94.4	1.12
Bicarbonate (mEq/l)	-	-	-	-	-	21.4	3.90
Phosphorus (mg/dl)	7.31	-	5.30	5.30-9.60	-	5.32	1.13
Calcium (mg/dl)	10.7	10.2-11.9	5.30	5.50-5.70	-	10.2	0.61
Magnesium (mg/dl)	-	-	-	1.90-3.20	-	2.41	0.26
Amylase (Somogyi unts/dl)	-	-	-	-	-	282	29.5
Alkaline Phosphatase (I.U./l)	87.6	-	-	-	-	70.0	6.19
Acid Phosphatase (I.U./l)	-	-	2.30	9.00-31.0	-	30.5	2.90
Alanine Transaminase (SGPT) (I.U./l) (ALT)	47.0	-	-	9.00-17.0	-	41.7	6.95
Aspartate Transaminase (SGOT) (I.U./l) (AST)	45.0	-	-	8.20-21.6	-	46.9	8.25

Table 16. Biochemical Reference Values of the Normal Guinea Pig

Components	Caisey & King (1980) Mean	Coles[2] (1980) Range	Benirschke[2] et al (1978) Mean	Benjamin[2] (1978) Range	Melby & Altman[2] (1974) Range	Mitruka & Rawnsley[3] (1977) Mean	S.D.
Creatine Phosphokinase (CPK) (I.U./l) (CK)	1.76	–	–	–	–	103	18.0
Lactic Dehydrogenase (LDH) (I.U./l)	103	–	–	96.0-160	–	49.5	10.4
Total Protein (g/dl)	4.60	–	5.40	7.90-8.90	5.00-5.60	5.20	0.31
Albumin (g/dl)	1.60	–	3.20	1.80-3.30	2.72-2.92	2.58	0.29
(%)	–	–	–	–	–	49.7	5.45
Alpha 1 – Globulin (g/dl)	–	–	–	0.32-0.44	–	0.11	0.03
(%)	–	–	–	–	–	2.05	0.29
Alpha 2 – Globulin (g/dl)	–	–	–	1.23-1.54	–	0.28	0.07
(%)	–	–	–	–	–	5.35	1.34
Beta – Globulin (g/dl)	1.00	–	–	1.39-2.01	–	0.98	0.19
(%)	–	–	–	–	–	18.8	3.85
Gamma – Globulin (g/dl)	0.60	–	–	2.24-2.46	–	1.26	0.21
(%)	–	–	–	–	–	24.3	3.93
Albumin/Globulin	0.55	–	–	0.37-0.51	–	0.99	0.17

Source: [1]Data from 10 Dunkin Hartley Guinea pigs, 3 weeks old. See text for methodology.
[2]Derived from literature.
[3]Data from 95 female and 110 male Hartley Guinea pigs weighing 500-800 g.
See text for methodology.

176

Table 17. Chemical Components in Serum of the Normal Rabbit

Components	Values in				Literature values*
	Male		Female		
	Mean	S.D.	Mean	S.D.	Range
Bilirubin (mg/dl)	0.32	0.04	0.30	0.04	0.00-0.74
Cholesterol (mg/dl)	46.7	12.9	44.5	11.2	10.0-80.0
Creatinine (mg/dl)	1.59	0.34	1.67	0.38	0.50-2.65
Glucose (mg/dl)	135.	12.0	128.	14.0	78.0-155.
Urea nitrogen (mg/dl)	19.2	4.93	17.6	4.36	13.1-29.5
Uric acid (mg/dl)	2.65	0.88	2.62	0.87	1.00-4.30
Sodium (mEq/l)	146.	1.15	141.	1.40	138.-155.
Potassium (mEq/l)	5.75	0.20	6.40	0.16	3.70-6.80
Chloride (mEq/l)	101.	1.45	105.	1.22	92.0-112.
Bicarbonate (mEq/l)	24.2	3.15	22.8	3.20	16.2-31.8
Phosphorus (mg/dl)	4.82	1.05	5.06	0.93	2.30-6.90
Calcium (mg/dl)	10.0	1.11	9.50	1.10	5.60-12.1
Magnesium (mg/dl)	2.52	0.24	3.20	0.22	2.00-5.40

*Source: Spector (1961); Burns and Lannoy (1966); Bonilla (1972);
Kaneko and Cornelius (1972); Laird (1972); Simesen (1972);
Byrd (1973); Moreland (1974); Harkness (1977).

Table 18. Enzyme Activities in Serum of the Normal Rabbit

Enzyme	Values in				Literature values*
	Male		Female		
	Mean	S.D.	Mean	S.D.	Range
Amylase (Somogyi units/dl)	132.	16.0	127.	12.0	90.0-170.
Alkaline Phosphatase (I.U./1)	10.4	2.28	9.96	3.10	4.10-16.2
Acid Phosphatase (I.U./1)	1.56	0.53	1.40	0.38	0.30-2.70
Alanine Transaminase (SGPT) (I.U./1) (ALT)	65.7	6.54	62.5	5.85	48.5-78.9
Aspartate Transaminase (SGOT) (I.U./1) (AST)	72.3	12.0	68.1	10.5	42.5-98.0
Creatine Phosphokinase (CPK) (I.U./1) (CK)	135	56.0	130	45.0	20-250
Lactic Dehydrogenase (LDH) (I.U./1)	84.4	20.5	78.5	22.0	33.5-129.

*Source: Albritton (1952); Altman and Dittmer (1974); Laird (1972); Byrd (1973); Nomura et al. (1975).

Table 19. Serum Proteins of the Normal Rabbit

Components	Values in				Literature values *
	Male		Female		
	Mean	S.D.	Mean	S.D.	Range
Total Protein (g/dl)	6.90	0.36	6.70	0.41	6.00-8.30
Albumin (g/dl)	3.39	0.29	3.04	0.26	2.42-4.05
(%)	49.2	4.60	45.3	3.80	35.5-63.5
α_1-Globulin (g/dl)	0.60	0.12	0.37	0.08	0.10-0.90
(%)	8.73	1.70	5.50	1.10	2.10-12.5
α_2-Globulin (g/dl)	0.43	0.09	0.21	0.05	0.15-0.75
(%)	6.21	1.30	3.20	0.94	1.50-11.8
β-Globulin (g/dl)	1.01	0.20	1.46	0.24	0.50-2.10
(%)	14.6	2.90	21.8	3.20	12.0-27.4
γ-Globulin (g/dl)	1.46	0.21	1.69	0.22	1.00-2.15
(%)	21.2	3.21	25.2	3.25	14.4-32.7
Albumin/Globulin	0.97	0.16	0.83	0.15	0.68-1.15

*Source: Spector (1961); Altman and Dittmer (1964, 1971); Burns and Lannoy (1966); Dimopoullos (1972); Laird (1972); Byrd (1973); Moreland (1974).

rabbits over the period of 1 year. The data compare well with the literature values reported by many investigators (Table 20).

Different breeds of rabbits have essentially similar levels of biochemical constituents in their sera (Table 21). The rabbit blood chemistry values reported by Spector (1956) were relatively lower than those found by Jones (1975) in Belgian Semi-lop rabbits. Jones (1975) also found that sodium, urea nitrogen, and total proteins were lower in his animals than in those tested by Spector (1956). Our data show that the differences due to species (breeds), age, or sex were not important.

Compared with the chemical values of normal humans, rabbits had lower bilirubin, cholesterol, uric acid, alkaline phosphatase, creatine phosphokinase, lactic dehydrogenase, and A/G ratio. However the BUN, inorganic phosphorus, amylase, alkaline phosphatase, acid phosphatase, AST, ALT, β-globulin, and α-globulin were considerably higher in normal rabbits than in normal humans. Cholesterol values were lower and inorganic phosphorus was higher in all strains of rabbits than in human sera. AST, ALT, and BUN varied widely both in the rabbits tested by us and in those reported in the literature.

Other Mammalian Species

Comparative biochemical values of certain small mammalian species are given in Table 22. There has been recent interest in these species as potential models for the study of specific human diseases. However there have been no extensive clinical biochemical studies in these animals. Mays (1969) reported significant differences between male and female gerbils in levels of inorganic phosphorus and uric acid. All other values for gerbils were not remarkable and fell generally within the ranges reported for other small rodents. Compared to normal humans, gerbils had lower bilirubin, calcium, cholesterol, and uric acid, and higher total protein, β-globulin, and inorganic phosphorus.

Blood chemistries of chinchillas were essentially similar to those of other small rodents reported in the literature. However lactic dehydrogenase and inorganic phosphorus were considerably higher, and bilirubin, total protein, and cholesterol were lower in the sera of chinchillas than in the sera of humans.

Physiological and biochemical studies of various species of opossums have been reported. However, because of the variety of internal parasites in clinically normal opossums, it has been difficult to establish correct normal values in this species (Timmons, and Marques, 1969). The data reported in the literature indicate that bilirubin, cholesterol, lactic dehydrogenase, albumin, and uric acid are lower, and that inorganic phosphorus, potassium, α_2-globulin, β-globulin, and urea nitrogen are higher in the sera of opossums than in the sera of normal humans. It is also interesting that Rothstein and Hunsaker (1972) observed that opossum chemistries are similar to those of primates.

The armadillo has recently been adopted as an experimental animal species for the study of reproductive biology, endocrinology, immunology, transplantation biology, genetic research, anatomical studies, adaptation to artificial environments outside their normal range, and for histological investigations (Mitruka et al., 1976; Anderson and Benirschke, 1966; Blair, 1970; Dunloo, 1969; Lewis and Doyle, 1964). The armadillo is the only mammal with ossified epidermal scales, and it exhibits certain unique features. The most important of these features are a relatively low body temperature (32–34°C), a long delay in the implantation of the blastocyst (3–4 months), extremely large adrenal glands in the embryo, and specific polyembryony resulting in monozygotic quadruplets (Talmage and Buchanan, 1954).

Table 20. Biochemical Reference Values of the Normal Rabbit

Components	Caisey & King (1980)[1] Mean	Yu et al[2] (1979) Mean	S.D.	Benirschke et al (1978)[3] Mean	S.D.	Melby and Altman (1974)[3] Mean	S.D.	Mitruka and Rawnsley (1977)[4] Mean	S.D.
Bilirubin (mg/dl)	–	0.28	0.13	0.07	0.06	0.21	0.26	0.31	0.04
Cholesterol (mg/dl)	77.6	38.8	20.6	20.5	4.36	24.6	9.60	25.6	12.1
Creatinine (mg/dl)	–	1.20	0.23	–	–	–	–	1.63	0.36
Glucose (mg/dl)	–	111	12.1	–	–	113	31.8	132	13.0
Urea nitrogen (mg/dl)	–	18.1	3.80	21.3	0.68	21.9	5.90	18.4	4.65
Uric acid (mg/dl)	–	0.99	0.32	0.94	0.04	1.18	0.29	2.64	0.88
Sodium (mEq/l)	156	142	8.30	158	–	–	–	144	1.28
Potassium (mEq/l)	6.00	3.92	0.36	4.10	–	4.60	–	6.08	0.18
Chloride (mEq/l)	108	105	2.80	105	–	109	–	103	1.34
Bicarbonate (mEq/l)	–	18.2	2.74	–	–	–	–	23.5	3.18
Phosphorus (mg/dl)	7.24	4.39	0.71	3.89	0.80	3.99	0.83	4.94	0.99
Calcium (mg/dl)	13.2	13.6	0.73	14.4	–	14.0	–	9.75	1.10
Magnesium (mg/dl)	–	2.00	0.20	–	–	2.86	0.23	3.20	0.59
Amylase (Somogyi units/dl)	–	262	57.5	–	–	–	–	130	14.0
Alkaline Phosphatase (I.U./l)	406	39.8	25.8	11.5	0.57	11.4	4.84	10.2	2.69
Acid Phosphatase (I.U./l)	–	0.86	0.30	–	–	–	–	1.48	0.46
Alanine Transaminase (ALT) (I.U./l)	79.0	33.3	13.4	50.8	10.6	49.1	20.2	64.1	6.20
Aspartate Transaminase (I.U./l) (AST)	47.0	16.1	6.80	89.9	11.2	61.3	22.1	70.2	11.3
Creatinine Phosphokinase (I.U./l) (CK)	544	311	169	114	13.9	174	49.0	133	51.0

181

Table (con't) 20. Biochemical Reference Values of the Normal Rabbit

Components	Caisey & King[1] (1980) Mean	Yu et al[2] (1979) Mean	S.D.	Benirschke et al[3] (1978) Mean	S.D.	Melby and Altman[3] (1974) Mean	S.D.	Mitruka and Rawnsley[4] (1977) Mean	S.D.
Lactic Dehydrogenase (LDH) (I.U./l)	209	77.5	32.7	62.4	2.66	99.6	29.9	81.4	21.3
Total Protein (g/dl)	5.90	6.44	0.57	6.12	0.38	6.14	0.54	6.80	0.39
Albumin (g/dl)	2.20	–	–	4.60	–	3.30	0.24	3.22	0.28
(%)	–	–	–	66.8	7.90	–	–	47.3	4.20
Alpha 1-Globulin (g/dl)	–	–	–	–	–			0.49	0.10
(%)	–	–	–	7.21	0.96			7.12	1.40
Alpha 2-Globulin (g/dl)	–	–	–	–	–			0.32	0.07
(%)	–	–	–	5.80	1.05			4.71	1.12
Beta-Globulin (g/dl)	1.20	–	–	–	–			1.24	0.22
(%)	–	–	–	–	–			18.2	3.05
Gamma Globulin (g/dl)	1.50	–	–	–	–			1.58	0.22
(%)	–	–	–	16.8	6.80			23.2	3.23
Albumin/Globulin	0.58	–	–	–	–	–	–	0.90	0.16

Source: [1] Data from 6 New Zealand White Rabbits, 9 to 10 weeks old. See text for methodology.

[2] Data from 45 purebred New Zealand White male rabbits 3 to 5 kg weight (age 1 to 2 years). See text for methodology.

[3] Derived from literature.

[4] Data from 80 female and 120 male New Zealand White rabbits weighing 3 to 5 kg. See text for methodology.

Table 21.   Biochemical Values in Sera of Certain Breeds of Rabbits

Components	New Zeland White	Dutch Belted	Polish
Alkaline Phosphatase	4.90-15.6 (K.A. units)	2.10-16.0 (K.A. units)	4.30-5.60 (K.A. units)
Chloride (mEq/1)	98.0-116.	102.-120.	89.0-112.
Cholesterol (mg/dl)	20.0-83.0	35.0-82.0	50.0-92.0
Creatinine (mg/dl)	0.80-2.57	0.80-1.70	1.30-2.90
Glucose (mg/dl)	50.0-93.2	69.0-159.	87.0-115.
Lactic Dehydrogenase (LDH) (I.U./1)	180.-300.	140.-250.	145.-292.
Magnesium (mg/dl)	2.10-4.30	1.98-4.10	2.00-4.50
Phosphorus, Inorganic (mg/dl)	4.40-7.80	3.10-8.15	4.00-5.10
Protein, Total (g/dl)	5.30-7.90	5.70-9.70	7.70-9.80
Protein, Electrophoresis Albumin (%)	47.0-67.2	52.5-62.4	46.8-65.5
α-Globulin (%)	7.40-10.4	15.2-19.5	12.0-18.2
β-Globulin (%)	7.94-11.8	15.3-21.4	11.7-17.6
γ-Globulin (%)	9.50-14.6	9.20-14.3	10.7-15.8
Potassium (mEq/1)	3.60-6.90	4.40-7.40	5.00-5.70
Sodium (mEq/1)	100.-145.	138.-156.	114.-146.
Transaminase, Aspartate (SGOT) (Karmen units) (AST)	14.0-113.	16.0-75.0	48.0-101.
Transaminase, Alanine (SGPT) (Wroblewski-LaDue units)(ALT)	16.0-67.0	12.0-37.0	25.0-46.0
Urea, nitrogen (mg/dl)	9.17-31.7	13.0-40.0	26.0-45.0

Source: Albritton (1952); Allen and Watson (1958); Burns and Lannoy (1966); Altman and Dittmer (1968); Fox _et al._ (1970); Houpt (1970); Weisbroth _et al._ (1974).

Table 22. Comparative Biochemical Values in Sera of Certain Mammalian Species

Component	Gerbil	Chinchilla	Opossum	Armadillo
Bilirubin, Total	0.20-0.60	0.00-0.23	0.30-0.80	0.00-0.19
Calcium (mg/dl)	3.70-6.20	4.40-10.0	9.60-11.2	9.10-13.3
Chloride (mEq/l)	93.0-118.	96.6-113.	85.0-117.	100.-115.
Cholesterol (mg/dl)	90.0-130.	96.0-147.	80.0-151.	70.0-210.
Creatinine (mg/dl)	0.50-1.40	---	---	1.00-1.12
Glucose (mg/dl)	40.0-141.	89.0-163.	99.0-145.	55.0-260.
Lactic Dehydrogenase (U/L)	----	406.-636.	60.0-185.	----
Phosphatase, Alkaline	12.3-37.0 (K.A. unit)	----	13.0-35.0 (K.A. unit)	20.0-80.0 (K.A. unit)
Phosphate, Inorganic (mg/dl)	3.70-7.20	4.70-7.00	2.90-8.20	4.40-7.10
Potassium (mEq/l)	3.30-6.30	3.00-4.60	4.00-7.00	4.10-6.30
Protein, Total (g/dl)	4.80-16.8	3.29-4.61	5.60-8.00	5.50-7.70
Protein, Electrophoresis				
Albumin (%)	22.7-64.1	55.8-73.5	9.50-24.8	38.2-53.5
α_1 - Globulin (%)	2.10-6.45	4.00-10.4	2.90-6.50	9.74-13.6
α_2 - Globulin (%)	2.85-9.75	5.30-9.70	14.8-27.9	10.2-14.3
β - Globulin (%)	8.46-22.1	4.60-16.6	11.4-32.0	13.5-18.9
γ - Globulin (%)	7.50-29.6	4.10-14.5	1.50-15.3	19.5-27.3
Urea, Nitrogen (mg/dl)	16.8-31.3	----	8.50-35.4	14.0-45.5
Uric Acid (mg/dl)	1.10-3.08	----	0.90-2.20	0.05-0.80

Source: Kritchevsky (1958); Altman and Dittmer (1964, 1971); Anderson and Benirscheke (1966); Mays (1969); Timmons and Marques (1969); Strozier et al. (1971); Peters and Storrs (1972); Rothstein and Hunsaker (1972).

Some aspects of morphological and physiological data of armadillos have been detailed by Szabuniewicz and McCrady (1969). Strozier *et al.* (1971) reported several serum chemical components of normal armadillos and found that there were no sex-related differences except in cholesterol value, which was considerably higher in males. The bilirubin, cholesterol, and uric acid of armadillos were considerably lower, and glucose, inorganic phosphorus, potassium, α-, β-, and γ-globulins, and blood urea nitrogen were higher than those of humans.

Chickens

Tables 23–25 present blood chemistry data of normal white leghorn chickens weighing 1.50–2.50 kg. The values were recorded from 52 male and 46 female laboratory-conditioned birds. When compared to measurements of human sera, the normal chicken's bilirubin, cholesterol, urea nitrogen, creatine phosphokinase, albumin, and A/G ratio were lower, and glucose, sodium, chloride, phosphorus, calcium, amylase, acid phosphatase, AST, ALT, α1-, α2-, and β-globulins were considerably higher.

Reptiles, Amphibians, and Fishes

Blood chemistries of birds are comparable to those of reptiles, amphibians, and fishes rather than to those of mammals. However there are many species and varieties of these animals, and therefore it is rather difficult to establish normal values for the group of animal species under experimental investigation. Table 26 compares the representative data of snakes, turtles, frogs, fishes, and birds. Although these data are not comprehensive, general comparisons may be made with human chemistry data. Snakes have lower bicarbonate, creatinine, glucose, protein, blood urea nitrogen, and alkaline phosphatase, and higher calcium, sodium, chloride, cholesterol, phosphate, α1-, β-, and γ-globulins than do humans.

When the values of turtle sera are compared with those of humans, cholesterol, creatinine, glucose, total protein, α2-globulin, γ-globulin, urea nitrogen, and uric acid are generally lower, and calcium, bicarbonate, magnesium, phosphate, and α1-globulin are higher in turtles.

In frogs, chloride, cholesterol, glucose, alkaline phosphatase, total protein, sodium, urea nitrogen, and bicarbonate are lower, and magnesium, inorganic phosphate, β- and γ-globulins are higher than in the sera of normal humans.

In fishes, cholesterol, creatinine, alkaline phosphatase, total protein, α1-globulin, urea nitrogen, and bicarbonate are reported to be lower, and calcium, sodium, chloride, glucose, phosphate, and α2-globulin are reported to be higher than in humans.

Cats

Tables 27–29 list the normal blood chemistries of conditioned mongrel cats weighing 2–4 kg. These data were established from a study of 32 male and 30 female cats over a period of 2 years. Variability in blood chemistry values due to age, sex, and nutritional condition were not important, but the values of an individual animal varied considerably over the study period, perhaps because of the easily excitable nature of the animal. However our data generally agreed well with those reported in the literature (Table 30).

In comparison with the normal blood chemistry values of humans, cats had lower cholesterol, uric acid, alkaline phosphatase, creatine phosphokinase, lactic dehydrogenase, albumin, and A/G ratio, and generally higher blood urea nitrogen, phosphorus, amylase, α1-globulin, α2-globulin, and β-globulin.

Table 23. Chemical Components in Serum of the Normal Chicken

Components	Values in Male		Values in Female		Literature values*
	Mean	S.D.	Mean	S.D.	Range
Bilirubin (mg/dl)	0.10	0.02	0.05	0.04	0.00-0.20
Cholesterol (mg/dl)	100.	25.0	92.0	21.6	52.0-148.
Creatinine (mg/dl)	1.38	0.27	1.10	0.30	0.90-1.85
Glucose (mg/dl)	162.	15.1	167.	16.2	152.-182.
Urea nitrogen (mg/dl)	1.95	0.75	1.80	0.80	1.50-6.30
Uric acid (mg/dl)	5.28	1.20	5.30	1.40	2.47-8.08
Sodium (mEq/l)	153.	2.35	158.	2.46	148.-163.
Potassium (mEq/l)	5.06	0.38	5.63	0.41	4.60-6.50
Chloride (mEq/l)	119.	1.38	117.	1.26	116.-140.
Bicarbonate (mEq/l)	23.0	2.10	24.6	2.30	17.6-29.8
Phosphorus (mg/dl)	7.05	0.80	6.85	0.91	6.20-7.90
Calcium (mg/dl)	14.4	5.20	19.6	4.86	9.0-23.7
Magnesium (mg/dl)	2.58	0.27	1.70	0.30	1.30-3.80

*Source: Albritton (1952); Spector (1961); Weiser et al. (1965); Schermer (1967); Lloyd et al. (1970); Swenson (1970); Altman and Dittmer (1974); Nomura et al. (1975).

Table 24. Enzyme Activities in Serum of the Normal Chicken

Enzyme	Values in				Literature values*
	Male		Female		
	Mean	S.D.	Mean	S.D.	Range
Amylase (Somogyi units/dl)	240.	35.0	265.	40.00	160.-345.
Alkaline Phosphatase (I.U./1)	36.4	3.78	32.5	3.30	24.5-44.4
Acid Phosphatase (I.U./1)	31.2	3.50	33.4	4.10	23.0-41.6
Alanine Transaminase (SGPT) (I.U./1) (ALT)	25.0	5.48	21.9	6.02	9.50-37.2
Aspartate Transaminase (SGOT) (I.U./1) (AST)	142.	24.5	154.	27.0	88.0-208.
Creatine Phosphokinase (CPK) (I.U./1) (CK)	490	110	560	120	240-810
Lactic Dehydrogenase (LDH) (I.U./1)	208.	51.1	190.	45.6	88.0-310.

*Source: Albritton (1952); Spector (1961); Swenson (1970); Altman and Dittmer (1974); Nomura et al. (1975).

Table 25. Serum Proteins of the Normal Chicken

Components	Values in				Literature values*
	Male		Female		
	Mean	S.D.	Mean	S.D.	Range
Total protein (g/dl)	6.10	0.32	5.91	0.30	5.20-6.90
Albumin (g/dl)	2.81	0.21	2.68	0.24	2.10-3.45
(%)	46.1	3.20	45.4	4.10	36.5-55.0
α_1 - Globulin (g/dl)	0.74	0.11	0.70	0.12	0.45-0.96
(%)	12.2	1.78	11.8	1.82	7.8-16.5
α_2 - Globulin (g/dl)	0.69	0.15	0.80	0.16	0.35-1.20
(%)	11.3	2.34	13.6	2.40	6.30-18.6
β-Globulin (g/dl)	1.19	0.24	1.03	0.26	0.50-1.80
(%)	19.5	4.25	17.4	4.10	8.40-29.0
γ-Globulin (g/dl)	0.66	0.07	0.70	0.07	0.46-1.00
(%)	10.9	1.20	11.8	1.50	7.90-15.0
Albumin/Globulin	0.85	0.22	0.83	0.23	0.58-1.30

*Source: Spector (1961); Schermer (1967); Samadieh et al. (1969); Altman and Dittmer (1974).

Table 26. Comparative Biochemical Values in the Sera of Reptiles, Amphibia, Fishes and Birds

Component	Snake[1]	Turtle[2]	Frog[3]	Fish[4]	Bird[5]
Calcium (mg/dl)	10.8-17.2	6.80-14.8	8.00-10.0	9.45-14.8	10.6-23.7
Chloride (mEq/l)	120.-138.	81.0-109.	55.0-95.0	225.-265.	106.-124.
Cholesterol (mg/dl)	266.-495.	24.0-36.5	117.-141.	20.0-60.0	52.0-200.
Creatinine (mg/dl)	0.50-0.60	0.34-0.92	---	0-0.72	0.86-1.85
Glucose (mg/dl)	25.0-108.	22.5-75.0	36.0-49.0	98.0-256.	117.-189.
Magnesium (mg/dl)	2.18-6.07	1.20-18.9	1.92-9.12	2.10-3.88	1.40-2.00
Phosphate, Inorganic (mg/dl)	2.17-9.30	2.79-10.8	5.89-10.9	8.79-40.0	6.20-7.90
Phosphatase, Alkaline (I.U./l)	9.90-19.8	----	6.20-13.5	8.10-27.0	24.0-45.0
Potassium (mEq/l)	4.10-7.20	2.40-6.70	2.50-5.00	4.50-6.80	5.30-6.41
Protein, Total (g/dl)	3.73-6.85	2.60-5.40	3.74-6.51	2.40-3.40	4.42-5.81
Albumin (%)	27.0-64.0	29.0-57.3	29.9-55.4	47.0-67.7	37.0-58.0
α_1 (%)	5.0-21.0	16.6-17.6	2.40-8.90	1.00-4.65	10.0-16.0
α_2 (%)	5.40-10.3	4.50-7.50	9.50-15.4	17.0-26.4	3.40-7.50
β(%)	10.1-22.6	1.60-18.6	29.9-47.5	11.3-16.9	9.00-18.0
γ(%)	27.4-41.4	9.50-14.5	17.8-28.6	15.0-26.1	9.00-46.9
Sodium (mEq/l)	152.-176.	120.-165.	88.-113.	235.-316.	140.-161.
Urea, Nitrogen (mg/dl)	5.00-6.00	0.90-1.35	0.50-1.10	1.04-1.60	4.60-6.80
Uric acid (mg/dl)	4.85-10.1	0.80-1.8	----	2.60-8.60	2.47-8.08
Bicarbonate (mEq/l)	2.00-14.0	23.0-52.0	14.0-25.0	6.0-12.0	17.5-30.0

Table 26 (con'td.)

Source: 1. Snake (Species of Bitis, Bothrops, Philohydrus and Crotalus): Prado (1946); Carmichael and Petcher (1948); Deutsch and McShan (1949); Dessauer et al. (1956); Hutton (1958); Altman and Dittmer (1971, 1974).

2. Turtle (Species of Caretta, Terrapene, Chrysemys, Coluber and Pseudemys Scripta): Drilhon and Marcoux (1942); Albritton (1952); Cohen (1954); Caumer and Goodnight (1957); Hutton and Goodnight (1957); Altman and Dittmer (1974).

3. Frog (Species of Rana Catesbeiana, R. areolata, R. pipiens and R. esculenta): Fenn (1936); Hashimoto and Nukata (1951); Albritton (1952) Ogaswara (1954); Gibbons and Kaplan (1959); Schermer (1967); Altman and Ditmer (1971, 1974).

4. Fish (Species of Salvelinus; Prosopium; Caysiodes and Ameiurus): Hartman et al. (1941); Field et al. (1943); Moore (1945); Deutsch and McShan (1949); Platner (1949); Ogasawara (1953).

5. Bird (Species of Columba, Coturnix, Meleagris, Anas Platyrhynchos):

See Table 21 for references.

190

Table 27. Chemical Components in Serum of the Normal Cat

Components	Values in Male		Values in Female		Literature values*
	Mean	S.D.	Mean	S.D.	Values
Bilirubin (mg/dl)	0.18	0.05	0.15	0.04	0.10-1.89
Cholesterol (mg/dl)	110.	28.0	101.	22.0	83.0-135.
Creatinine (mg/dl)	1.50	0.50	1.40	0.45	0.40-2.60
Glucose (mg/dl)	120.	14.0	114.	15.0	60.0-145.
Urea nitrogen (mg/dl)	25.0	5.00	27.5	4.50	14.0-32.5
Uric acid (mg/dl)	1.45	0.22	1.30	0.20	0.00-1.85
Sodium (mEq/1)	150.	1.15	152.	1.20	147.-156.
Potassium (mEq/1)	4.25	0.24	5.30	0.31	4.00-6.00
Chloride (mEq/1)	120.	1.10	112.	1.00	110.-123.
Bicarbonate (mEq/1)	20.4	2.40	21.8	2.80	14.5-27.4
Phosphorus (mg/dl)	6.20	1.07	6.40	1.17	4.50-8.10
Calcium (mg/dl)	10.1	0.85	11.2	0.92	8.10-13.3
Magnesium (mg/dl)	2.64	0.25	2.54	0.21	2.00-3.00

*Source: Albritton (1952); Altman and Dittmer (1974); Kaneko and Cornelius (1972); Laird (1972); Moreland (1974).

Table 28. Enzyme Activities in Serum of the Normal Cat

Enzyme	Values in				Literature values*
	Male		Female		
	Mean	S.D.	Mean	S.D.	Range
Amylase (Somogyi units/dl)	152.	32.0	138.	27.0	68.-222.
Alkaline Phosphatase (I.U./l)	14.3	3.54	10.4	3.20	3.40-21.3
Acid Phosphatase (I.U./l)	2.67	1.15	2.48	1.27	0.10-5.20
Alanine Transaminase (SGPT) (I.U./l) (ALT)	19.6	4.95	18.5	3.60	8.50-29.6
Aspartate Transaminase (SGOT) (I.U./l) (AST)	19.0	4.80	17.0	4.90	7.00-29.0
Creatine Phosphokinase (CPK) (I.U./l) (CK)	170	110	190	130	50-450
Lactic Dehydrogenase (LDH) (I.U./l)	67.5	10.6	87.4	11.5	34.5-110.

*Source: Altman and Dittmer (1974); Kaneko and Cornelius (1972); Laird (1972).

Table 29. Serum Proteins of the Normal Cat

Components	Values in				Literature values*
	Male		Female		
	Mean	S.D.	Mean	S.D.	Range
Total protein (g/dl)	6.10	0.32	5.91	0.30	4.30-7.5
Albumin (g/dl)	2.81	0.21	2.68	0.24	2.20-3.20
(%)	46.1	3.20	45.4	4.10	44.0-56.0
α_1 - Globulin (g/dl)	0.74	0.11	0.70	0.12	0.40-1.00
(%)	12.2	1.78	11.8	1.82	7.80-16.5
α_2 - Globulin (g/dl)	0.69	0.15	0.80	0.16	0.30-1.30
(%)	11.3	2.34	13.6	2.40	6.30-18.6
β-Globulin (g/dl)	1.19	0.24	1.03	0.26	0.43-1.80
(%)	19.5	4.25	17.4	4.10	8.40-28.5
γ-Globulin (g/dl)	0.66	0.07	0.70	0.07	0.46-1.00
(%)	10.9	1.20	11.8	1.50	7.50-15.1
Albumin/Globulin	0.85	0.18	0.83	0.19	0.60-1.20

*Source: Albritton (1952); Altman and Dittmer (1974); Kaneko and Cornelius (1972); Moreland (1974).

Table 30. Biochemical Reference Values of the Normal Cat

Components	Coles[2] (1980) Range	Caisey[1] & King (1980) Mean	Benirschke[2] et al (1978) Mean	Benirschke[2] et al (1978) S.D.	Benjamin[2] (1978) Range	Tasker[3] (1978) Mean	Tasker[3] (1978) S.D.	Mitruka & Rawnsley[4] (1977) Mean	Mitruka & Rawnsley[4] (1977) S.D.
Bilirubin (mg/dl)	-	-	0.31	0.44	0.15-0.20	0.30	0.15	0.27	0.05
Cholesterol (mg/dl)	95 -130	159	120	32.8	-	134	4.60	106	25.0
Creatinine (mg/dl)	1.0-2.0	-	-	-	-	1.10	0.30	1.45	0.48
Glucose (mg/dl)	50 - 75	-	149	42.1	-	84.0	30.0	117	14.5
Urea nitrogen (mg/dl)	20 - 30	-	22.2	8.19	-	24.0	4.52	26.3	4.75
Uric acid (mg/dl)	-	-	0.80	0.45	-	-	-	1.38	0.21
Sodium (mEq/l)	147-156	-	5.70	0.10	147 - 156	156	2.98	151	1.18
Potassium (mEq/l)	4.0-4.5	4.80	16.7	0.30	4.00-4.50	4.60	0.45	4.78	0.28
Chloride (mEq/l)	117-123	118	-	-	-	119	2.74	116	1.05
Bicarbonate (mEq/l)	-	-	-	-	-	-	-	21.1	2.60
Phosphorus (mg/dl)	2.6-4.7	5.70	4.75	1.03	-	5.60	0.80	6.30	1.12
Calcium (mg/dl)	6.3-10	10.7	8.22	0.97	-	9.80	0.68	10.7	0.89
Magnesium (mg/dl)	-	-	-	-	-	-	-	2.59	0.23
Amylase (Somogyi units/dl)	-	-	-	-	-	-	-	145	29.5
Alkaline Phosphatase (I.U./l)	-	291	8.34	4.84	2.00-7.00	15.0	8.00	12.4	3.37
Acid Phosphatase (I.U./l)	-	-	-	-	-	-	-	2.58	1.21
Alanine Transaminase (ALT) (I.U./l)	-	27.0	76.0	70.2	1.70-14.0	-	-	19.0	4.28
Aspartate Transaminase (AST) (I.U./l)	-	11.0	160	141	6.70-11.0	46.0	11.0	18.0	4.85
Creatine Phosphokinase (CK) (I.U./l)	-	137	-	-	-	69.0	31.0	180	120

Table 30. Biochemical Reference Values of the Normal Cat

Components	Coles[2] (1980) Range	Caisey[1] & King (1980) Mean	Benirschke[2] et al (1978) Mean	S.D.	Benjamin[2] (1978) Range	Tasker[3] (1978) Mean	S.D.	Mitruka & Rawnsley[4] (1977) Mean	S.D.
Lactic Dehydrogenase (LDH) (I.U./l)	-	137	152	143	16.0-69.0	61.0	30.0	77.5	11.1
Total Protein (g/dl)	-	7.00	6.43	0.65	5.40-7.30	6.50	0.50	6.05	0.31
Albumin (g/dl)	-	3.00	2.99	-	2.10-3.30	2.60	0.30	2.75	0.23
(%)	-	-	-	-	-	-	-	45.8	3.65
Alpha 1 - Globulin (g/dl)	-	-	4.53	1.40	0.20-1.10	-	-	0.72	0.12
(%)	-	-	-	-	-	-	-	12.0	1.80
Alpha 2 - Globulin (g/dl)	-	1.07	-	-	0.40-0.90	-	-	0.75	0.16
(%)	-	-	-	-	-	-	-	12.5	2.37
Beta - Globulin (g/dl)	-	1.30	1.08	-	1.00-1.90	-	-	1.11	0.25
(%)	-	-	-	-	-	-	-	18.5	4.18
Gamma - Globulin (g/dl)	-	0.70	1.96	-	1.70-4.40	-	-	0.68	0.07
(%)	-	-	-	-	-	-	-	11.4	1.35
Albumin/Globulin	-	0.74	-	-	0.45-1.19	-	-	0.80	0.19

Source:
[1] Data from 6 crossbred cats 5 month old. See text for methodology.
[2] Derived from literature.
[3] Data from 27 adult animals, 5 blood donors and 22 privately owned, male and female cats. See text for methodology.
[4] Data from 30 female and 32 male mongrel cats weighing 2-4 kg. See text for methodology.

Dogs

Tables 31–33 present chemical values of 18 male and 10 female mongrel dogs weighing 12–16 kg. Bilirubin, uric acid, alkaline phosphatase, CK, LDH, albumin, γ-globulin, and the A/G ratio of dogs were lower than those of humans. However, amylase acid phosphatase, ALT, AST, and β-globulin were higher in dogs than in normal humans. Larger variations were observed in certain chemical constituents, particularly in enzymes and other proteins (Table 34) perhaps due to methodological differences and other test conditions. However differences due to the sex or age of the animal were not important.

Michaelson *et al.* (1966) found no differences due to sex in chemistry of beagles. Differences due to age appeared specifically in the serum proteins between 6 months to 1 year and 1- to 2-year age groups in which there was an increase in total protein, albumin, and globulins. There was a significant decrease in transaminase levels between these two age groups. There was also a significant increase in venous CO_2 content between 1- and 3-year-old females. In general, however, there were no significant differences in various components among dogs over 2 years old.

McKelvie *et al.* (1966) reported the results of tests on serum from young (1-year-old) beagles and from older (10-year-old) beagles. The older dogs had higher glucose and bilirubin values and lower BUN. Analysis of serum taken at 120-day intervals for the first year of life revealed that the values were within the normal ranges reported by others. Changes with age were apparent in glucose, inorganic phosphorus, total protein, albumin, and globulin. Other variable factors included the effect of climate and the time of blood sample collection relative to the feeding of the dogs.

McKelvie (1971) determined the electrophoretic protein fractions in the sera of several hundred clinically normal beagles until they were 8 years old. Total protein increased until 8 months of age because of an increase in all but the α fraction of globulins, which decreased slightly. Thereafter the values remained quite consistent. A slight fluctuation in the albumin concentration was reflected in changes in total protein values. The mean values of serum proteins from 1-year-old beagles varied considerably from those reported by Van Stewart and Langwell (1969).

Robinson and Ziegler (1968) reported the clinical laboratory values of a large number of newly purchased beagles with minimum stabilization. The animals were young and had been acquired from several sources. Considerable variation was found; for example, AST and ALT were higher in females than in males, and phosphorus, potassium, and sodium were higher in males. Adaptation of dogs from the environment to laboratory conditions caused statistically significant differences in serum proteins, calcium, phosphorus, and lactic dehydrogenase. Although there were statistical differences between the normal values of conditioned male and female dogs, the differences were small and had little practical significance.

Differences in the clinical chemistries of different breeds of dogs are probably caused by differences in the dogs' size, genetic factors, nutrition, and metabolism. Table 35 lists the serum chemical values of four breeds of dogs. It shows that in German shepherds, glucose, LDH, alkaline phosphatase, blood urea nitrogen (BUN), and uric acid are lower, and that amylase and transaminases are higher than those of humans. In other breeds, LDH, alkaline phosphatase, BUN, and uric acid have been reported to be lower than those values in man. However amylase and AST are higher in all breeds of dogs, and inorganic phosphorus and ALT are considerably higher in beagles and Labrador breeds of dogs than in humans.

Lane and Robinson (1970) studied sev-

Table 31. Chemical Components in Serum of the Normal Dog

Components	Values in				Literature values*
	Male		Female		
	Mean	S.D.	Mean	S.D.	Range
Bilirubin (mg/dl)	0.25	0.11	0.21	0.10	0.00-0.50
Cholesterol (mg/dl)	211.	12.0	150.	11.0	137.-275.
Creatinine (mg/dl)	1.35	0.35	1.08	0.15	0.80-2.05
Glucose (mg/dl)	132.	10.4	110.	8.5	80.0-165.
Urea nitrogen (mg/dl)	15.0	1.90	13.9	1.20	5.00-23.9
Uric acid (mg/dl)	0.55	0.10	0.42	0.10	0.20-0.90
Sodium (mEq/l)	147.	2.20	146.	1.90	139.-153.
Potassium (mEq/l)	4.54	0.10	4.42	0.20	3.60-5.20
Chloride (mEq/l)	114.	1.15	111.	1.20	103.-121.
Bicarbonate (mEq/l)	21.8	1.60	22.2	1.91	14.6-29.4
Phosphorous (mg/dl)	4.40	1.00	3.70	0.50	2.70-5.70
Calcium (mg/dl)	10.2	0.42	9.40	0.50	9.30-11.7
Magnesium (mg/dl)	2.10	0.30	2.20	0.28	1.50-2.80

*Source: Spector (1961); Crawley and Swenson (1965); Fletcher et al. (1966); McKlevie (1966, 1970); Michaelson (1966); Coles (1967); Robinson and Siegler (1968); Brunden et al. (1970); Lane and Robinson (1970); Altman and Dittmer (1974); Kaneko and Cornelius (1972); Laird (1972) Ralston Purina Co. Bulletin (1973); Moreland (1974).

Table 32. Enzyme Activities in Serum of the Normal Dog

Enzyme	Values in				Literature values*
	Male		Female		
	Mean	S.D.	Mean	S.D.	Range
Amylase (Somogyi units/dl)	178.	30.0	175.	26.0	140.-480.
Alkaline Phosphatase (I.U./1)	17.7	3.80	16.5	4.30	7.90-26.3
Acid Phosphatase (I.U./1)	3.54	1.20	3.30	1.10	0.80-6.00
Alanine Transaminase (SGPT) (I.U./1) (ALT)	40.6	7.70	44.0	7.80	24.5-60.0
Aspartate Transaminase (SGOT) (I.U./1) (AST)	54.8	8.90	59.2	9.00	36.0-77.5
Creatine Phosphokinase (CPK) (I.U./1) (CK)	120	30.0	80.0	11.0	20-200
Lactic Dehydrogenase (LDH) (I.U./1)	76.5	17.8	67.0	15.0	30.0-112.

*Source: Bloom (1960); McKelvie (1966); Coles (1967); Robinson and Ziegler (1968); Brunden et al. (1970); Lane and Robinson (1970); Nomura (1975); Ayers et al. (1975).

Table 33. Serum Proteins of the Normal Dog

Components	Values in				Literature values*
	Male		Female		
	Mean	S.D.	Mean	S.D.	Range
Total protein (g/dl)	7.11	0.20	6.30	0.50	4.90-9.6
Albumin (g/dl)	3.68	0.22	3.10	0.14	2.12-4.00
(%)	51.7	2.70	49.2	2.20	43.5-57.8
α_1 - Globulin (g/dl)	0.27	0.03	0.23	0.04	0.16-0.35
(%)	3.80	0.26	3.70	0.24	2.72-7.68
α_2 - Globulin (g/dl)	0.66	0.08	0.62	0.08	0.45-0.85
(%)	9.30	1.15	9.90	1.20	4.64-15.6
β-Globulin (g/dl)	1.72	0.20	1.83	0.22	1.25-2.30
(%)	24.2	2.60	29.0	2.52	14.1-36.2
γ-Globulin (g/dl)	0.77	0.08	0.52	0.08	0.35-0.95
(%)	10.9	1.19	8.20	1.32	3.75-12.9
Albumin/Globulin	1.07	0.23	0.97	0.22	0.50-1.68

*Source: Hahn et al. (1956); Spector (1961); McKelvie (1966); Bulgin et al. (1971); Dimopoullos (1972); Tumbleson and Hutcheson (1972); Moreland (1975).

Table 34. Biochemical Reference Values of the Normal Dog

Components	Caisey & King (1980)[1] Mean	Coles (1980)[4] Mean	Coles (1980)[4] S.D.	Coles (1980)[4] Range	Lumsden et al (1979)[2] Mean	Lumsden et al (1979)[2] S.D.	Benirschke et al (1978)[4] Mean	Benirschke et al (1978)[4] S.D.	Benirschke et al (1978)[4] Range
Bilirubin (mg/dl)	-	-	-	-	0.30	0.10	0.51	0.09	0.07-0.61
Cholesterol (mg/dl)	128	110	28.0	125 - 250	198.8	40.9	143	29.0	90.0-280
Creatinine (mg/dl)	-	-	-	1.00-1.70	0.80	0.20	-	-	1.00-2.00
Glucose (mg/dl)	-	-	-	70.0-110	84.0	9.01	107	2.35	-
Urea nitrogen (mg/dl)	-	-	-	10.0-20.0	31.9	12.0	17.0	4.00	12.0-18.0
Uric acid (mg/dl)	-	-	-	0.10-1.00	-	-	0.67	0.04	0.00-10.0
Sodium (mEq/l)	-	146	3.43	-	150	3.70	160	8.00	-
Potassium (mEq/l)	4.40	4.20	0.39	-	4.20	0.20	5.40	0.60	-
Chloride (mEq/l)	107	112	4.56	-	110	3.10	106	-	-
Bicarbonate (mEq/l)	-	-	-	-	-	-	-	-	-
Phosphorus (mg/dl)	4.61	5.60	-	-	6.01	1.39	4.11	0.14	2.20-4.00
Calcium (mg/dl)	10.0	9.90	0.80	-	11.0	0.52	11.6	0.50	8.40-11.2
Amylase (Somogyi units/dl)	-	-	-	-	56.0	10.0	-	-	-
Alkaline Phosphatase (I.U./l)	173	-	-	-	32.0	11.7	-	-	-
Acid Phosphatase (I.U./l)	-	-	-	-	-	-	21.7	0.56	-
Alanine Transaminase (ALT) (I.U./l)	60.0	-	-	-	14.3	6.57	64.4	27.5	-
Aspartate Transaminase (AST) (I.U./l)	32.0	-	-	-	18.6	4.14	46.0	14.0	-
Creatine Phosphokinase (CK) (I.U./l)	118	-	-	-	-	-	30.5	10.8	-

Table 34. Biochemical Reference Values of the Normal Dog

Components	Caisey[1] & King (1980)			Coles[5] (1980)			Lumsden[2] et al (1979)			Benirschke[4] et al (1978)		
	Mean	S.D.	Range	Mean	S.D.	Range	Mean	S.D.	Range	Mean	S.D.	Range
Lactic Dehydrogenase (LDH) (I.U./1)	112			-	-	-	77.1	50.0		48.7	2.54	-
Total Protein (g/dl)	5.70			-	-	-	6.00	0.40		7.70	0.80	6.00-8.00
Albumin (g/dl)	2.60			-	-	-	3.00	0.30		-	-	3.00-4.80
(%)	-			-	-	-	-	-		-	-	-
Alpha 1 - Globulin (g/dl)	-			-	-	-	0.70	0.10		-	-	-
(%)	-			-	-	-	-	-		-	-	-
Alpha 2 - Globulin (g/dl)	-			-	-	-	0.60	0.10		-	-	-
(%)	-			-	-	-	-	-		-	-	-
Beta - Globulin (g/dl)	1.50			-	-	-	1.30	0.30		-	-	-
(%)	-			-	-	-	-	-		-	-	-
Gamma - Globulin (g/dl)	0.40			-	-	-	0.50	0.10		-	-	-
(%)	-			-	-	-	-	-		-	-	-
Albumin/Globulin	0.81			-	-	-	10.0	2.00		-	-	-

Source: [1]Data from 10 beagle dogs, 10 to 11 months old. See text for methodology.

[2]Data from 51 clinically healthy dogs (26 female and 25 male), approximately 6 to 24 months age and of mixed breed. See text for methodology.

[3]Data from 146 dogs in two laboratory colonies, one beagles and one Labrador retrievers. Assays were performed with Coulter Chemistry Analyzer.

[4]Derived from literature.

[5]Data from 10 female and 18 male mongrel dogs weighing 12-16 kg. See text for methodology.

Table 34. Biochemical Reference Values of the Normal Dog

Components	Benjamin (1978)[4]			Tasker (1978)[3]		Melby and Altman (1974)[4]		Mitruka & Rawnsley (1977)[5]	
	Mean	S.D.	Range	Mean	S.D.	Mean	S.D.	Mean	S.D.
Bilirubin (mg/dl)	0.25	0.10	0.07-0.61	0.50	0.10	0.75	0.68	0.23	0.11
Cholesterol (mg/dl)	-	-	-	182	50.0	195	71.6	181	11.5
Creatinine (mg/dl)	-	-	-	0.90	0.18	0.81	0.20	1.22	0.25
Glucose (mg/dl)	-	-	-	84.0	12.0	105	28.5	121	9.5
Urea nitrogen (mg/dl)	-	-	-	13.8	2.90	16.7	8.90	14.5	1.60
Uric acid (mg/dl)	-	-	-	-	-	1.10	1.43	0.49	0.10
Sodium (mEq/l)	143	-	137 - 149	147	2.10	163	17.4	147	2.05
Potassium (mEq/l)	4.40	-	3.30-5.80	4.90	0.40	3.95	0.46	4.50	0.15
Chloride (mEq/l)	106	-	99.0-110	109	3.20	106	8.50	113	1.18
Bicarbonate (mEq/l)	20.5	-	18.0-24.0	-	-	-	-	22.0	1.80
Phosphorus (mg/dl)	5.60	-	-	4.00	0.70	4.54	1.87	4.05	0.75
Calcium (mg/dl)	4.90	-	4.20-5.60	10.6	0.40	5.08	1.08	9.80	0.46
Amylase (Somogyi units/dl)	-	-	-	-	-	-	-	177	28.0
Alkaline Phosphatase (I.U./l)	-	-	3.00-16.0	29.0	13.0	19.7	15.4	17.1	4.05
Acid Phosphatase (I.U./l)	-	-	-	-	-	-	-	3.42	1.15
Alanine Transaminase (ALT) (I.U./l)	13.0	5.30	4.80-24.0	-	-	82.0	54.5	42.3	7.75
Aspartate Transaminase (AST) (I.U./l)	9.60	3.40	6.20-13.0	41.0	27.0	81.2	53.9	57.0	8.95
Creatine Phosphokinase (CK) (I.U./l)	-	-	-	48.0	18.0	23.0	27.0	100	70

Table 34. Biochemical Reference Values of the Normal Dog

Components	Benjamin[4] (1978)			Tasker[3] (1978)		Melby and[4] Altman (1974)		Mitruka & Rawnsley (1977)[5]	
	Mean	S.D.	Range	Mean	S.D.	Mean	S.D.	Mean	S.D.
Lactic Dehydrogenase (LDH) (I.U./l)	-	-	10.0-35.0	29.0	18.0	65.7	41.5	71.8	16.4
Total Protein (g/dl)	-	-	5.40-7.10	6.00	0.44	5.90	0.73	6.71	0.35
Albumin (g/dl)	-	-	2.30-3.20	3.30	0.25	3.72	0.29	3.40	0.18
(%)	-	-	-	-	-	-	-	50.5	2.45
Alpha 1 - Globulin (g/dl)	-	-	0.20-0.50					0.25	0.04
(%)	-	-	-					3.75	0.25
Alpha 2 - Globulin (g/dl)	-	-	0.30-1.10					0.64	0.08
(%)	-	-	-					9.60	1.18
Beta - Globulin (g/dl)	-	-	0.60-1.20					1.78	0.21
(%)	-	-	-					27.0	2.56
Gamma - Globulin (g/dl)	-	-	0.90-2.20					0.65	0.08
(%)	-	-	-					9.55	1.26
Albumin/Globulin	-	-	0.59-1.11					1.02	0.23

Source: [3]Data from 146 dogs in two laboratory colonies, one beagles' and one labrador retrievers'. Assays were performed with Coulter Chemistry Analyzer.
[4]Derived from literature
[5]Data from 10 female and 18 male mongrel dogs weighing 12-16 kg. See text for methodology.

203

Table 35. Comparative Biochemical Values in Sera of Different Breeds of Dogs

Components	Alsatian German Shepherd	Beagle	Labrador Retriever	Mixed Breed
Amylase (Somogyi Units)	130.-370.	79.5-265.	92.0-274.	105.-295.
Bilirubin, Total (mg/dl)	0.10-0.95	0.12-0.45	0.07-1.00	0.05-1.10
Calcium (mg/dl)	9.30-11.0	9.61-11.7	9.80-11.4	9.30-10.8
Chloride (mEq/l)	103.-112.	110.-114.	105.-116.	108.-116.
Cholesterol (mg/dl)	125.-250.	147.-275.	95.0-193.	138.-240.
Creatinine (mg/dl)	0.80-1.40	0.98-1.56	1.00-2.00	1.07-1.93
Glucose (mg/dl)	64.0-90.0	92.1-127.	70.0-115.	72.0-120.
Lactic Dehydrogenase (I.U./l)	45.0-113.	47.0-125.	51.0-176.	93.0-120.
Phosphatase, Alkaline (I.U./l)	7.40-23.2	9.60-24.5	7.50-24.6	7.10-24.7
Phosphorus, Inorganic (mg/dl)	2.70-4.70	3.34-6.52	4.00-9.00	3.50-4.80
Potassium (mEq/l)	4.19-4.55	4.42-5.00	4.10-4.68	4.32-4.74
Protein, Total (g/dl)	5.70-7.30	6.76-7.68	6.90-7.80	6.70-7.80
Albumin (%)	45.6-55.9	43.0-53.8	43.3-52.2	42.2-54.2
Globulin (%)	41.9-52.9	44.5-54.5	46.3-57.5	40.8-53.5
Sodium (mEq/l)	142.-150.	143.-151.	143.-151.	143.-152.
Transaminase, Alanine (SGPT) (I.U./l) (ALT)	24.2-66.5	17.5-47.3	18.0-62.0	16.5-45.0
Transaminase, Aspartate (SGOT) (I.U./l) (AST)	35.8-55.0	40.0-68.4	38.5-74.6	33.2-75.0
Urea, Nitrogen (mg/dl)	10.1-19.9	7.58-21.0	9.00-20.0	9.80-22.2
Uric Acid (mg/dl)	0.20-0.60	0.21-0.64	0.20-0.75	0.22-0.74

Source: Albritton (1952); Hoffman (1961); Spector (1961); Pick (1965); McKelvie et al. (1966); Coles (1967); Schermer (1967); Robinson and Ziegler (1968); Von Stewart and Longwell (1969); Altman and Dittmer (1974); Kaneko and Cornelius (1972); Secord and Russel (1973); Moreland (1974); Nomura et al. (1975).

eral breeds of dogs including Alsatians, Labrador retrievers, golden retrievers, boxers, and collies (all castrated or spayed). Serum urea values were found to be markedly affected by the animals' protein intakes. The normal ranges were generally narrower than those reported by earlier workers because of the precautions taken to study all the dogs under comparable conditions and because the animals formed a more homogeneous population than many other workers had studied.

Secord and Russell (1973) reported the values of 150 conditioned mongrel dogs and 89 Labrador retrievers. Significant but not large differences reflected the greater homogeneity, less stressed prior experience, and metabolic characteristics of the animals. The mongrel dogs had much higher sodium concentrations and lower chlorides than the Labrador retrievers. Sodium was influenced by seasonal variation in that it was higher in colder months because of stress. Blood glucose was found to be lower and LDH higher in Labrador retrievers, as reported previously by Crawley and Swenson (1965).

Monkeys

Tables 36–38 present the chemical components, enzyme activities, and electrophoretically separated proteins in the sera of normal Rhesus monkeys weighing 2–5 kg. The values are those of 38 male and 30 female laboratory-conditioned monkeys recorded over a period of 2 years. The data are in good agreement with those reported in the literature (Tables 39–40). Compared to human chemistries, monkeys had lower serum bilirubin, cholesterol, uric acid, alkaline phosphatase, CK, albumin, and A/G ratio, and higher sodium, chloride, phosphorus, amylase, acid phosphatase, ALT, AST, β- and γ-globulins.

There are many species of monkeys and their blood chemistries differ on the bases of age, sex, nutritional habits and environmental factors. It is beyond the scope of this book to present comprehensive data on the blood chemistry of every species of monkey used in experimental medicine and biology; however data on seven species are compared in Table 41. In general, bilirubin, cholesterol, uric acid, and A/G ratios in all species of monkeys are lower than in humans, and AST, ALT, β-, and γ-globulin are generally higher in monkeys than in humans. Other general comparisons with human serum chemistry may be made in the values of creatinine (higher in baboons and squirrel monkeys); glucose (lower in stump-tailed monkeys, sooty mangabeys, and Indian langurs); and urea nitrogen (higher in squirrel monkeys and Indian langurs).

A number of investigators have reported normal values in *Macaca mulatta* (Anderson, 1966; King and Fargus, 1967; Rao and Shipley, 1970; Robinson and Ziegler, 1968; Turbyfill *et al.*, 1968, 1970). Altshuler *et al.* (1971) compared biochemical values of *M. arctoides, M. fascicularis,* and *M. radiata* and found highly significant differences between the normal ranges of the different species and between the normal ranges of males and females of the same species. In the same comparison, glucose, total bilirubin, SGPT, and α_1-globulin did not show any significant differences. In general, their results agreed with those of other investigators in this field.

Group and sex-related differences were observed in a number of parameters of the baboon (*Papio* species) by De la Pena *et al.* (1970, 1972). The data showed definite intergroup and intersex differences in glucose, inorganic phosphorus, uric acid, cholesterol, ALT, AST, lactic dehydrogenase, albumin, and A/G ratio. In longitudinally studied animals, variations between animals and within individual animals were indistinguishable. However definite trends were observed in the mean values of some parameters. They reported that

Table 36. Chemical Components in Serum of the Normal Rhesus Monkey

Components	Values in				Literature values*
	Male		Female		
	Mean	S.D.	Mean	S.D.	Range
Bilirubin (mg/dl)	0.38	0.18	0.57	0.20	0.05-1.32
Cholesterol (mg/dl)	129.	6.6	150.	5.0	100.-220.
Creatinine (mg/dl)	1.50	0.09	1.28	0.06	1.00-2.00
Glucose (mg/dl)	88.7	3.6	77.0	2.0	43.0-148.
Urea nitrogen (mg/dl)	10.1	0.90	12.8	0.18	7.00-23.0
Uric acid (mg/dl)	0.90	0.10	1.29	0.10	1.10-1.50
Sodium (mEq/l)	155.	0.60	154.	0.80	143.-164.
Potassium (mEq/l)	5.10	0.12	5.10	0.12	3.79-667.
Chloride (mEq/l)	112.	0.80	114.	0.70	103.-118.
Bicarbonate (mEq/l)	29.9	1.60	30.3	1.40	21.5-38.6
Phosphorus (mg/dl)	5.92	0.15	5.79	0.11	2.80-6.70
Calcium (mg/dl)	11.2	0.66	10.8	0.72	9.40-12.0
Magnesium (mg/dl)	1.65	0.32	1.82	0.41	1.00-2.70

*Source: Krise and Wald (1958); Spector (1961); Anderson (1966); Petery (1967); Robinson and Ziegler (1968) Rollins et al. (1970); Vogin and Oser (1971); Laird (1972); Olsen and Burns (1972); McClure et al. (1973); Moreland (1974); Vondruska and Greco (1974); Nomura et al. (1975).

Table 37. Enzyme Activities in Serum of the Normal Rhesus Monkey

Enzyme	Values in				Literature values*
	Male		Female		
	Mean	S.D.	Mean	S.D.	Range
Amylase (Somogyi units/dl)	190.	25.0	165.	23.0	110.-250.
Alkaline Phosphatase (I.U./1)	9.60	5.10	8.70	4.95	3.00-29.0
Acid Phosphatase (I.U./1)	31.5	3.32	33.2	3.50	24.5-41.0
Alanine Transaminase (SGPT) (I.U./1) (ALT)	24.5	9.50	23.6	10.2	3.50-45.0
Aspartate Transaminase (SGOT) (I.U./1) (AST)	27.9	6.60	28.7	7.60	12.5-44.2
Creatine Phosphokinase (CPK) (I.U./1) (CK)	115	14.0	63.0	13.0	33-150
Lactic Dehydrogenase (LDH) (I.U./1)	175.	70.0	180.	71.0	30.0-320.

*Source: Majmundar and Das-Gupta (1944); Anderson (1966); Robinson and Ziegler (1968); Rollins et al. (1970); Vogin and Oser (1971); Laird (1972); Nomura et al. (1975).

Table 38.　Serum Proteins of the Normal Rhesus Monkey

Components	Values in				Literature values*
	Male		Female		
	Mean	S.D.	Mean	S.D.	Range
Total protein (g/dl)	7.20	0.44	6.86	0.56	5.90-8.70
Albumin (g/dl)	3.50	0.32	3.53	0.33	1.80-4.60
(%)	48.6	5.21	51.5	4.70	47.5-62.5
α_1 - Globulin (g/dl)	0.35	0.08	0.23	0.04	0.20-0.55
(%)	4.90	0.50	3.30	0.62	2.90-7.50
α_2 - Globulin (g/dl)	0.56	0.08	0.63	0.07	0.40-0.80
(%)	7.80	1.10	9.20	1.20	5.70-11.5
β-Globulin (g/dl)	1.40	0.15	1.17	0.11	0.80-2.00
(%)	19.5	1.85	17.1	1.72	12.0-25.0
γ-Globulin (g/dl)	1.38	0.15	1.30	0.14	1.00-1.80
(%)	19.2	2.20	18.9	2.15	13.8-24.2
Albumin/Globulin	0.95	0.20	1.06	0.21	0.61-1.55

*Source:　Robinson and Ziegler (1968); Altman and Dittmer (1974); Dimopoullos (1972); Moreland (1974).

Table 39. Biochemical reference values of the normal animal (cont.)

Components	Caisey and King (1980)[1] Mean	Benirschke et al (1978)[2] Mean	S.D.	Melby and Altman (1974)[2] Mean	S.D.	Mitruka and Rawnsley (1977)[3] Mean	S.D.
Bilirubin (mg/dl)	–	0.36	0.21	0.44	0.39	0.48	0.19
Cholesterol (mg/dl)	121	128	34.6	137.9	24.6	140	5.8
Creatinine (mg/dl)	–	–	–	0.79	0.14	1.39	0.08
Glucose (mg/dl)	–	70.1	17.3	70.1	17.2	82.9	2.8
Urea nitrogen (mg/dl)	–	16.9	2.70	16.9	2.70	11.5	0.85
Uric acid (mg/dl)	–	0.88	0.11	0.66	0.11	1.10	0.10
Sodium (mEq/l)	154	158	13	152.7	4.5	154.5	0.70
Potassium (mEq/l)	4.1	4.7	0.8	5.30	0.73	5.10	0.12
Chloride (mEq/l)	113	114	9	106.2	4.0	113	0.75
Bicarbonate (mEq/l)	–	–	–	–	–	30.1	1.50
Phosphorus (mg/dl)	2.23	5.00	0.95	5.30	1.40	5.86	0.13
Calcium (mg/dl)	10.1	4.86 (mEq/l)	0.81	5.33 (mEq/l)	0.60	11.0	0.69
Magnesium (mg/dl)	–	–	–	–	–	1.74	0.37
Amylase (Somogyi units/dl)	–	–	–	–	–	177.5	24
Alkaline Phosphatase (I.U./l)	1134 (U/l)	9.5	5.15	12.8	6.5	9.20	5.03
Acid Phosphatase (I.U./l)	–	–	–	–	–	32.4	3.41
Alanine Transaminase (SGPT) (I.U./l) (ALT)	94 (U/l)	42.1	21.2	46.6	23.3	24.1	9.85

Table 39. (con't)   Biochemical Reference Values of the Normal Rhesus Monkey

Components	Caisey and King (1980)[1] Mean	Benirschke et al (1978)[2] Mean	Benirschke et al (1978)[2] S.D.	Melby and Altman (1974)[2] Mean	Melby and Altman (1974)[2] S.D.	Mitruka and Rawnsley (1977)[3] Mean	Mitruka and Rawnsley (1977)[3] S.D.
Aspartate Transaminase (SGOT) (AST) (I.U./l)	31.0	27.0	6.50	23.3	6.60	28.3	7.10
Creatine Phosphokinase (CPK) (I.U./l)	125 (U/l)	8.90	1.40 (U)	172.8	69.6	89.0	13.5
Lactic Dehydrogenase (LDH) (I.U./l)	397 (U/l)	186.9	68.4	186.9	68.4	177.5	70.5
Total protein (g/dl)	8.00	6.61	0.46	6.61	0.46	7.03	0.50
Albumin (g/dl)	3.20	4.43	0.95	4.04	0.55	3.52	0.33
Albumin (%)	–	–	–	–	–	50.1	4.96
Alpha-1 Globulin (g/dl)	0.90	0.90	0.21	2.73	0.45	0.29	0.06
Alpha-1 Globulin (%)	–	–	–			4.10	0.56
Alpha-2 Globulin (g/dl)						0.60	0.08
Alpha-2 Globulin (%)						8.50	1.15
Beta Globulin (g/dl)	1.60	1.20	0.26			1.29	0.13
Beta Globulin (%)	–	–	–			18.30	1.79
Gamma Globulin (g/dl)	2.30	–	–			1.34	0.15
Gamma Globulin (%)	–	–	–			19.05	2.18
Albumin/Globulin	0.63	–	–	–	–	1.01	0.21

Source: [1] Data from 16 Macaca fascicularis, 5 to 9 years old Monkeys. See text for methodology.

[2] Derived from Literature.

[3] Data from 30 female and 38 male, laboratory conditioned, normal Rhesus Monkeys weighing 2 - 5 Kg. See text for methodology.

Table 40. Biochemical Reference Values of Squirrel Monkey (Saimiri sciureus)

Components	Beland et al[1] (1979)		Benirschke et al[2] (1978)		Mitruka and Rawnsley[3] (1977)	
	Mean	S.D.	Mean	S.D.	Mean	S.D.
Bilirubin (mg/dl)	0.30	0.23	0.50	0.20	0.35	0.35
Cholesterol (mg/dl)	161	40.0	155	30.8	155	30.8
Creatinine (mg/dl)	—	—	—	—	7.55	1.74
Glucose (mg/dl)	80.0	28.0	75.5	17.4	86.9	27.2
Urea nitrogen (mg/dl)	31.0	8.00	25.3	7.50	26.4	9.20
Uric acid (mg/dl)	1.20	0.40	1.00	0.30	1.00	0.30
Sodium (mEq/l)	148	4.00	148	0.10	150	5.85
Potassium (mEq/l)	3.60	0.70	6.30	1.10	4.40	0.90
Chloride (mEq/l)	110	4.00	113.6	5.20	114	6.65
Bicarbonate (mEq/l)	—	—	12.4	4.00	—	—
Phosphorus (mg/dl)	4.50	2.20	4.90	1.70	4.90	1.70
Calcium (mg/dl)	9.00	0.70	9.60	0.50	9.75	0.60
Alkaline Phosphatase (I.U./l)	117	53.0	22.4	15.0	—	—
Alanine Transaminase (SGPT) (I.U./l) (ALT)	—	—	174	57.0	121	50.0
Aspartate Transaminase (SGOT) (I.U./l) (AST)	87.0	31.0	196	53.0	143	31.2

Table 40.(con't) Biochemical Reference Values of Squirrel Monkey (Saimiri sciureus)

Components	Beland et al[1] (1979)		Benirschke et al[2] (1978)		Mitruka and[3] Rawnsley (1977)	
	Mean	S.D.	Mean	S.D.	Mean	S.D.
Lactic Dehydrogenase (LDH) (I.U./l)	43.0	28.0	382	111	–	–
Total protein (g/dl)	7.20	1.30	7.80	0.70	7.42	0.62
Albumin (%)	3.20	0.50			2.28	0.22
Alpha Globulin (%)	–	–	1.70	1.20	1.96	0.48
Beta Globulin (%)	–	–	6.40	1.40	0.82	0.19
Gamma Globulin (%)	–	–			1.87	0.47

*Source: [1]From 129 (65 females and 64 males) animals, the average age of the group 3 years, Chemistry analysis performed with SMA analyzer.

[2]Derived from Literature

[3]Derived from Literature

212

Table 41. Comparative Biochemical Values in Sera of Different Species of Monkeys

Components	Sex	Macaca mulatta[1] (Rhesus) Mean S.D.	Macaca arctoides[2] (Stump-tailed) Mean S.D.	Papio species[3] (Baboon) Mean S.D.	Saimiri sciureus[4] (Squirrel monkey) Mean S.D.	Cercopithecus aethiops[5] (African Green) Mean S.D.	Cercocebus atys[6] (Sooty Mangabey) Mean S.D.	Presbytis entellus[7] (Indian Langur) Mean S.D.
Bilirubin, Total (mg/dl)	M	0.38±0.28	0.35±0.10	0.33±0.16	0.50±0.40	0.27±0.13	0.37±0.10	0.32±0.18
	F	0.57±0.60	----	----	0.20±0.30	0.24±0.10	0.36±0.09	0.26±0.11
Calcium (mg/dl)	M	9.61±0.33	10.3±0.54	9.40±0.60	10.0±0.50	10.4±1.00	10.4±0.80	11.4±1.40
	F	10.9±0.70	10.1±0.68	9.60±1.20	9.50±0.70	10.9±0.30	10.1±0.70	11.6±0.90
Chloride (mEq/l)	M	115.±12.5	118.±2.10	108.±3.30	115.±7.90	110.±1.40	105.±3.80	103.±3.60
	F	110.±27.6	114.±2.10	100.±6.00	113.±5.40	113.±3.50	105.±4.10	107.±3.80
Cholesterol (mg/dl)	M	128.±34.6	129.±24.4	89.0±19.0	155.±30.8	222.±13.5	155.±32.3	158.±30.7
	F	219.±52.4	154.±36.9	102.±24.0	----	226.±5.40	139.±29.3	183.±26.0
Creatinine (mg/dl)	M	1.50±0.09	0.85±0.14	4.90±0.50	7.55±17.4	1.00±0.10	----	----
	F	1.28±0.06	0.57±0.10	4.60±0.60	----	1.00±0.10	----	----
Glucose (mg/dl)	M	91.0±14.0	39.6±8.10	95.9±20.4	100.±37.0	102.±13.0	86.3±31.6	99.2±17.0
	F	71.8±10.6	44.8±12.2	----	73.8±17.4	107.±19.2	78.7±29.2	112.±21.2
Phosphate, Inorganic (mg/dl)	M	5.16±1.00	5.51±1.08	5.20±1.40	4.90±1.70	4.70±0.80	5.50±1.40	5.20±1.40
	F	5.25±1.20	4.55±1.58	3.60±1.40	----	3.20±3.20	5.00±1.30	----
Potassium (mEq/l)	M	4.70±0.80	4.12±0.69	3.80±0.50	4.30±0.90	4.40±0.50	6.30±0.80	5.90±0.70
	F	4.08±0.65	4.39±0.47	3.30±0.70	4.50±0.90	4.13±0.50	5.00±0.70	5.70±0.70
Transaminase, Aspartate (SGOT) (I.U./l) (AST)	M	33.4±8.00	43.2±11.5	30.3±6.70	88.6±19.4	61.0±9.80	51.6±15.3	36.1±15.6
	F	33.8±10.9	39.1±15.8	36.2±8.50	196.±53.0	54.0±9.90	46.4±12.9	----
Transaminase, Alanine (SGPT) (I.U./l) (ALT)	M	20.3±5.80	28.6±10.3	14.0±6.00	68.6±43.1	38.0±9.20	28.8±13.7	23.3±9.10
	F	23.4±9.80	25.7±13.2	20.0±13.0	174.±57.0	47.0±13.3	31.7±17.0	----
Sodium (mEq/l)	M	153.±7.50	150.±38.0	143.±2.40	152.±6.80	150.±3.40	151.±4.40	154.±5.10
	F	150.±6.30	152.±2.60	145.±5.00	148.±4.90	151.±4.00	149.±4.50	154.±4.40
Urea Nitrogen (mg/dl)	M	12.3±1.90	16.6±3.80	9.50±2.30	24.0±11.5	12.0±1.50	17.9±370	25.1±6.20
	F	13.0±1.10	16.6±3.60	12.4±2.70	28.7±6.95	12.0±1.30	17.3±3.80	24.9±4.20
Uric Acid (mg/dl)	M	0.90±0.11	----	0.70±0.30	1.00±0.30	----	----	----
	F	1.29±0.14	----	0.40±0.20	----	----	----	----

Table 41 (continued).

Components	Sex	Macaca mulatta[1] (Rhesus) Mean±S.D.	Macaca arctoides[2] (Stump-tailed) Mean±S.D.	Papio[3] species (Baboon) Mean±S.D.	Saimiri[4] sciureus (Squirrel monkey) Mean±S.D.	Cercopithecus[5] aethiops (African Green) Mean±S.D.	Cercoebus[6] atys (Sooty Mangabey) Mean±S.D.	Presbytis[7] entellus (Indian Langur) Mean±S.D.
Protein, Total (g/dl)	M	7.80±0.78	8.21±0.56	7.10±0.52	7.63±0.63	7.60±0.80	8.50±0.70	7.70±0.70
	F	7.21±0.56	8.44±0.24	7.40±0.40	7.20±0.60	7.40±0.80	8.50±0.80	7.70±0.60
Albumin (g%)	M	4.73±0.52	3.17±0.36	3.65±0.52	4.11±0.23	3.70±1.00	4.20±1.00	4.40±0.90
	F	4.44±0.32	2.98±0.60	3.10±0.60	4.05±0.25	3.40±1.00	3.80±0.60	3.90±0.70
α_1-Globulin	M	0.90±0.21	0.28±0.14	0.20±0.04	0.51±0.20	amorphous band	0.25±0.12	0.37±0.14
	F	----	0.29±0.16	----	0.54±0.18		0.21±0.09	0.32±0.09
α_2-Globulin	M	----	0.54±0.25	0.70±0.14	1.35±0.25	amorphous band	0.62±0.20	0.42±0.18
	F	----	0.61±0.27	----	1.25±0.37		0.58±0.21	0.50±0.17
β-Globulin	M	1.20±0.26	2.15±0.52	1.10±0.40	0.85±0.19	2.40±0.47	2.17±0.60	1.66±0.41
	F	----	2.30±0.54	----	0.79±0.19	2.50±0.43	2.00±0.70	1.67±0.37
γ-Globulin	M	1.20±0.33	1.69±0.73	1.60±0.27	1.89±0.53	1.44±0.44	1.71±0.53	1.72±0.49
	F	----	1.74±0.72	----	1.85±0.43	1.53±0.47	2.03±0.45	1.80±0.50
Albumin/Globulin	M	1.84±0.61	0.82±0.66	1.16±0.40	1.20±0.30	----	----	----
	F	1.68±0.39	0.83±0.41	0.91±0.28	1.40±0.40	----	----	----

Source:

1. Macaca mulatta (Rhesus): Krise and Wald (1958); Hensley and Langham (1964); Robinson et al. (1964); Anderson (1966); King and Gargus (1967); Petery (1967); Robinson and Ziegler (1968); Turbyfill et al. (1968, 1970); Oser et al. (1970); Rao and Shipley (1970); Rollins et al. (1970); Vogin and Oser (1971); Spiller et al. (1975).
2. Macaca arcticodes (Stump-tailed): Hensley and Langham (1964); Oser et al. (1970); Vondruska (1970); Altshuler et al. (1971); Vogin and Oser (1971).
3. Papio Species (Baboon): De la Pena and Goldzieher (1965, 1967); Burns et al. (1967); De la Pena et al. (1970, 1972); Steyn (1975).
4. Saimiri sciureus (Squirrel Monkey): Garcia and Hunt (1966); New (1968); Manning et al. (1969); Vogin and Oser (1971).
5. Cercopithecus aethiops (African Green): Pridgen (1967); Altshuler and Stowell (1972); Hambelton et al. (1979).
6. Cercoebus atys (Sooty Mangabey): Altshuler and Stowell (1972).
7. Presbytis entellus (Indian Langur): Altshuler and Stowell (1972).
8. General References: Albritton (1952); Schermer (1967); Morrow and Terry (1968); Swenson et al. (1970); Laird (1972); Altman and Dittmer (1974).

the normal values for uric acid, cholesterol, and certain enzyme activities of baboons were quite different from those in man.

Vondruska (1970) reported that glucose and BUN values were significantly lower in *M. arctoides* than in *M. mulatta* species. A very wide range in the blood values of *M. arctoides* was reported by Oser *et al.* (1970). However values found in *M. arctoides* were within the normal range reported for *M. mulatta*. They concluded that as far as hematological and biochemical parameters were concerned, there were no significant differences between the two species.

Pridgen (1967) reported the physiological parameters of African green monkeys (*Cercopethicus aethiops*). He found that the mean values for blood glucose, blood urea nitrogen, plasma glutamic oxalacetic transaminase, ALT, plasma sodium, and plasma chloride were similar to the range of these blood components in *M. mulatta*. Alkaline phosphatase in male monkeys was twice the plasma levels of the enzyme in females.

Vogin and Oser (1971) compared the blood values of four species of monkeys. Rhesus, Galago, and Cynomolgus monkeys had essentially similar values for all of the parameters examined. However, squirrel monkeys often showed marked deviations from the mean values of the other species. Blood glucose, AST, and ALT were elevated in the squirrel monkey.

Manning *et al.* (1969) reported that the serum enzyme activities of AST, ALT, and alkaline phosphatase were considerably higher in the squirrel monkey (*Saimiri sciureus*, a new world primate) than values reported for old-world nonhuman primates such as the Rhesus monkey. Sodium and chloride values in the squirrel monkey have been reported to be higher than in other mammals (Manning *et al.*, 1969; Holmes *et al.*, 1967). Serum calcium, AST, albumin, and γ-globulin values in *Ceropethicus aethiops* (African green monkey), *Cercocebus atys* (sooty mangabey), and *Presbytis entillus* (Indian langur) were significantly different in these species and in the males and females within the same species. In the same comparison between sexes, serum carbon dioxide, α_1-globulin, α_2-globulin, urea nitrogen, and ALT were not significantly different.

Almost all clinical biochemical studies of nonhuman primates have found a fairly high degree of variability in the serum enzymes (Morrow and Terry, 1968a, b, c; Altshuler and Stowell, 1972). Wellde *et al.* (1971) reported the results of serum chemical tests in the night monkey (*Aotus trivirgatus*). Both AST and ALT were elevated in the night monkey when compared to values in man or old-world primates. Other authors have noted increased levels of transaminases in the serum of *Aotus* as well as other new world monkeys (Holmes *et al.*, 1967; Malinow, *et al.*, 1968; Manning *et al.*, 1969: Morrow and Terry, 1968a, b, c; Neu, 1968; Schnell *et al.*, 1969; Sheldon and Ross, 1966). The higher enzyme levels in monkeys may be the result of a genetic trait, an infection, or other stress factors brought about by captivity.

Age-related differences in the species of monkeys have been reported by Rao and Shipley (1970). Marked differences in alkaline phosphatase values distinguished different age groups; they were high in young and small monkeys and low in old and large monkeys. Young adult monkeys had significantly higher AST values than adults. No age-related differences were found in other blood chemistries of monkeys.

Pigs

Tables 42–44 show clinical chemical components, enzymatic activities, and electrophoretically separated proteins of normal Hormel mini-pigs (miniature swine) weighing 70–100 kg. The values were recorded from blood samples of 12 male and 10 female pigs over a 2-year period. In

Table 42. Chemical Components in Serum of the Normal Pig

| Components | Values in | | | | Literature values* |
| | Male | | Female | | |
	Mean	S.D.	Mean	S.D.	Range
Bilirubin (mg/dl)	0.20	0.20	0.15	0.10	0.00-0.60
Cholesterol (mg/dl)	154.	9.00	151.	7.00	76.0-174.
Creatinine (mg/dl)	1.85	0.26	1.70	0.30	1.00-2.60
Glucose (mg/dl)	85.0	12.5	90.0	14.0	60.0-136.
Urea nitrogen (mg/dl)	16.0	2.65	14.5	3.10	8.00-24.0
Uric acid (mg/dl)	1.22	0.52	1.15	0.46	0.10-1.90
Sodium (mEq/l)	142.	1.20	149.	1.32	135.-152.
Potassium (mEq/l)	5.60	0.20	6.20	0.23	4.90-7.10
Chloride (mEq/l)	100.	0.45	105.	0.62	94.0-106.
Bicarbonate (mEq/l)	30.6	1.54	29.8	2.20	24.5-35.0
Phosphorus (mg/dl)	7.50	2.10	8.40	1.85	5.30-11.0
Calcium (mg/dl)	9.65	0.99	11.1	1.08	7.50-13.2
Magnesium (mg/dl)	1.60	0.24	2.80	0.20	1.20-3.70

*Source: Spector (1961); Coles (1967); Tumbleson et al. (1971); Kaneko and Cornelius (1972); Tumbleson (1972); Tumbleson (1972 a); Tumbleson et al. (1972 b); Tumbleson et al. (1973); Altman and Dittmer (1974); Nomura et al. (1975).

Table 43. Enzyme Activities in Serum of the Normal Pig

Enzyme	Values in				Literature values*
	Male		Female		
	Mean	S.D.	Mean	S.D.	Range
Amylase (Somogyi units/dl)	110.	14.0	96.0	12.5	68.0-140.
Alkaline Phosphatase (I.U./1)	65.0	14.7	62.0	14.5	35.0-110.
Acid Phosphatase (I.U./1)	26.0	4.80	23.0	4.45	16.0-36.0
Alanine Transaminase (SGPT) (I.U./1) (ALT)	27.6	8.10	28.5	8.50	10.5-45.0
Aspartate Transaminase (SGOT) (I.U./1) (AST)	36.2	11.5	35.4	10.7	30.0-61.0
Creatine Phosphokinase (CPK) (I.U./1) (CK)	127.	21.1	119.	15.4	85.0-170.
Lactic Dehydrogenase (LDH) (I.U./1)	65.0	15.0	62.0	14.8	32.0-100.

*Source: Wretlind et al. (1959); Coles (1967); Kaneko and Cornelius (1972); Tumbleson et al. (1972 a, 1972 b); Nomura et al. (1975).

Table 44. Serum Proteins of the Normal Pig

Components	Values in				Literature values*
	Male		Female		
	Mean	S.D.	Mean	S.D.	Range
Total protein (g/dl)	8.90	0.48	8.45	0.42	4.80-10.3
Albumin (g/dl)	4.19	0.40	3.59	0.34	1.80-5.60
(%)	47.1	4.60	42.5	4.10	32.0-57.0
α_1 - Globulin (g/dl)	0.55	0.06	0.62	0.07	0.40-0.80
(%)	6.20	0.81	7.40	0.75	4.50-9.80
α_2 - Globulin (g/dl)	1.17	0.13	1.42	0.16	0.80-1.85
(%)	13.2	1.50	16.8	1.45	10.0-20.0
β-Globulin (g/dl)	1.77	0.20	1.83	0.19	1.25-2.30
(%)	19.9	2.15	21.7	2.10	14.8-26.7
γ-Globulin (g/dl)	1.25	0.31	0.98	0.25	0.35-1.90
(%)	14.1	3.48	11.6	3.22	4.60-21.5
Albumin/Globulin	0.89	0.12	0.74	0.11	0.47-1.19

*Source: Altman and Dittmer (1971, 1974); Dimopoullos (1972); Kaneko and Cornelius (1972); Tumbleson et al. (1972 a, 1972 b); Tumbleson et al. (1973).

comparison to human serum biochemical constituents, the pigs had lower bilirubin, cholesterol, uric acid, lactic dehydrogenase, and A/G ratio, and higher amounts of inorganic phosphorus, alkaline phosphatase, acid phosphatase, creatine phosphokinase, and total protein α_2-, β-, and γ-globulins.

Selected normal biochemical parameters of miniature swine have been reported previously (Hill, 1966; McClellan et al., 1966; Tegeris et al., 1966). Other investigators studied the biochemical parameters of swine from birth to maturity using crossbred, conventional swine (Jakobsen and Monstgard, 1950; Miller et al., 1961a, b; Ulrey et al., 1967). Pond et al. (1968) reported differences in the concentration of blood components between conventional and miniature swine. Tumbelson et al. (1969) evaluated age- and sex-related differences in the blood hematological and serum biochemical parameters of 4–9-month-old Hormel miniature swine. They found that with increasing age, total protein, albumin β-globulin, A/G ratio, and calcium increased, and that alkaline phosphatase, inorganic phosphorus, sodium, potassium, and chloride decreased. Also, alkaline phosphatase and inorganic phosphorus concentrations were greater in serum from males than from females. Serum cholesterol, total protein, and calcium values were higher in serum from females than from males.

Earl et al. (1971) reported the biochemical values of neonatal and weaning miniature pigs and the range of values in animals from 1 to 112 days old. They found that alkaline phosphatase was considerably higher in the cesarean-derived pigs, and that serum alkaline phosphatase values were generally higher for younger pigs than for older swine. Blood urea nitrogen did not vary greatly from birth to 112 days of age. Fasting blood glucose levels were not as high at birth as at day 28, and they reached their peak (124 mg/dl) at 42 days of age.

The LDH values were lowest at birth; little variation was found in the values at other ages. AST was very high at birth but decreased rapidly to the normal range in the older pigs. In our studies data show that the sex-related differences in biochemical parameters were not important, and our findings were within the range of values reported in the literature. Benjamin (1978) and Coles (1980) reported blood chemistry values on miniature swine and other breeds of adult animals.

Goats

Data on the blood chemistry of goats are sparse. We measured the serum chemical constituents of 12 male and 8 female goats (domestic breed) weighing 20–25 kg. The normal values presented in Tables 45–47 show that bilirubin, cholesterol, glucose, uric acid, potassium, creatine phosphokinase, lactic dehydrogenase, albumin, and A/G ratio are lower than in humans. However, blood urea nitrogen, phosphorus, amylase, alkaline phosphatase, acid phosphatase, AST, ALT, β-globulin, and γ-globulin were higher in goats than in humans.

The differences in serum biochemical constituents between male and female goats were not important. No age-related or breed-related differences have been reported in blood chemistry of goats. Recent blood chemistry values reported in literature are compared with our values in Table 48.

Sheep

Blood chemistry values in sheep (Tables 49–51) were rather similar to those in goats. Thirty-two male and 26 female Dorset-Delane sheep weighing 40–60 kg were used over a 1-year period to establish normal values. The data show that the sera from

Table 45. Chemical Components in Serum of the Normal Goat

Components	Values in				Literature values*
	Male		Female		
	Mean	S.D.	Mean	S.D.	Range
Bilirubin (mg/dl)	0.05	0.01	0.05	0.01	0.00-0.10
Cholesterol (mg/dl)	70.0	12.0	68.0	14.0	55.0-210.
Creatinine (mg/dl)	1.36	0.46	1.15	0.42	0.20-2.21
Glucose (mg/dl)	83.5	15.0	72.0	16.5	43.0-100.
Urea nitrogen (mg/dl)	20.5	3.80	17.4	3.62	13.0-44.0
Uric acid (mg/dl)	0.67	0.33	0.60	0.30	0.20-1.10
Sodium (mEq/1)	147.	3.52	149.	4.10	141.-157.
Potassium (mEq/1)	3.61	0.18	2.95	0.24	2.45-4.11
Chloride (mEq/1)	103.	0.52	106.	0.46	98.0-111.
Bicarbonate (mEq/1)	24.6	2.10	26.1	2.20	19.6-31.1
Phosphorus (mg/dl)	10.9	0.98	7.87	1.42	5.00-13.7
Calcium (mg/dl)	10.3	0.70	10.7	0.62	8.80-12.2
Magnesium (mg/dl)	2.50	0.36	3.20	0.35	1.80-3.95

*Source: Albritton (1952); Spector (1961); Altman and Dittmer (1971, 1974); Kaneko and Cornelius (1971; 1972).

Table 46. Enzyme Activities in Serum of the Normal Goat

Enzyme	Values in				Literature values*
	Male		Female		
	Mean	S.D.	Mean	S.D.	Range
Amylase (Somogyi units/dl)	225.	32.0	200.	29.0	135.-290.
Alkaline Phosphatase (I.U./l)	85.4	15.1	82.9	16.2	45.0-125.
Acid Phosphatase (I.U./l)	37.4	8.85	35.9	8.90	17.1-57.4
Alanine Transaminase (SGPT) (I.U./l) (ALT)	22.1	11.2	24.5	9.60	0.50-47.0
Aspartate Transaminase (SGOT) (I.U./l) (AST)	67.5	26.8	66.5	26.7	12.0-122.
Creatine Phosphokinase (CPK) (I.U./l) (CK)	31.1	4.85	30.9	4.90	20.0-41.5
Lactic Dehydrogenase (LDH) (I.U./l)	92.0	27.0	89.0	26.2	35.0-137.

*Source: Albritton (1952); Coles (1967); Altman and Dittmer (1971, 1974); Kaneko and Cornelius (1972).

Table 47. Serum Proteins of the Normal Goat

Components	Values in				Literature values
	Male		Female		
	Mean	S.D.	Mean	S.D.	Range
Total protein (g/dl)	6.75	0.35	6.90	0.38	5.90-7.80
Albumin (g/dl)	3.46	0.41	3.35	0.42	2.45-4.35
(%)	51.3	7.40	48.5	6.30	33.5-66.5
α_1 - Globulin (g/dl)	0.45	0.05	0.40	0.04	0.30-0.60
(%)	6.70	0.74	5.80	0.62	4.20-8.30
α_2 - Globulin (g/dl)	0.51	0.06	0.68	0.07	0.30-0.90
(%)	7.50	0.90	9.80	1.10	5.00-12.5
β-Globulin (g/dl)	1.33	0.14	1.61	0.15	1.00-2.00
(%)	19.8	2.15	23.4	2.30	14.8-28.5
γ-Globulin (g/dl)	1.05	0.15	0.86	0.14	0.50-1.50
(%)	15.5	2.58	12.5	2.30	7.00-21.0
Albumin/Globulin	1.05	0.11	0.95	0.12	0.71-1.26

[*]Source: Altman and Dittmer (1971, 1974); Kaneko and Cornelius (1972).

Table 48. Biochemical Reference Values of the Normal Goat

Components	Coles[1] (1980)		Benjamin[1] (1978)		Melby and Altman[1] (1974)		Mitruka and Rawnsley[2] (1977)	
	Mean	Range	Mean	Range	Mean	Range	Mean	S.D.
Bilirubin (mg/dl)	-	-	-	0.00-0.10	-	0.00-0.10	0.05	0.01
Cholesterol (mg/dl)	-	80.0-130	-	-	-	80.0-130	69.0	13.0
Creatinine (mg/dl)	-	0.90-1.82	-	-	1.60	1.00-2.20	1.26	0.43
Glucose (mg/dl)	-	50.0-75.0	-	-	62.8	48.6-77.0	78.8	15.8
Urea nitrogen (mg/dl)	-	13.0-28.0	-	-	-	21.0-44.0	19.0	3.71
Uric acid (mg/dl)	-	0.33-1.00	-	-	-	0.30-1.00	0.64	0.32
Sodium (mEq/l)	-	142 - 155	-	-	150	144 - 157	148	3.81
Potassium (mEq/l)	-	3.50-6.70	-	-	-	3.50-6.70	3.28	0.21
Chloride (mEq/l)	-	99.0-110	-	-	105	99.5-111	105	0.49
Bicarbonate (mEq/l)	-	-	-	-	-	-	25.4	2.15
Phosphorus (mg/dl)	3.10	2.39-3.81	-	-	6.50	-	9.39	1.20
Calcium (mg/dl)	10.3	-	-	-	5.15	4.80-5.50	10.5	0.66
Magnesium (mg/dl)	3.20	2.85-3.55	-	-	3.20	2.80-3.60	2.90	0.36
Amylase (Somogyi units/dl)	-	-	-	7.00-30.0	-	-	213	30.5
Alkaline Phosphatase (I.U./l)	-	-	-	-	17.0	11.0-23.0	84.2	15.7
Acid Phosphatase (I.U./l)	-	-	-	-	-	-	36.7	8.88
Alanine Transaminase (ALT) (I.U./l)	-	-	-	7.00-24.0	-	7.00-24.0	23.3	10.4
Aspartate Transaminase (AST) (I.U./l)	-	-	-	43.0-132	-	43.0-132	67.0	26.8
Creatine Phosphokinase (CK) (I.U./l)	-	-	-	-	4.50	1.70-7.30	31.0	4.88

Table 48. Biochemical Reference Values of the Normal Goat

Components	Coles[1] (1980)		Benjamin[1] (1978)		Melby and Altman[1] (1974)		Mitruka and Rawnsley[2] (1977)	
	Mean	Range	Mean	Range	Mean	Range	Mean	S.D.
Lactic Dehydrogenase (LDH) (I.U./1)	-	-	-	31.0-99.0	71.0	7.00-24.0	90.5	26.6
Total Protein (g/dl)	-	-	-	6.40-7.90	6.90	6.42-7.38	6.83	0.37
Albumin (g/dl)	-	-	-	2.70-3.90	6.40	5.44-7.36	3.41	0.41
(%)	-	-	-	-	-	-	49.9	68.5
Alpha-1 Globulin (g/dl)	-	-	-	0.50-0.70			0.43	0.05
(%)	-	-	-	-			6.25	0.68
Alpha-2 Globulin (g/dl)	-	-	-	-			0.60	0.07
(%)	-	-	-	-			8.65	1.00
Beta Globulin (g/dl)	-	-	-	1.00-1.80			1.47	0.15
(%)	-	-	-	-			21.6	2.23
Gamma Globulin (g/dl)	-	-	-	0.90-3.00			0.96	0.15
(%)	-	-	-	-			14.0	2.44
Albumin/Globulin	-	-	-	0.63-1.26			1.00	0.12

Source: [1]Derived from Literature.
[2]Data from 8 female and 12 male goats (domestic, young adult, weighing 20-25 kg. See text for methodology.

Table 49. Chemical Component in Serum of the Normal Sheep

Components	Values in				Literature values*
	Male		Female		
	Mean	S.D.	Mean	S.D.	Range
Bilirubin (mg/dl)	0.29	0.09	0.15	0.05	0.00-0.60
Cholesterol (mg/dl)	110.	24.1	90.0	18.2	50.0-140.
Creatinine (mg/dl)	1.56	0.36	2.20	0.40	0.70-3.00
Glucose (mg/dl)	96.0	17.0	80.8	18.0	55.0-131.
Urea nitrogen (mg/dl)	24.0	2.55	28.0	4.10	15.0-36.0
Uric acid (mg/dl)	1.22	0.70	1.15	0.72	0.00-1.90
Sodium (mEq/1)	149.	4.25	155.	3.56	140.-164.
Potassium (mEq/1)	4.70	0.91	5.40	0.62	4.40-6.70
Chloride (mEq/1)	120.	0.60	116.	0.74	115.-121.
Bicarbonate (mEq/1)	26.2	1.80	27.1	2.20	21.2-32.1
Phosphorus (mg/dl)	5.90	0.11	4.40	0.21	4.00-7.00
Calcium (mg/dl)	11.4	0.32	12.2	0.28	10.4-14.0
Magnesium (mg/dl)	2.27	0.25	2.50	0.30	1.80-2.40

*Source: Coles (1967); Tumbleson et al. (1970); Tumbleson and Hutcheson (1971); Kaneko and Cornelius (1972); Tumbleson et al. (1973).

Table 50. Enzyme Activities in Serum of the Normal Sheep

Enzyme	Values in				Literature values*
	Male		Female		
	Mean	S.D.	Mean	S.D.	Range
Amylase (Somogyi units/dl)	205.	28.0	210.	30.0	140.-270.
Alkaline Phosphatase (I.U./l)	98.5	12.6	94.8	12.2	69.5-125.
Acid Phosphatase (I.U./l)	10.8	4.75	8.80	4.20	0.20-21.0
Alanine Transaminase (SGPT) (I.U./l)	41.4	7.60	42.7	6.20	25.0-70.0
Aspartate Transaminase (SGOT) (I.U./l) (ALT)	82.5	18.5	80.8	19.6	40.0-123.
Creatine Phosphokinase (CPK) (I.U./l) (AST)	15.8	2.40	14.5	2.30	9.50-21.4
Lactic Dehydrogenase (LDH) (I.U./l)	80.0	15.5	75.0	14.0	44.0-112.

*Source: Cornelius et al. (1959); Boyd (1962); Coles (1967); Baetz (1970); Kaneko and Cornelius (1972).

Table 51. Serum Proteins of the Normal Sheep

Components	Values in				Literature values*
	Male		Female		
	Mean	S.D.	Mean	S.D.	Range
Total protein (g/dl)	6.80	0.30	7.20	0.31	5.70-9.10
Albumin (g/dl)	3.70	0.35	3.81	0.33	2.70-4.55
(%)	54.4	5.20	52.9	4.80	41.5-65.5
α_1 - Globulin (g/dl)	0.33	0.08	0.38	0.06	0.15-0.50
(%)	4.80	1.20	5.30	0.80	2.40-7.70
α_2 - Globulin (g/dl)	0.96	0.13	0.73	0.12	0.45-1.25
(%)	14.1	2.60	10.2	1.85	5.00-20.0
β-Globulin (g/dl)	0.52	0.10	0.91	0.13	0.25-1.20
(%)	7.70	1.75	12.6	1.80	3.70-16.6
γ-Globulin (g/dl)	1.33	0.20	1.37	0.25	0.82-1.90
(%)	19.5	3.50	19.0	4.20	12.0-27.5
Albumin/Globulin	1.19	0.20	1.12	0.21	0.70-1.60

*Source: Spector (1961); Tumbleson et al. (1970); Altman and Dittmer (1971, 1974); Kaneko and Cornelius (1972).

sheep had lower values of bilirubin, cholesterol, uric acid, CK, LDH, and A/G ratio
than sera from normal humans. However,
blood urea nitrogen, chloride, sodium, inorganic phosphorus, amylase, alkaline
phosphatase, acid phosphatase, ALT, AST,
α_2-globulin, and γ-globulin were higher in sheep than in humans. Sex-related differences in the sheep were not considerable. The blood chemistry values of different breeds and age groups of sheep have
been reported and discussed in many
textbooks on veterinary clinical pathology
(Miller, 1960; Kaneko and Cornelius, 1972;
Medway *et al.*, 1969; Coles, 1967, 1980),
some of which are compared in Table 52.

Cattle

Large animals such as cattle are used in
experimental medicine as the source of tissues and fluids for experimental purposes
or media preparation, for environmental
studies, and for some aspects of genetic
research. It is rather difficult to maintain
these animals under laboratory conditions
so the animals are usually kept on the farm
and samples are taken as required to study
clinical biochemical and hematological parameters. Therefore greater variation in the
blood chemistries of cattle is expected.

In this study normal blood chemistry values for cattle were established from blood
samples taken over a 2-year period from 6
castrated, young adult bulls and 10 Holstein
cows weighing 500–600 kg. Also included in
these data were samples from other breeds
of cattle on various farms and veterinary
clinics and from commercial sera (Grand
Island Biological Company, Grand Island,
New York). The blood chemistries of the
cattle (Tables 53–55) are very similar to the
values reported in the literature (Table 56).
No sex-related differences were remarkable
in the blood components tested.

In comparison to human blood chemical
values, cattle had lower serum bilirubin,
cholesterol, uric acid, albumin, and A/G
ratios, and higher creatinine, phosphorus,
calcium, amylase, alkaline phosphatase,
acid phosphatase. ALT, AST, CK, α_1-
globulin, and γ-globulin.

Horses

Tables 57–59 list the blood chemicals
values of 15 male and 6 female light horses
of mixed breed, weighing 350–400 kg. Also
included in these values are results found
with commercial horse sera (Grand Island
Biological). The data show that cholesterol,
uric acid, albumin, and A/G ratios are lower
in horses than in humans. However, bilirubin, creatinine, calcium, alkaline phosphatase, acid phosphatase, AST, ALT,
CK, LDH, total protein, and all globulin
fractions were higher in the sera of horses
than in human sera. Table 60 presents data
on thoroughbred horses reported by Lumsden *et al.* (1979) & Sato *et al.* (1978).

Our data fell within the normal range of
values reported in the literature (Table 61).
Medway (1970) reported the blood chemistry studies of 60 light horses and 10 ponies
and found nothing remarkable except some
differences in the values of urea nitrogen,
total protein, and chlorides from those reported earlier (Tasker, 1966). Wolf *et al.*
(1970) assayed sera from 102 horses for 20
different components. The animals were
classified according to clinical condition (no
abnormalities, infectious diseases, or metabolic diseases), age (<2½ years, 2½–4½
years, 4½–8 years, or >8 years) and sex
(male, gelding, or female). Mean serum albumin values were consistently lower and
there were differences between the normal
and diseased populations, but sex- or age-
related differences were not remarkable.

Humans

The blood chemistries of human sera are
presented in Tables 62–66. The data repre-

Table 52. Biochemical Reference Values of the Normal Sheep

Components	Coles[1] (1980) Mean	Range	Benjamin[1] (1978) Mean	Range	Smith et al[2] (1978) Mean	S.D.
Bilirubin (mg/dl)	-	-	0.19	0.00-0.39	0.20	0.10
Cholesterol (mg/dl)	64.0	52.0-76.0	-	-	57.0	8.00
Creatinine (mg/dl)	-	1.20-1.93	-	-	1.00	0.20
Glucose (mg/dl)	-	50.0-80.0	-	-	61.0	10.0
Urea nitrogen (mg/dl)	-	8.00-20.0	-	-	28.0	4.00
Uric acid (mg/dl)	-	0.50-1.93	-	-	less than 3.5	
Sodium (mEq/l)	147	142 - 152	153	146 - 161	145	3.00
Potassium (mEq/l)	4.85	4.46-5.24	4.80	-	5.20	0.40
Chloride (mEq/l)	108	96.5-113	116	-	104	2.00
Bicarbonate (mEq/l)	-	-	25.0	-	26.0	3.00
Phosphorus (mg/dl)	6.40	6.00-6.80	6.90	-	4.70	0.90
Calcium (mg/dl)	9.80	6.90-10.0	5.70	-	9.22	0.40
Magnesium (mg/dl)	2.50	2.20-2.80	2.30	2.20-2.40	-	-
Amylase (Somogyi units/dl)	-	-	-	-	-	-
Alkaline Phosphatase (I.U./l)	-	-	-	5.00-30.0	64.0	34.0
Acid Phosphatase (I.U./l)	-	-	-	-	-	-
Alanine Transaminase (ALT) (I.U./l)	-	-	11.0	10.0-12.0	-	-
Aspartate Transaminase (AST) (I.U./l)	-	-	79.0	68.0-90.0	71.0	26.0
Creatine Phosphokinase (CK) (I.U./l)	-	-	-	-	93.0	70.0
Lactic Dehydrogenase (LDH) (I.U./l)	-	-	-	60.0-111	311	55.0

Table 52. Biochemical Reference Values of the Normal Sheep

Components	Coles[1] (1980)		Benjamin[1] (1978)		Smith et al[2] (1978)	
	Mean	Range	Mean	Range	Mean	S.D.
Total protein (g/dl)	-	-	-	6.00-7.90	6.90	C.70
Albumin (g/dl)	-	-	-	2.70-3.90	4.10	0.60
(%)	-	-	-	-	-	-
Alpha-1 Globulin (g/dl)	-	-	-	0.30-0.60		
(%)	-	-	-	-		
Alpha-2 Globulin (g/dl)	-	-	-	-		
(%)	-	-	-	-		
Beta Globulin (g/dl)	-	-	-	1.10-2.60		
(%)	-	-	-	-		
Gamma Globulin (g/dl)	-	-	-	0.90-3.30		
(%)	-	-	-	-		
Albumin/Globulin	-	-	-	0.42-0.76		

Source: [1]Derived from literature

[2]Data from 18 female sheep (Suffolk & Hampshire), 3-5 years old. Serum Chemistry values were obtained with SMAC analyzer.

Table 52. (con't) Biochemical Reference Values of the Normal Sheep

Components	Benirschke[1] et al (1978)		Melby and Altman[1] (1974)		Mitruka and Rawnsley[3] (1977)	
	Mean	S.D.	Mean	S.D.	Mean	S.D.
Bilirubin (mg/dl)	-	-	0.23	0.10	0.22	0.07
Cholesterol (mg/dl)	110	20.5	64.0	12.0	100	21.2
Creatinine (mg/dl)	3.00	0.40	-	-	1.88	0.38
Glucose (mg/dl)	96.0	37.0	68.4	6.00	88.4	17.5
Urea nitrogen (mg/dl)	28.0	4.00	28.0	4.00	26.0	3.33
Uric acid (mg/dl)	-	-	0.00	1.90	1.19	0.71
Sodium (mEq/l)	149	4.00	149	4.00	152	3.91
Potassium (mEq/l)	5.00	0.90	4.80	0.40	5.05	0.77
Chloride (mEq/l)	120	3.00	120	3.00	118	0.67
Bicarbonate (mEq/l)	-	-	-	-	26.7	2.00
Phosphorus (mg/dl)	8.40	2.10	6.40	0.20	5.15	0.16
Calcium (mg/dl)	11.4	0.80	5.70	0.40	11.8	0.30
Magnesium (mg/dl)	-	-	2.50	0.30	2.39	0.28
Amylase (Somogyi units/dl)	-	-	-	-	208	29.0
Alkaline Phosphatase (I.U./l)	32 (KU)	21.0	14.0	8.00	96.7	12.4
Acid Phosphatase (I.U./l)	-	-	-	-	9.80	4.48
Alanine Transaminase (ALT) (I.U./l)	-	-	11.0	1.00	42.1	6.90
Aspartate Transaminase (AST) (I.U./l)	100	21.0 (KU)	79.0	11.0	81.7	19.1
Creatine Phosphokinase (CK) (I.U./l)	-	-	10.3	1.60	15.2	2.35

Table 52. (con't)   Biochemical Reference Values of the Normal Sheep

Components	Benirschke[1] et al (1978)		Melby and Altman (1974)[1]		Mitruka and Rawnsley (1977)[3]	
	Mean	S.D.	Mean	S.D.	Mean	S.D.
Lactic Dehydrogenase (LDH) (I.U./l)	290	62.0(WU)	89.0	15.0	77.5	14.8
Total protein (g/dl)	8.50	0.60	7.20	0.52	7.00	0.31
Albumin (g/dl)	2.60	0.30	2.70	0.19	3.76	0.34
(%)	-	-	-	-	53.7	5.00
Alpha-1 Globulin (g/dl)	0.95	0.18			0.36	0.07
(%)					5.05	1.00
Alpha-2 Globulin (g/dl)					0.85	0.13
(%)					12.2	2.23
Beta Globulin (g/dl)	1.98	0.38			0.72	0.12
(%)					10.2	1.78
Gamma Globulin (g/dl)	2.92	0.53			1.35	0.23
(%)					19.3	3.85
Albumin/Globulin	-	-	-	-	1.16	0.21

Source: [1] Derived from literature

[3] Data from 26 female and 32 male Pors et-Déélane Sheep weighing 40-60 kg.
See text for methodology.

232

Table 53. Chemical Components in Serum of the Normal Cattle

Components	Values in				Literature values*
	Male		Female		
	Mean	S.D.	Mean	S.D.	Range
Bilirubin (mg/dl)	0.37	0.15	0.31	0.17	0.10-0.70
Cholesterol (mg/dl)	117.	24.5	110.	22.0	80.0-170.
Creatinine (mg/dl)	1.54	0.52	1.25	0.45	0.20-2.60
Glucose (mg/dl)	85.0	20.1	89.0	22.0	45.0-130.
Urea nitrogen (mg/dl)	16.5	3.56	15.9	3.55	6.00-27.0
Uric acid (mg/dl)	1.29	0.70	1.10	0.68	0.00-2.70
Sodium (mEq/l)	146.	4.40	142.	2.67	132.-152.
Potassium (mEq/l)	5.63	0.26	4.90	0.22	03.9-6.80
Chloride (mEq/l)	108.	0.62	104.	0.71	97.0-111.
Bicarbonate (mEq/l)	27.4	3.56	26.6	4.12	18.5-36.4
Phosphorus (mg/dl)	56	0.86	6.00	0.45	4.50-9.30
Calcium (mg/dl)	10.6	1.05	10.8	1.40	9.40-12.2
Magnesium (mg/dl)	2.04	0.28	2.19	0.25	1.20-3.50

*Source: Albritton (1952); Spector (1961); Coles (1967); Altman and Dittmer (1971, 1974); Tumbleson and Hutcheson (1971); Kaneko and Cornelius (1972); Laird (1972); Pierce and Laird (1972); Tumbleson et al. (1973).

Table 54. Enzyme Activities in Serum of the Normal Cattle

Enzyme	Values in				Literature values*
	Male		Female		
	Mean	S.D.	Mean	S.D.	Range
Amylase (Somogyi units/dl)	190.	27.0	180.	25.0	126.-250.
Alkaline Phosphatase (I.U./l)	133.	18.2	131.	17.5	94.0-170.
Acid Phosphatase (I.U./l)	9.45	3.60	8.50	3.50	1.20-17.5
Alanine Transaminase (SGPT) (I.U./l) (ALT)	47.5	12.5	49.6	13.6	20.0-76.8
Aspartate Transaminase (SGOT) (I.U./l) (AST)	52.8	17.5	50.5	21.0	8.50-93.0
Creatine Phosphokinase (CPK) (I.U./l) (CK)	91.5	11.2	95.0	9.00	66.0-120.
Lactic Dehydrogenase (LDH) (I.U./l)	160.	70.0	150.	71.0	8.00-302.

*Source: Boyd (1962); Cardeilhac and CArdeilhac (1962); Coles (1967); Baetz (1970); Kaneko and Cornelius (1972).

Table 55. Serum Proteins of the Normal Cattle

| Components | Values in | | | | Literature values* |
| | Male | | Female | | |
	Mean	S.D.	Mean	S.D.	Range
Total protein (g/dl)	6.97	0.45	7.56	0.50	5.90-8.60
Albumin (g/dl)	3.24	0.36	3.40	0.32	2.45-4.20
(%)	46.6	5.10	45.0	4.75	43.5-57.0
α_1 - Globulin (g/dl)	0.98	0.12	0.85	0.12	0.60-1.23
(%)	14.0	2.10	11.2	1.95	7.20-18.0
α_2 - Globulin (g/dl)	0.72	0.08	0.57	0.06	0.35-0.95
(%)	10.4	1.20	7.54	0.82	7.00-13.0
β-Globulin (g/dl)	0.62	0.10	0.88	0.10	0.40-1.10
(%)	8.90	1.40	11.6	1.22	7.10-14.5
γ-Globulin (g/dl)	1.44	0.29	1.86	0.36	0.72-2.60
(%)	20.6	4.10	24.6	4.50	11.6-33.7
Albumin/Globulin	0.87	0.11	0.83	0.12	0.70-1.15

*Source: Spector (1972); Altman and Dittmer (1971, 1974); Dimopoullos (1972); Kaneko and Cornelius (1972); Laird (1972); Pierce and Laird (1972); Tumbleson et al. (1973)

Table 56. Biochemical Reference Values of the Normal Cow

Components	Coles[1] (1980) Range	Lumsden[2] et al (1980) Range	Benjamin[1] (1978) Range	Tasker[3] (1978) Mean	S.D.	Melby and Altman[1] (1974) Mean	S.D.	Mitruka &[4] Rawnsley (1977) Mean	S.D.
Bilirubin (mg/dl)	-	0.00-0.50	0.01-0.47	0.20	0.07	0.19	0.10	0.34	0.16
Cholesterol (mg/dl)	78.0-142	77.2-239	-	159	23.0	136	28.2	114	23.3
Creatinine (mg/dl)	1.00-2.07	0.70-1.40	-	1.20	0.25	-	-	1.40	0.48
Glucose (mg/dl)	45.0-70.0	45.1-68-5	-	-	-	51.3	9.80	87.0	21.1
Urea Nitrogen (mg/dl)	6.00-27.0	3.92-43.9	-	13.7	5.20	9.12	1.76	16.2	3.56
Uric acid (mg/dl)	0.50-2.08	-	-	-	-	1.27	0.17	1.20	0.69
Sodium (mEg/l)	132-152	135-145	132-152	136	3.10	-	-	144	3.54
Potassium (mEg/l)	3.90-5.80	3.60-5.10	3.90-5.80	4.60	0.33	-	-	5.27	0.24
Chloride (mEg/l)	97.0-111	96.0-105	97.0-111	100	1.60	-	-	106	0.67
Bicarbonate (mEg/l)	-	-	-	-		-	-	27.0	3.84
Phosphorus	4.70-6.30	4.09-8.20	5.60-6.50	5.20	1.00	-	-	5.78	0.66
Calcium (mg/dl)	6.60-8.20	8.40-10.7	4.70-6.10	9.80	0.50	-	-	10.7	1.23
Magnesium (mg/dl)	1.73-2.37	1.58-2.38	1.00-2.90	-	-	-	-	2.12	0.27
Amylase (Somogyi units/dl)	-	-	0.00-38.0	-	-	-	-	185	25.0
Alkaline Phosphatase (I.U./l)	-	3.00-46.0	-	16.0	8.00	15.0	10.0	132	17.9
Acid Phosphatase (I.U./l)	-	-	-	-	-	-	-	8.98	3.55
Alanine Transaminase (SGPT) (I.U./l) (ALT)	-	7.14-25.7	4.00-11.0	-	-	68.7	17.5	48.6	13.1
Asparate Transaminase (SGOT) (I.U./l) (AST)	-	34.3-64.3	20.0-34.0	133	21.0	56.8	17.9	51.7	19.3

Table 56. Biochemical Reference Values of the Normal Cow

Components	Coles[1] (1980) Range	Lumsden[2] et al (1980) Range	Benjamin[1] (1978) Range	Tasker[3] (1978) Mean	S.D.	Melby and Altman[1] (1974) Mean	S.D.	Mitruka & Rawnsley[4] (1977) Mean	S.D.
Creatine Phosphokinase (CPK) (I.U./l) (CK)	-	24.3-84.3	-	37.0	11.0	49.8	31.3	93.3	10.1
Lactic Dehydrogenase (LDH) (I.U./l)	-	406-730	176-365	234	330	270	560	155	70.5
Total protein (g/dl)	-	5.90-8.10	6.74-7.46	7.10	0.59	7.10	0.18	7.27	0.48
Albumin (g/dl)	-	2.90-3.90	3.03-3.55	3.00	0.22	3.29	0.13	3.32	0.34
(%)	-	-	-	-		-		45.8	4.93
Alpha-1 Globulin (g/dl)	-		0.75-0.83					0.92	0.12
(%)	-							12.6	2.03
Alpha-1 Globulin (g/dl)	-		0.80-1.12					0.65	0.07
(%)	-		-					8.97	1.01
Beta Globulin (g/dl)	-		-					0.75	0.10
(%)	-		-					10.3	1.31
Gamma Globulin (g/dl)	-		1.69-2.25					1.65	0.33
(%)	-		-					2.26	4.30
Albumin/Globulin	-		0.84-0.94					0.85	0.12

Source:
[1] Derived from Literature.
[2] Data from 172 female Holstein animals, 4 age groups. See text for methodology.
[3] Data from 146 Dairy adult cows. See text for methodology.
[4] Data from 10 Holstein cows weighing 500-600 kg. See text for methodology.

Table 57. Chemical Components in Serum of the Normal Horse

Components	Values in				Literature values*
	Male		Female		
	Mean	S.D.	Mean	S.D.	Range
Bilirubin (mg/dl)	1.70	0.70	1.40	0.70	0.20-6.20
Cholesterol (mg/dl)	86.8	12.8	77.0	10.6	50.0-130.
Creatinine (mg/dl)	2.50	0.41	2.10	0.40	1.60-3.35
Glucose (mg/dl)	104.	29.0	106.	23.0	65.0-135.
Urea nitrogen (mg/dl)	16.9	3.30	17.0	4.00	12.9-29.5
Uric acid (mg/dl)	0.96	0.10	0.90	0.10	0.85-1.00
Sodium (mEq/l)	139.	2.23	154.	2.40	126.-158.
Potassium (mEq/l)	3.60	0.24	5.60	0.26	2.60-6.20
Chloride (mEq/l)	104.	0.50	108.	0.64	95.0-119.
Bicarbonate (mEq/l)	28.1	3.90	27.6	3.50	19.5-36.2
Phosphorus (mg/dl)	3.35	1.20	4.90	1.10	2.00-6.50
Calcium (mg/dl)	12.3	0.72	11.6	0.74	9.50-14.4
Magnesium (mg/dl)	2.20	0.40	1.54	0.16	1.30-3.50

*Source: Spector (1961); Tasker (1966); Cole (1967); Medway (1970); Wolff et al. (1970); Kaneko and Cornelius (1972); Laird (1972); Pierce and Laird (1972).

Table 58. Enzyme Activities in Serum of the Normal Horse

Enzyme	Values in				Literature values*
	Male		Female		
	Mean	S.D.	Mean	S.D.	Range
Amylase (Somogyi units/dl)	75.0	12.0	60.0	10.6	35.0-100.
Alkaline Phosphatase (I.U./1)	88.6	12.5	87.5	11.6	60.5-115.
Acid Phosphatase (I.U./1)	5.40	2.50	4.65	2.30	0.50-10.5
Alanine Transaminase (SGPT) (I.U./1) (ALT)	70.1	17.8	76.2	18.5	33.9-113.
Aspartate Transaminase (SGOT) (I.U./1) (AST)	175.	20.2	174.	19.5	133.-216.
Creatine Phosphokinase (CPK) (I.U./1) (CK)	113.	12.9	111.	12.5	86.0-140.
Lactic Dehydrogenase (LDH) (I.U./1)	365.	115.	310.	110.	80.0-600.

*Source: Jennings and Mulligan (1953); Cornelius et al. (1959); Freedland et al. (1965); Cardinet et al. (1967); Baetz (1970); Wolff et al. (1970); Altman and Dittmer (1971, 1974).

Table 59. Serum Proteins of the Normal Horse

Components	Values in				Literature values*
	Male		Female		
	Mean	S.D.	Mean	S.D.	Range
Total protein (g/dl)	7.61	0.38	7.46	0.78	5.20-9.90
Albumin (g/dl)	3.10	0.34	2.99	0.28	1.66-3.99
(%)	40.7	3.85	30.7	3.10	21.5-48.8
α_1 - Globulin (g/dl)	0.70	0.12	0.72	0.11	0.64-1.67
(%)	5.26	0.75	9.65	1.10	5.00-11.9
α_2 - Globulin (g/dl)	0.86	0.08	1.07	0.12	0.38-2.06
(%)	11.3	1.10	14.3	1.20	8.70-18.5
β-Globulin (g/dl)	1.03	0.13	1.04	0.17	0.35-1.64
(%)	22.7	1.85	24.7	2.10	17.7-32.5
γ-Globulin (g/dl)	1.52	0.19	1.54	0.20	0.95-2.75
(%)	20.0	2.90	20.6	2.80	14.0-27.8
Albumin/Globulin	0.69	0.20	0.66	0.21	0.26-0.91

*Source: Albritton (1952); Gilman (1952); Jennings and Mulligan (1953); Hort (1968); Medway (1970); Altman and Dittmer (1971, 1974).

Table 60. Biochemical Reference Values of the Thoroughbred Horses

| Components | Lumsden et al[1] (1980) | | | | Sato et al[2] (1978) | |
| | Mares | | Foals | | | |
	Mean	S.D.	Mean	S.D.	Mean	S.D.
Bilirubin (mg/dl)	1.30	0.40	1.40	0.20	1.06	0.11
Cholesterol (mg/dl)	105	16.6	145	20.1	–	–
Creatinine (mg/dl)	1.30	0.20	1.40	0.20	–	–
Glucose (mg/dl)	94.9	27.8	121	10.7	116	3.20
Urea nitrogen (mg/dl)	49.6	9.24	29.9	8.96	15.1	1.31
Uric acid (mg/dl)	–	–	–	–	–	–
Sodium (mEq/l)	138	2.20	138	1.50	141	1.10
Potassium (mEq/l)	3.50	0.50	4.00	0.10	4.23	0.20
Chloride (mEq/l)	99.5	3.10	96.0	1.80	99.3	0.57
Bicarbonate (mEq/l)	–	–	–	–	–	–
Phosphorus (mg/dl)	2.69	0.80	6.19	0.50	5.46	1.20
Calcium (mg/dl)	11.6	0.40	11.4	0.32	10.4	0.40
Magnesium (mg/dl)	1.64	0.16	1.48	0.08	–	–
Amylase (Somogyi units/dl)	590	350 (caraway units/l 37°)	–	–	–	–
Alkaline Phosphatase (I.U./l)	59.0	16.7	157	20.0	27.1	3.10
Acid Phosphatase (I.U./l)	–	–	–	–	–	–
Alanine Transaminase (ALT) (I.U./l)	4.00	2.00	4.86	0.71	10.6	0.50
Aspartate Transaminase (AST) (I.U./l)	236	48.0	207	24.3	195	45.0
Creatinine Phosphokinase (CK) (I.U./l)	63.6	31.9	68.6	9.43	–	–

Table 60(con't).   Biochemical Reference Values of the Thoroughbred Horses

Components	Lumsden et al[1] (1980)				Sato et al[2] (1978)	
	Mares		Foals			
	Mean	S.D.	Mean	S.D.	Mean	S.D.
Lactic Dehydrogenase (LDH) (I.U./l)	218	52.4	351	49.1	633	61.1
Total Protein (g/dl)	6.5	0.40	5.60	0.30	5.83	0.15
Albumin (g/dl)	3.20	0.20	3.10	0.10	3.40	0.60
(%)	-	-	-	-	-	-
Alpha-1 Globulin (g/dl)	1.00	0.20	0.90	0.10	-	-
(%)	-	-	-	-		
Alpha-2 Globulin (g/dl)					2.40	0.34
(%)	-	-	-	-		
Beta-Globulin (g/dl)	1.30	0.20	1.10	0.10		
(%)	-	-	-	-		
Gamma Globulin (g/dl)	1.00	0.20	0.60	0.10		
(%)	-	-	-	-		
Albumin/Globulin	1.00	0.10	1.20	0.10	1.46	1.22

Source: [1]Data from 60 thoroughbred mares in early to late pregnancy and 12 foals between 2 to 8 weeks of age. See text for methodology.
[2]Data from 5 thoroughbred foals during the first six months of life. Blood samples were analyzed with manual methods.

Table 61. Biochemical Reference Values of Normal Standardbred Horses

Components	Lumsden[2] et al (1980) Mean	S.D.	Benjamin[1] (1978) Mean	Range	Tasker[3] (1978) Mean	S.D.	Melby and[1] Altman (1974) Mean	S.D.	Mitruka & Rawnsley (1977)[4] Mean	S.D.
Bilirubin (mg/dl)	1.90	0.50	1.68	0.68–3.28	1.50	0.56	1.25	0.07	1.55	0.70
Cholesterol (mg/dl)	86.9	13.3	–	–	84.0	12.0	111	18.0	81.9	11.7
Creatinine (mg/dl)	1.30	0.20	–	–	1.20	0.16	–	–	2.30	0.41
Glucose (mg/dl)	81.9	9.37	–	–	100	18.0	95.6	8.50	105	26.0
Urea nitrogen (mg/dl)	31.1	4.76	–	–	16.0	3.00	11.0	3.00	17.0	3.35
Uric acid (mg/dl)	–	–	–	–	–	–	–	–	0.93	0.10
Sodium (mEq/l)	140	1.50	149	146–152	137	2.50	139	3.50	147	2.32
Potassium (mEq/l)	3.60	0.40	3.30	2.70–3.50	3.80	0.61	3.51	0.57	4.60	0.25
Chloride (mEq/l)	99.0	1.00	102	98 – 106	102	1.90	104	2.60	106	0.57
Bicarbonate (mEq/l)	–	–	–	–	–	–	–	–	27.9	3.70
Phosphorus (mg/dl)	3.19	0.50	–	3.10–5.60	3.70	0.78	3.38	1.36	4.12	1.15
Calcium (mg/dl)	11.9	0.52	12.2	11.2–13.4	13.0	1.00	10.4	0.58	12.0	0.73
Magnesium (mg/dl)	1.64	0.08	2.20	1.60–2.50	–	–	2.50	0.31	1.85	0.28
Amylase (Somogyi units/dl)	810	200	–	–	–	–	–	–	63.0	11.3
Alkaline Phosphatase (I.U./l)	45.6	10.8	–	97.0– 209	93.0	28.0	19.0	8.00	88.0	12.0
Acid Phosphatase (I.U./l)	–	–	–	–	–	–	–	–	5.00	2.40
Alanine Transaminase (ALT) (I.U./l)	6.14	5.14	4.00	1.00–6.70	–	–	11.0	3.80	73.2	18.2
Aspartate Transaminase (AST) (I.U./l)	310	200	76.0	58.0–94.0	428	109	76.0	18.0	175	19.9
Creatine Phosphokinase (CK) (I.U./l)	55.3	57.14	–	–	53.0	11.0	12.9	5.30	112	12.7

Table (con't) 61. Biochemical Reference Values of Normal Standardbred Horses

Components	Lumsden et al (1980)[2] Mean	S.D.	Benjamin (1978)[1] Mean	Range	Tasker[3] (1978) Mean	S.D.	Melby and Altman (1974)[1] Mean	S.D.	Mitruka & Rawnsley (1977)[4] Mean	S.D.
Lactic Dehydrogenase (LDH)(I.U./l)	200	47.9	-	41.0-104	88.0	28.0	64.0	16.0	338	113
Total Protein (g/dl)	6.40	0.50	7.10	6.30-7.90	6.30	0.44	6.47	0.28	7.54	0.58
Albumin (g/dl)	3.10	0.20	3.10	2.40-3.80	3.40	0.28	3.14	0.24	3.05	0.31
(%)	-		-		-		-		35.7	3.48
Alpha 1-Globulin (g/dl)	-		0.70	0.58-0.82					0.71	0.12
(%)									7.00	0.92
Alpha 2-Globulin (g/dl)			1.31	1.00-1.62					0.97	0.10
(%)			-						12.8	1.15
Beta Globulin (g/dl)	1.70	0.40	1.58	1.18-1.98					1.04	0.15
(%)	-		0.89	0.60-1.18					23.7	1.98
Gamma Globulin (g/dl)	1.10	0.20	1.90	1.35-2.45					1.53	0.20
(%)	-		-						20.3	2.85
Albumin/Globulin	1.00	0.20	1.46	0.62	-		-		0.68	0.21

Source: [1] Derived from literature

[2] Data from 50 animals in light to heavy training. See text for methodology.

[3] Data from 50 horses maintained for experimental and teaching purposes. Assays were performed with Coulter Chemistry II (Coulter Electronics, Hialeah, Florida) Analyzer.

[4] Data from 6 female and 15 male light horses of mixed breed weighing 350-450 kg. See text for methodology.

244

Table 62. Chemical Components in Sera of Normal Humans

Components	Values in				Literature values*
	Male		Female		
	Mean	S.D.	Mean	S.D.	Range
Bilirubin (mg/dl)	0.76	0.24	0.72	0.22	0.09-1.10
Cholesterol (mg/dl)	197.	8.00	186.	10.0	170.-328.
Creatinine (mg/dl)	1.20	0.20	1.00	0.20	0.8-1.50
Glucose (mg/dl)	100.	10.0	95.0	10.0	80.0-120.
Urea nitrogen (mg/dl)	14.3	2.21	13.8	2.10	8.00-26.0
Uric acid (mg/dl)	5.40	0.60	4.10	0.50	2.80-8.80
Sodium (mEq/l)	141.	2.82	142.	2.45	135.-155.
Potassium (mEq/l)	4.10	0.37	4.40	0.42	3.60-5.50
Chloride (mEq/l)	104.	0.90	102.	0.80	98.0-109.
Bicarbonate (mEq/l)	27.0	2.00	28.0	2.20	22.0-33.0
Phosphorus (mg/dl)	3.50	0.10	3.40	0.10	2.50-4.80
Calcium (mg/dl)	9.80	0.44	9.50	0.40	8.50-10.7
Magnesium (mg/dl)	2.12	0.50	2.32	0.60	1.80-2.90

*Source: Donum and King (1965); Best et al. (1969); Frankel et al. (1970); Misanik (1970); Wolf et al. (1972); Bauer et al. (1974); Henry et al. (1974); Rawnsley (1976); Tietz (1976); Werner (1976).

Table 63. Enzyme Activities in Sera of Normal Humans

Enzyme	Values in				Literature values*
	Male		Female		
	Mean	S.D.	Mean	S.D.	Range
Amylase (Somogyi units/dl)	110.	30.0	105.	28.0	40.0-140.
Alkaline Phosphatase (I.U./1)	25.1	12.0	24.8	11.0	19.0-69.0
Acid Phosphatase (I.U./1)	1.25	0.90	1.01	0.50	1.00-1.35
Alanine Transaminase (SGPT) (I.U./1) (ALT)	9.92	5.32	9.50	5.30	6.00-40.0
Aspartate Transaminase (SGOT) (I.U./1) (AST)	13.6	5.40	13.2	5.30	8.00-55.0
Creatine Phosphokinase (CPK) (I.U./1) (CK)	55.0	40.0	45.0	35.0	6.00-57.0
Lactic Dehydrogenase (LDH) (I.U./1)	205.	65.0	190.	60.0	85.0-283.

*Source: Donum and King (1965); Reece (1968); Altman and Dittmer (1971, 1974); Wolf et al. (1973); Henry et al. (1974); Rawnsley (1976); Tietz (1976); Werner (1976).

Table 64. Enzyme Activities in Sera of Human Adults[*]

	Values Mean ± S.D.	Representative Literature Values
Alkaline Phosphatase (I.U./l)	81 ± 7	20 - 90
Aminotransferase (I.U./l)		
Alanine (ALT)	10 ± 5	5 - 25
Aspartate (AST)	14 ± 5	8 - 30
Amylase (Somogyi units)	110 ± 30	60 - 180
Creatine Kinase (I.U./l)	50 ± 38	30 - 150
Lactate Dehydrogenase (I.U./l)	205 ± 65	90 - 310

[*]Values are from manual methods described in Section V.
Representative literature values are from similar methods.

Table 65 (a). Serum Protein Values of Normal Humans

| Components | Values in | | | | Literature values* |
| | Male | | Female | | |
	Mean	S.D.	Mean	S.D.	Range
Total protein (g/dl)	6.80	0.40	6.62	0.35	6.60-8.30
Albumin (g/dl)	4.28	0.31	4.10	0.32	3.50-4.70
(%)	63.0	4.50	62.0	4.70	53.0-65.0
α_1 - Globulin (g/dl)	0.34	0.04	0.28	0.03	0.17-0.40
(%)	5.00	0.62	4.25	0.53	2.5-8.00
α_2 - Globulin (g/dl)	0.46	0.07	0.54	0.08	0.42-0.87
(%)	6.80	1.10	8.20	1.50	5.50-13.0
β-Globulin (g/dl)	0.88	0.11	0.93	0.10	0.52-1.25
(%)	12.9	1.72	14.1	1.65	8.00-14.0
γ-Globulin (g/dl)	0.82	0.20	0.76	0.19	0.50-1.45
(%)	12.0	3.40	11.5	3.20	8.50-22.0
Albumin/Globulin	1.69	0.22	1.62	0.24	1.13-2.17

*Source: Altman and Dittmer (1971, 1974); Bauer et al. (1974); Henry et al. (1974); Tietz (1976).

Table 65b. Representative Serum Protein Values in Human Adults

	Percent of Total Protein	g/dl
Total Protein	-	6.00-8.00
Albumin	53.0-65.0	3.50-5.00
Alpha - 1	2.50-5.00	0.10-0.40
Alpha - 2	7.00-13.0	0.40-1.00
Beta	8.00-14.0	0.50-1.10
Gamma Globulin	12.0-22.0	0.60-1.65
Ig G	-	0.80-1.50
A	-	0.10-0.50
M	-	0.05-0.25
D	-	0.003
E	-	0.00003

Source: Henry et al (1979); Tietz (1976)

Table 66. Comparative Reference Values in Various Age Groups
         of Humans

CONSTITUENT	NORMAL VALUE RANGE
Acid Phosphatase (37°C)	
Newborn	7.4 - 19.4 IU/l
2-13 years	6.4 - 15.2 IU/l
Adult	0.2 - 11.0 IU/l
Aldolase	
Children	2x to 4x Adult value
Adult	< 11 IU/l
Alkaline Phosphatase (37°C)	
Newborn	20 - 266 IU/l
1 month - 1 year	50 - 260 IU/l
1 - 2 years	146 - 477 IU/l
2 - 6 years	70 - 160 IU/l
7 - 10 years	45 - 273 IU/l
Puberty	56 - 258 IU/l
Adult	13 - 40 IU/l
Ammonia	
Newborn	90 - 150 ug/dl
0 - 2 weeks	70 - 129 ug/dl
> 1 month	20 - 79 ug/dl
Infant-child	29 - 80 ug/dl
Adult	13 - 48 ug/dl
Amylase (37°C)	60 - 160 Somogyi Units/dl
Bicarbonate	20 - 30 meq/l
Bilirubin, Total	
Cord	< 1.8 mg/dl
24 hours	Premature 1 - 6 mg/dl
	Full Term 2 - 6 mg/dl
48 hours	Premature 6 - 8 mg/dl
	Full Term 6 - 7 mg/dl
3 - 5 days	Premature 10 - 15 mg/dl
	Full Term 4 - 12 mg/dl
1 month - adult	1 mg/dl
Conjugated Bilirubin, Direct	< 0.4 mg/dl

Table 66.(con't) Comparative Reference Values in Various
Age Groups of Humans

CONSTITUENT	NORMAL VALUE RANGE
Calcium (Total)	
Premature 1 week	6 - 10 mg/dl
Full Term 1 week	7 - 12 mg/dl
Child	8 - 11 mg/dl
Adult	8.5 - 11 mg/dl
Calcium (ionized)	4.4 - 5.4 mg/dl
Chloride	94 - 106 meq/l
Cholesterol (Total)	
Full Term	50 - 120 mg/dl
1 - 2 years	70 - 190 mg/dl
2 years	135 - 250 mg/dl
Adult	130 - 270 mg/dl
Creatine Kinase (Creatine Phosphokinase)	
Newborn	30 - 100 U/l
Child	15 - 50 U/l
Adult	5 - 40 U/l
Creatinine	
1 - 18 months	0.2 - 0.5 mg/dl
2 - 12 years	0.3 - 0.8 mg/dl
13 - 20 years	0.5 - 1.2 mg/dl
Adult	0.8 - 1.5 mg/dl
Fibrinogen	200 - 400 mg/dl
Folic Acid (Folate)	5 - 21 nanograms/ml
Glucose	
Children	60 - 110 mg/dl
Premature	> 30 mg/dl
Full Term	> 40 mg/dl
Iron	
Newborn	110 - 270 ug/dl
4 - 10 months	30 - 70 ug/dl
3 - 10 years	53 - 119 ug/dl
Adult	87 - 279 ug/dl

Table 66. (con't)   Comparative Reference Values in Various
Age Groups of Humans

CONSTITUENT	NORMAL VALUE RANGE
Iron Binding Capacity (TIBC)	
Newborn	59 - 175 ug/dl
4 - 10 months	250 - 400 ug/dl
3 - 10 years	250 - 400 ug/dl
Adult	250 - 400 ug/dl
Lactate	
Venous Blood	0.5 - 1.6 meq/l
Arterial Blood	0.3 - 1.0 meq/l
Lactate Dehydrogenase (LDH)(37°C)	
Birth	290 - 501 U/l
1 day - 1 month	188 - 404 U/l
1 month - 2 years	110 - 244 U/l
4 years	60 - 170 U/l
3 - 17 years	85 - 165 U/l
Adult	30 - 90 U/l
LDH Isozymes (37°C)	
LD_1 (Heart)	24 - 34%
LD_2 (Heart, Erythrocytes)	35 - 45%
LD_3 (Muscles)	15 - 25%
LD_4 Liver,(Trace Muscle)	4 - 10%
LD_5 Liver, Muscle	1 - 9%
Lipids (Total)	450 - 1000 mg/dl
Lipoproteins	
Alpha (HDL)	286 - 450 mg/dl
Pre Beta (VLDL)	22 - 72 mg/dl
Beta (LDL)	276 - 438 mg/dl
Phospholipids	
Cord Blood	0.48 - 1.6 g/l
2 - 13 years	1.66 - 2.47 g/l
3 - 20 years	1.93 - 3.38 g/l

Table 66.(con't) Comparative Reference Values in Various
Age Groups of Humans

CONSTITUENT	NORMAL VALUE RANGE
Phosphorus (Inorganic)	
Newborn	4 - 10.5 mg/dl
1 year	4.0 - 6.8 mg/dl
5 years	3.6 - 6.5 mg/dl
Adult	3.0 - 4.5 mg/dl
Potassium	
10 days of age	3.5 - 7.0 meq/l
10 days of age	3.5 - 5.5 meq/l
Proteins (Total)	
Premature	5.5 (4.0 - 7.0) g/dl
Newborn	6.4 (5.0 - 7.1) g/dl
1 - 3 months	6.6 (4.7 - 7.4) g/dl
3 - 12 months	6.8 (5.0 - 7.5) g/dl
1 - 15 years	7.4 (6.5 - 8.6) g/dl
Proteins (Albumin)	
Premature	3.7 (2.5 - 4.5) g/dl
Newborn	3.4 (2.5 - 5.0) g.dl
1 - 3 months	3.8 (3.0 - 4.2) g/dl
3 - 12 months	3.9 (2.7 - 5.0) g/dl
1 - 15 years	4.0 (3.2 - 5.0) g/dl
Gamma Globulin	
Premature	0.7 (0.5 - 0.9) g/dl
Newborn	0.8 (0.7 - 0.9) g/dl
1 - 3 months	0.3 (0.1 - 0.5) g/dl
3 - 12 months	0.6 (0.4 - 1.2) g.dl
1 15 years	0.9 (0.6 - 1.2) g/dl
Pyruvate	0.05 - 0.14 meq/l
Sodium	
Premature	130 - 140 meq/l
Adult	135 - 145 meq/l

Table 66. (con't)   Comparative Reference Values in Various Age Groups of Humans

CONSTITUENT	NORMAL VALUE RANGE
Aspartate Transaminase (AST) (37°C)	
1 - 3 days	16 - 74 IU/1
6 months	20 - 43 IU/1
6 months to 1 year	16 - 35 IU/1
1 year - 5 years	6 - 30 IU/1
5 years - adult	19 - 28 IU/1
Adult Male	8 - 46 IU/1
Female	7 - 34 IU/1
Alanine Transaminase (ALT) (37°C)	
Infants	54 IU/1
Children	1 - 30 IU/1
Adults	0 - 19 IU/1
Triglycerides	29 - 154 mg/dl
Urea Nitrogen (BUN)	6 - 23 mg/dl
Uric Acid	2.5 - 8.0 mg/dl
Vitamin A (Retinol)	
6 months	20 - 90 ug/dl
1 - 5 years	30 - 100 ug/dl
5 - 16 years	60 - 100 ug/dl
Adult	20 - 80 ug/dl
Vitamin B-12	330 - 1025 ug/dl
Vitamin C	0.4 - 1.5 mg/dl
Vitamin E	
Newborn	0.3 ug/dl
Child - Adult	0.5 - 1.2 ug/dl
Zinc	
0 - 1 year	74 - 146 ug/dl
2 - 10 years	72 - 128 ug/dl
11 - 18 years	65 - 125 ug/dl
Adult	60 - 120 ug/dl

Source: Henry et al (1979), Tietz (1976), Rawnsley (1981)

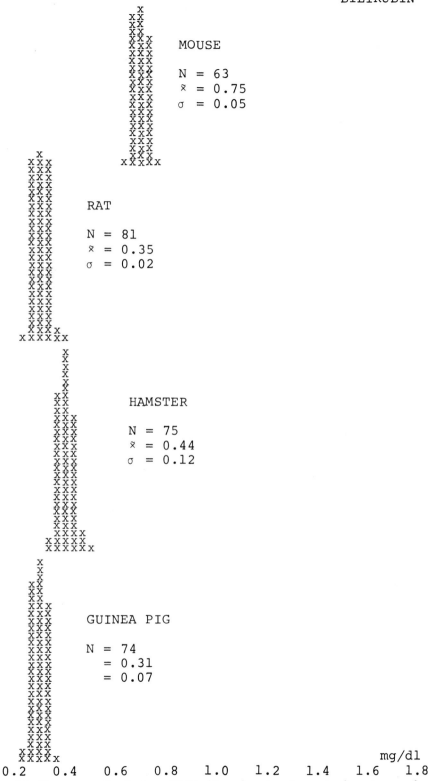

Figure 1A. *Frequency distribution of bilirubin values in mice, rats, hamsters, and guinea pigs.*

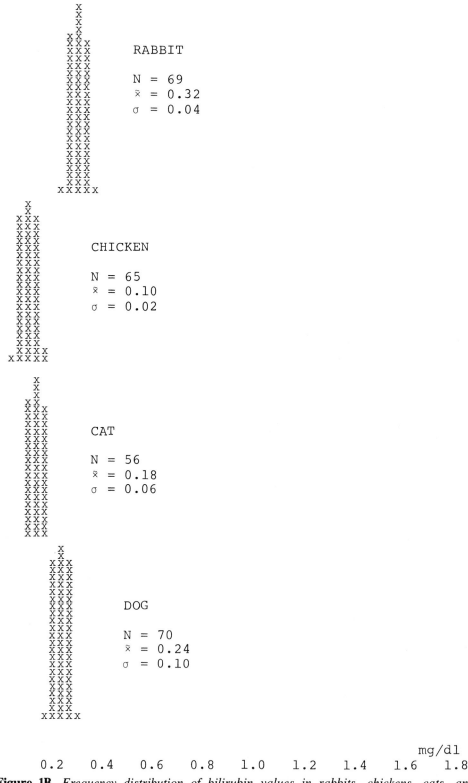

Figure 1B. *Frequency distribution of bilirubin values in rabbits, chickens, cats, and dogs.*

256

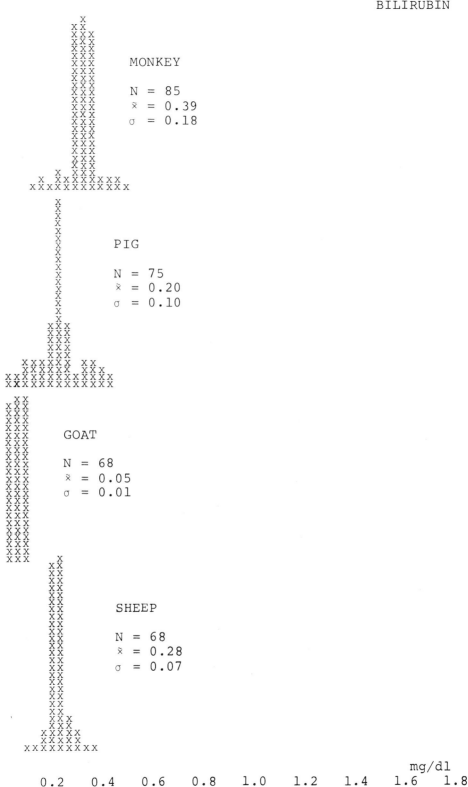

Figure 1C. *Frequency distribution of bilirubin values in monkeys, pigs, goats, and sheep.*

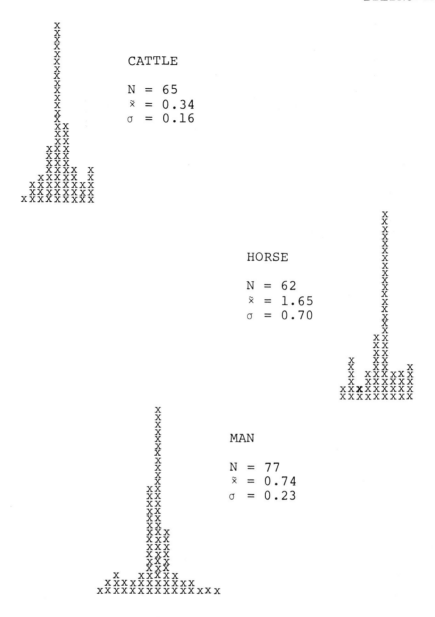

Figure 1D. *Frequency distribution of bilirubin values in cattle, horses, and man.*

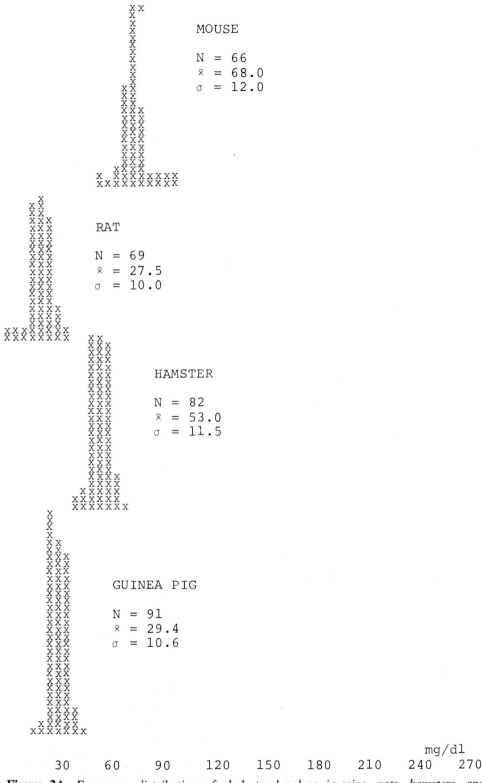

Figure 2A. *Frequency distribution of cholesterol values in mice, rats, hamsters, and guinea pigs.*

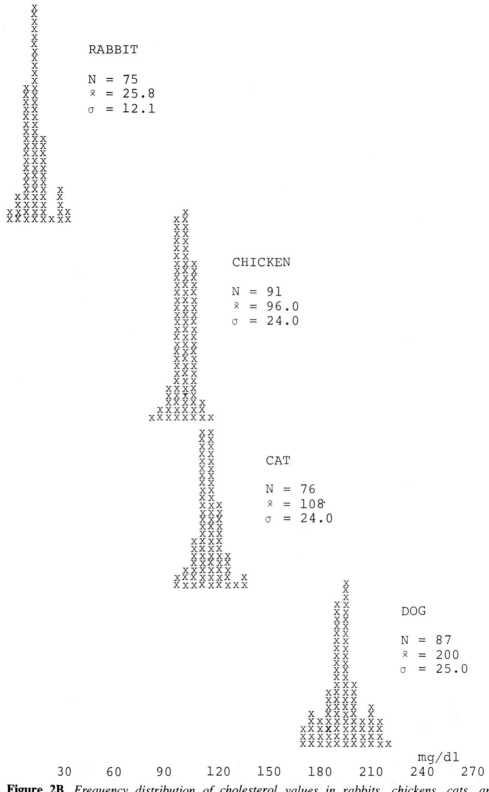

Figure 2B. *Frequency distribution of cholesterol values in rabbits, chickens, cats, and dogs.*

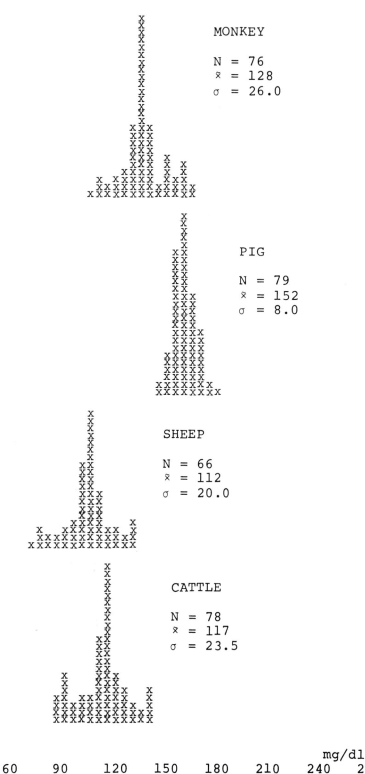

Figure 2C. *Frequency distribution of cholesterol values in monkeys, pigs, sheep, and cattle.*

261

CHOLESTEROL

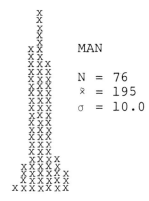

```
              X
              X
              X
              X
            X X        MAN
            X X X
            X X X      N = 76
            X X X      x̄ = 195
            X X X      σ = 10.0
            X X X
            X X X
            X X X
            X X X
            X X X
            X X X
            X X X X
          X X X X X
        X X X X X X X
        X X X X X X
    X X X X X X X X
```

mg/dl

  30   60   90   120   150   180   210   240   270

Figure 2D. *Frequency distribution of cholesterol values in man.*

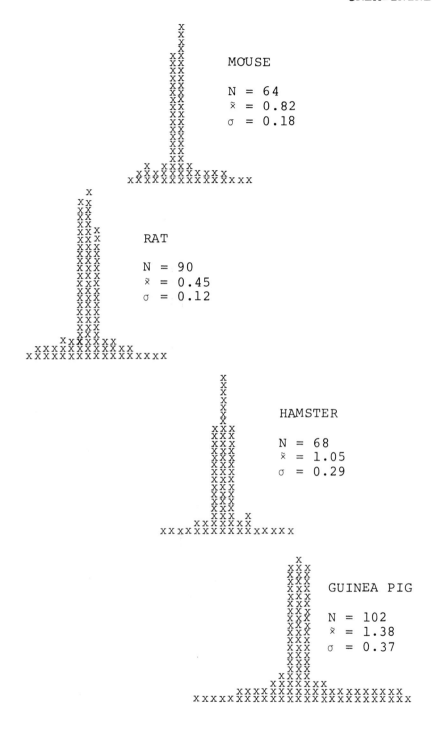

Figure 3A. *Frequency distribution of creatinine values in mice, rats, hamsters, and guinea pigs.* 263

CREATININE

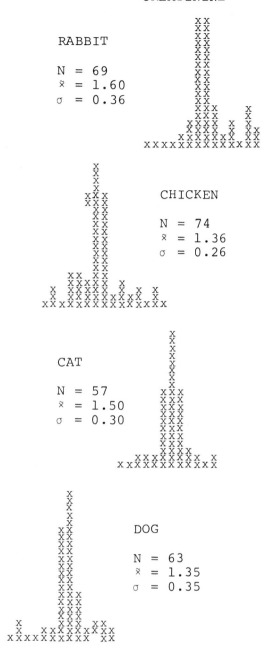

RABBIT

N = 69
x̄ = 1.60
σ = 0.36

CHICKEN

N = 74
x̄ = 1.36
σ = 0.26

CAT

N = 57
x̄ = 1.50
σ = 0.30

DOG

N = 63
x̄ = 1.35
σ = 0.35

mg/dl
0.2   0.4   0.6   0.8   1.0   1.2   1.4   1.6   1.8

Figure 3B. *Frequency distribution of creatinine values in rabbits, chickens, cats, and dogs.*

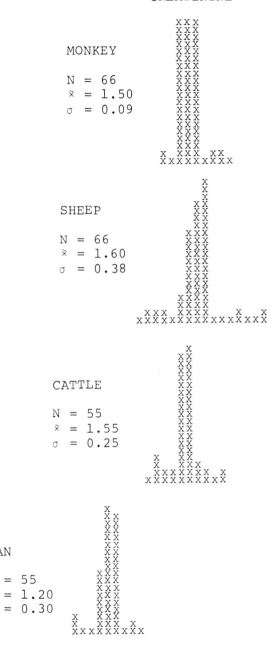

CREATININE

MONKEY

N = 66
x̄ = 1.50
σ = 0.09

SHEEP

N = 66
x̄ = 1.60
σ = 0.38

CATTLE

N = 55
x̄ = 1.55
σ = 0.25

MAN

N = 55
x̄ = 1.20
σ = 0.30

mg/dl

0.2   0.4   0.6   0.8   1.0   1.2   1.4   1.6   1.8

Figure 3C. *Frequency distribution of creatinine values in monkeys, sheep, cattle, and man.*

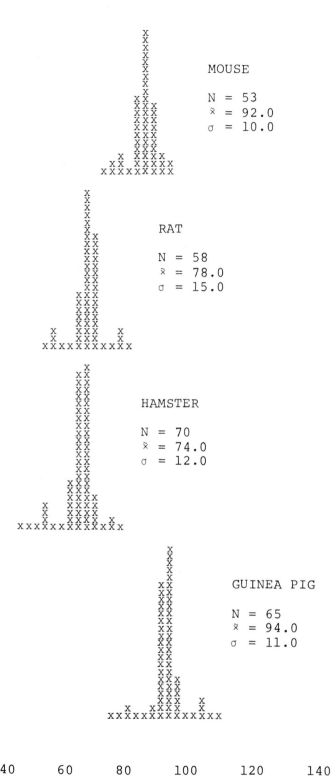

Figure 4A. *Frequency distribution of glucose values in mice, rats, hamsters, and guinea pigs.*

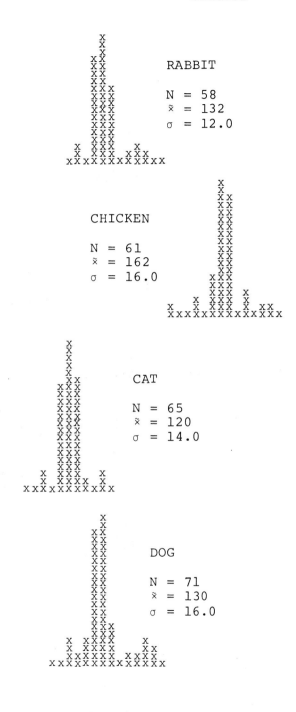

Figure 4B. *Frequency distribution of glucose values in rabbits, chickens, cats, and dogs.*

267

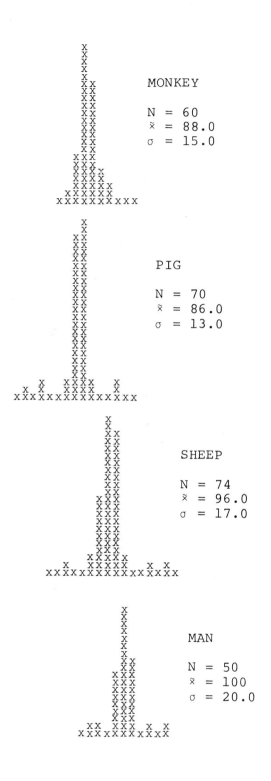

Figure 4C. *Frequency distribution of glucose values in monkeys, pigs, sheep, and man.*

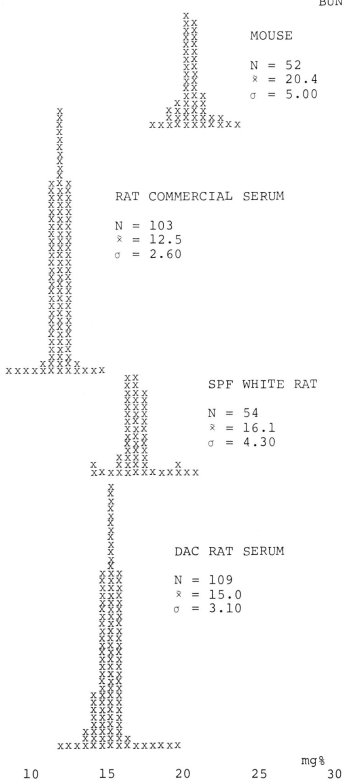

MOUSE

N = 52
x̄ = 20.4
σ = 5.00

RAT COMMERCIAL SERUM

N = 103
x̄ = 12.5
σ = 2.60

SPF WHITE RAT

N = 54
x̄ = 16.1
σ = 4.30

DAC RAT SERUM

N = 109
x̄ = 15.0
σ = 3.10

mg%

5   10   15   20   25   30

Figure 5A. *Frequency distribution of BUN values in mice and rats.*

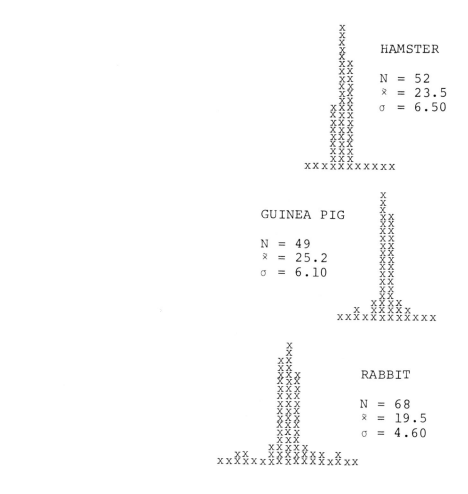

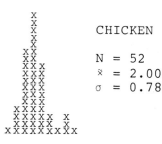

Figure 5B. *Frequency distribution of BUN values in hamsters, guinea pigs, rabbits, and chickens.*

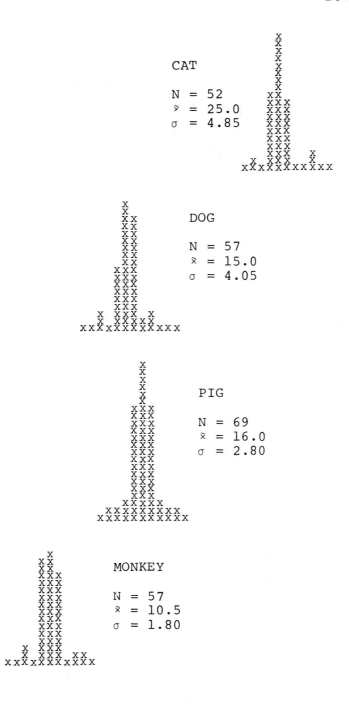

Figure 5C. *Frequency distribution of BUN values in cats, dogs, pigs, and monkeys.*

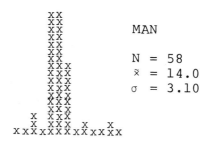

```
                 XX
                 XX
                 XX            MAN
                 XX
                 XX            N = 58
                 XX            x̄ = 14.0
                XXX            σ = 3.10
                XXX
                XXX
                XXX
                XXX
            X   XXX
            X   XXX   X     X
        X XX XXXXXX XX XX XX
```

mg%

Figure 5D. *Frequency distribution of BUN values in man.*

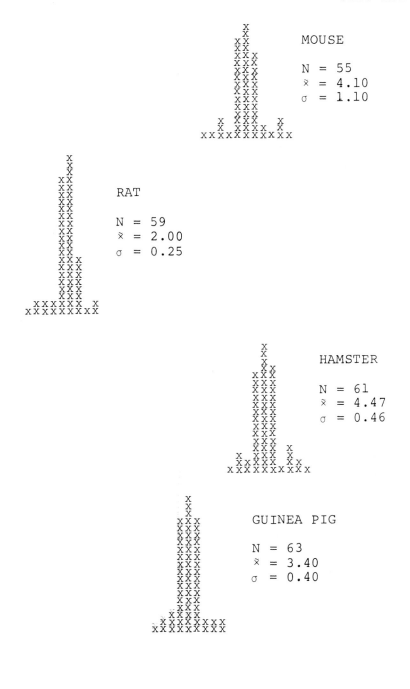

Figure 6A. *Frequency distribution of uric acid values in mice, rats, hamsters, and guinea pigs.*

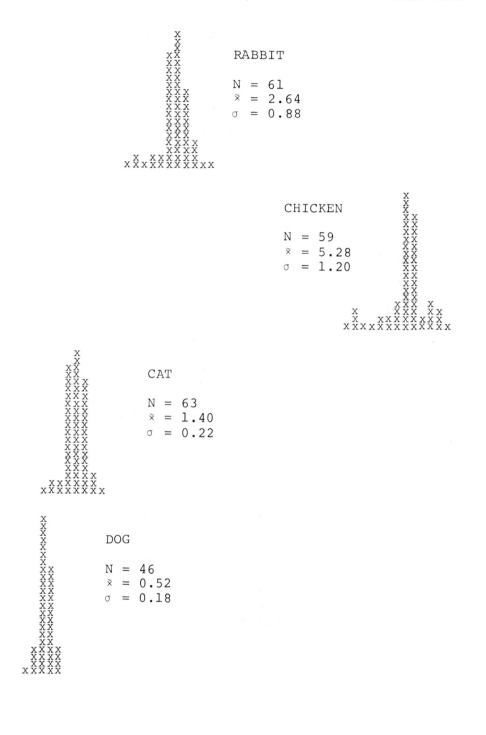

Figure 6B. *Frequency distribution of uric acid values in rabbits, chickens, cats, and dogs.*

274

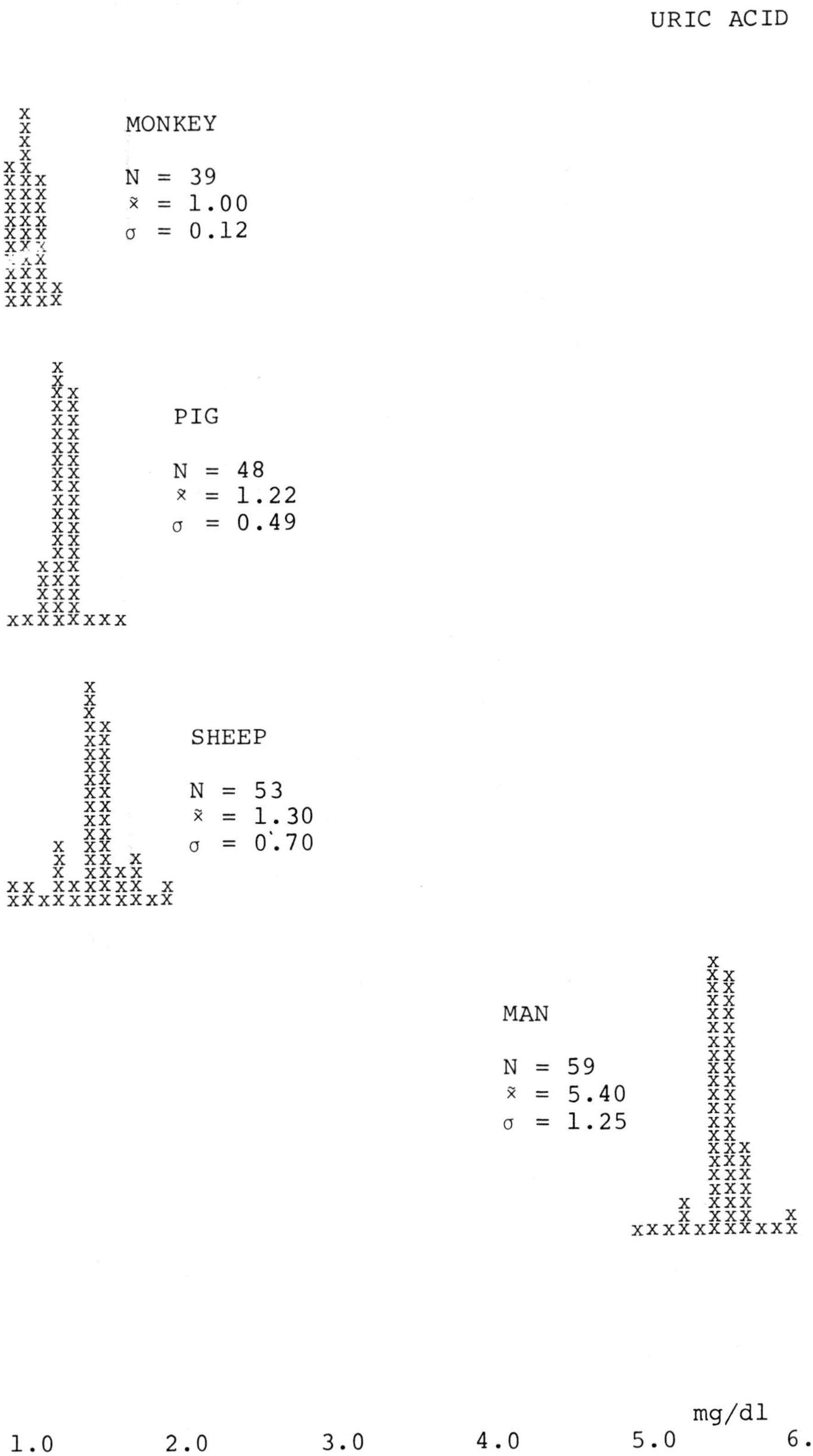

Figure 6C. *Frequency distribution of uric acid values in monkeys, pigs, sheep, and man.*

275

TOTAL PROTEIN

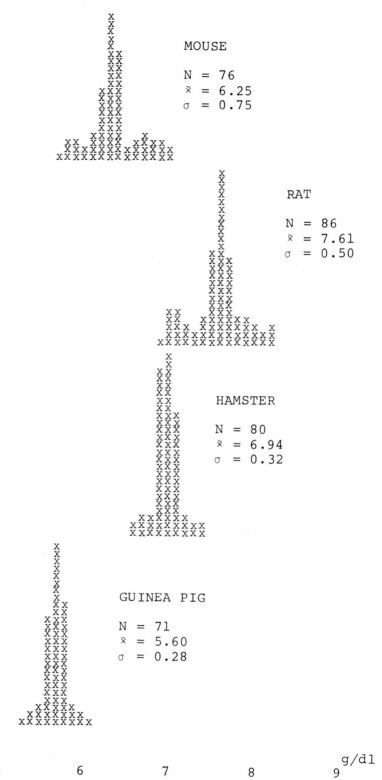

MOUSE

N = 76
x̄ = 6.25
σ = 0.75

RAT

N = 86
x̄ = 7.61
σ = 0.50

HAMSTER

N = 80
x̄ = 6.94
σ = 0.32

GUINEA PIG

N = 71
x̄ = 5.60
σ = 0.28

g/dl

5   6   7   8   9

Figure 7A. *Frequency distribution of total protein values in mice, rats, hamsters, and guinea pigs.*
276

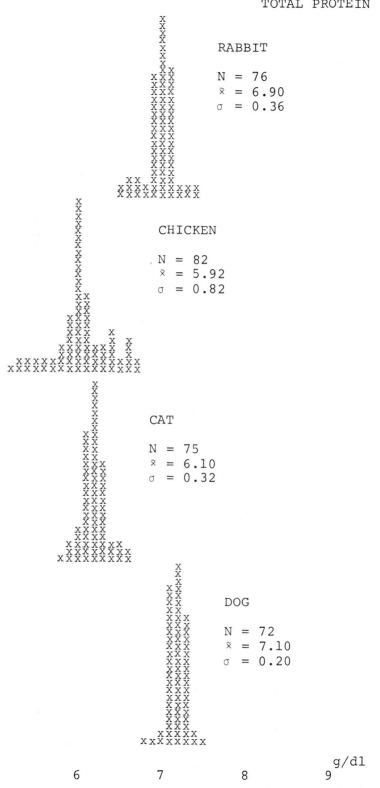

Figure 7B. *Frequency distribution of total protein values in rabbits, chickens, cats, and dogs.*

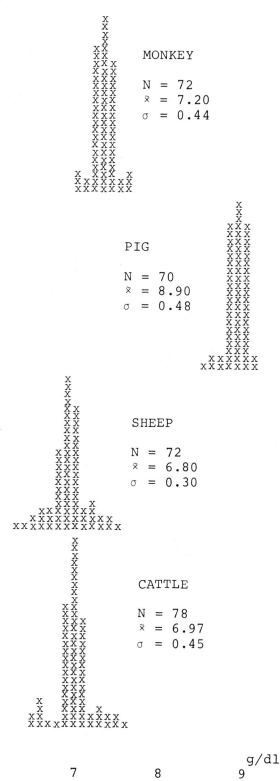

Figure 7C. *Frequency distribution of total protein values in monkeys, pigs, sheep, and cattle.*

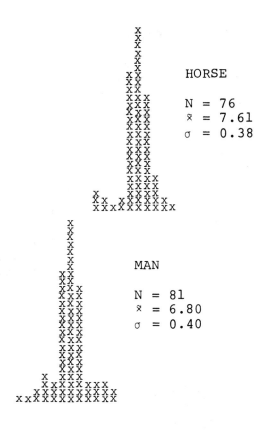

HORSE

N = 76
x̄ = 7.61
σ = 0.38

MAN

N = 81
x̄ = 6.80
σ = 0.40

g/dl

5   6   7   8   9

Figure 7D. *Frequency distribution of total protein values in horses and man.*

279

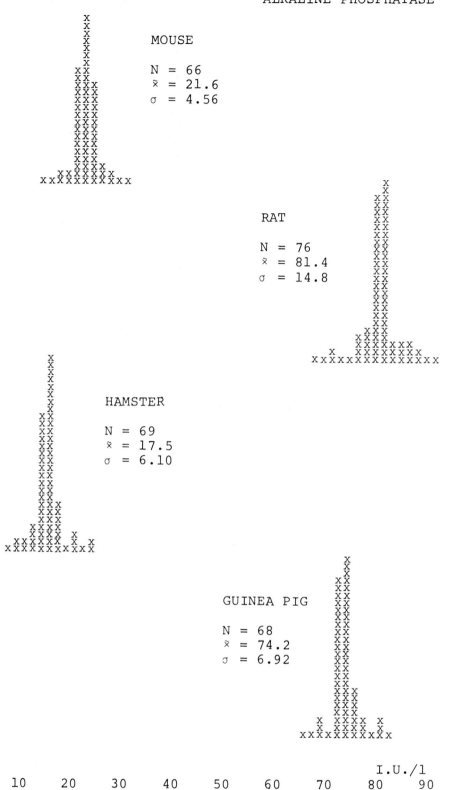

Figure 8A. *Frequency distribution of alkaline phosphatase values in mice, rats, hamsters, and guinea pigs.*

280

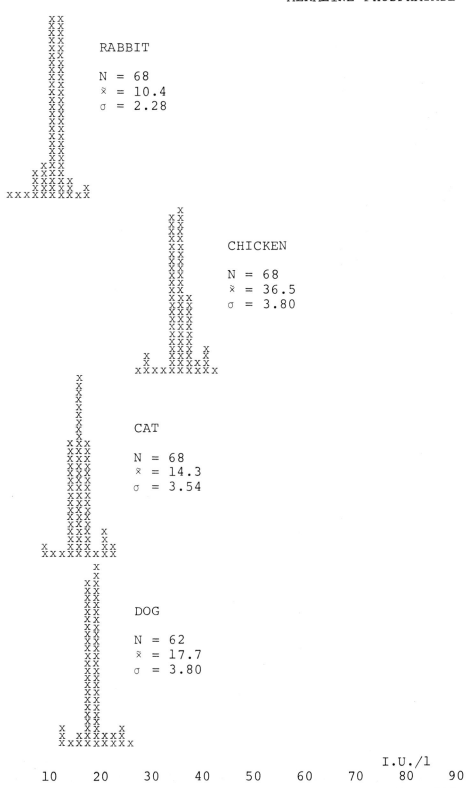

Figure 8B. *Frequency distribution of alkaline phosphatase values in rabbits, chickens, cats, and dogs.*

281

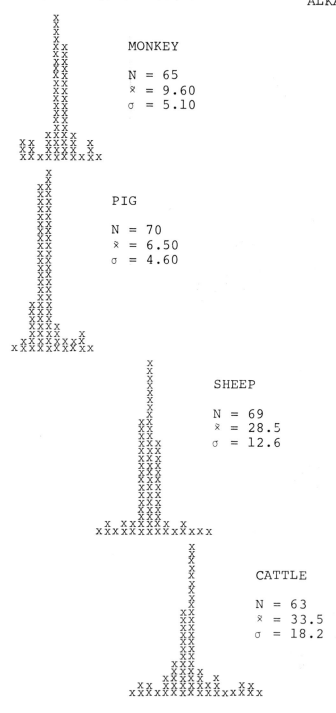

Figure 8C. *Frequency distribution of alkaline phosphatase values in monkeys, pigs, sheep, and cattle.*

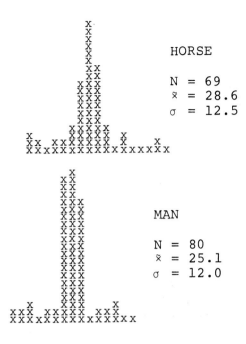

Figure 8D. *Frequency distribution of alkaline phosphatase values in horses and man.*

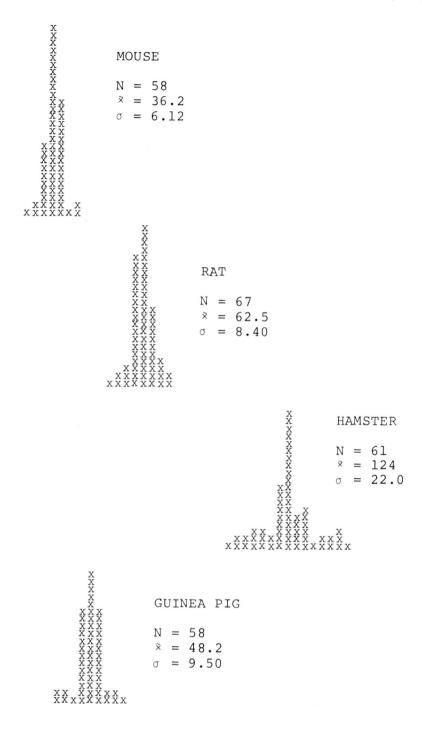

Figure 9A. *Frequency distribution of SGOT values in mice, rats, hamsters, and guinea pigs.*

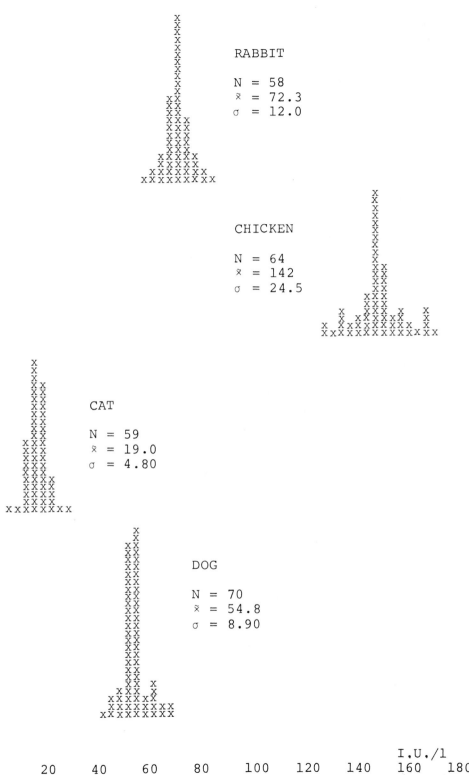

Figure 9B. *Frequency distribution of SGOT values in rabbits, chickens, cats, and dogs.*

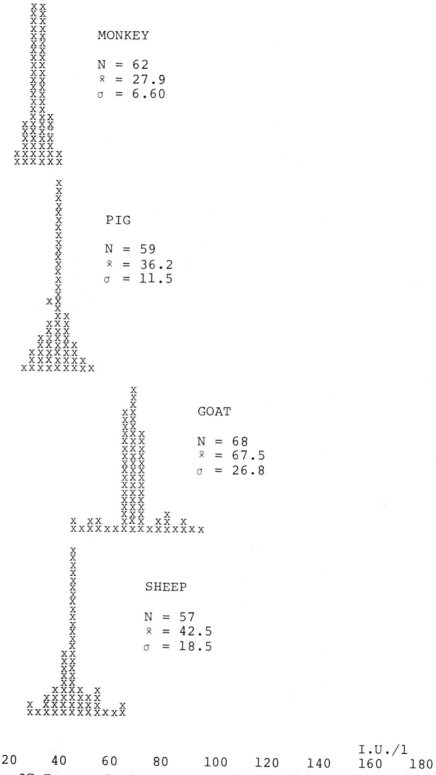

Figure 9C. *Frequency distribution of SGOT values in monkeys, pigs, goats, and sheep.*

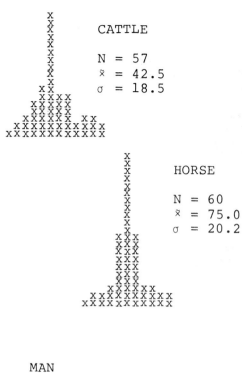

Figure 9D. *Frequency distribution of SGOT values in cattle, horses, and man.*

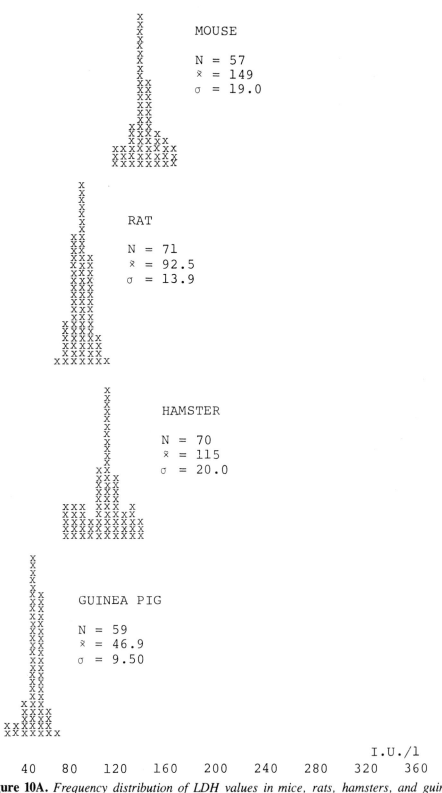

Figure 10A. *Frequency distribution of LDH values in mice, rats, hamsters, and guinea pigs.*

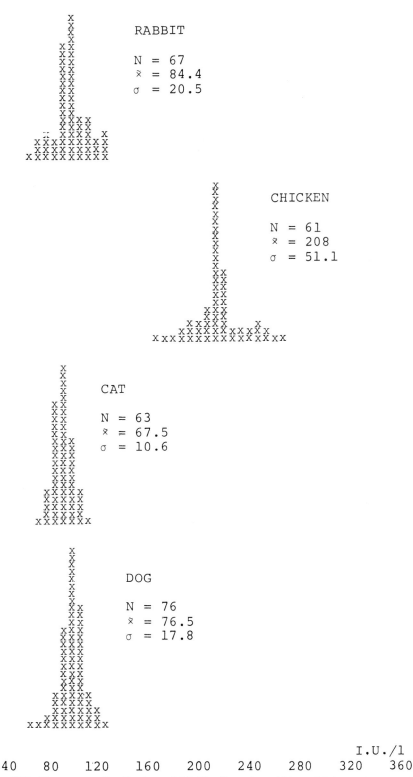

Figure 10B. *Frequency distribution of LDH values in rabbits, chickens, cats, and dogs.*

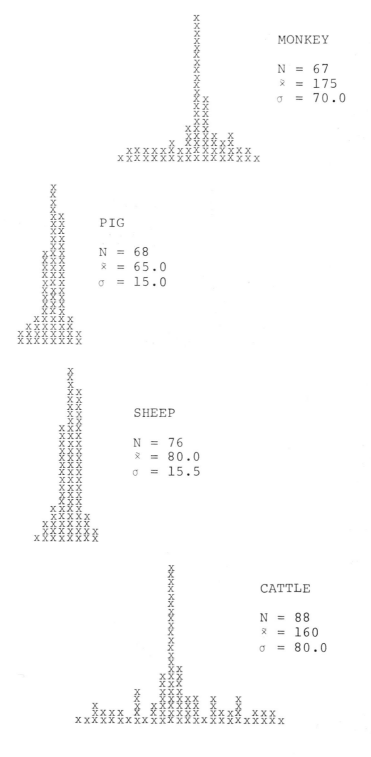

Figure 10C. *Frequency distribution of LDH values in monkeys, pigs, sheep, and cattle.*

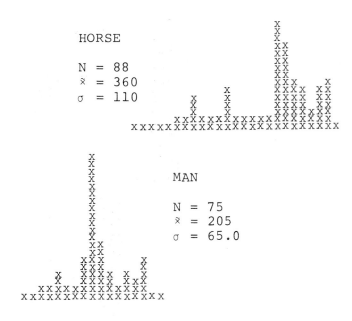

HORSE

N = 88
x̄ = 360
σ = 110

MAN

N = 75
x̄ = 205
σ = 65.0

I.U./l

40   80   120   160   200   240   280   320   360

Figure 10D. *Frequency distribution of LDH values in horses and man.*

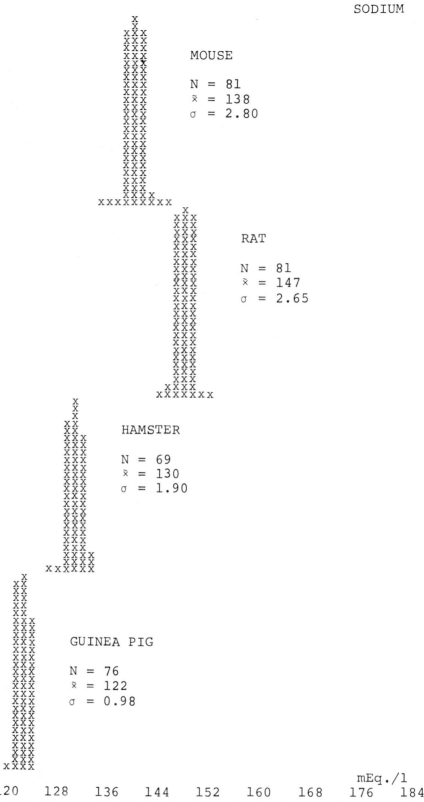

Figure 11A. *Frequency distribution of sodium values in mice, rats, hamsters, and guinea pigs.*

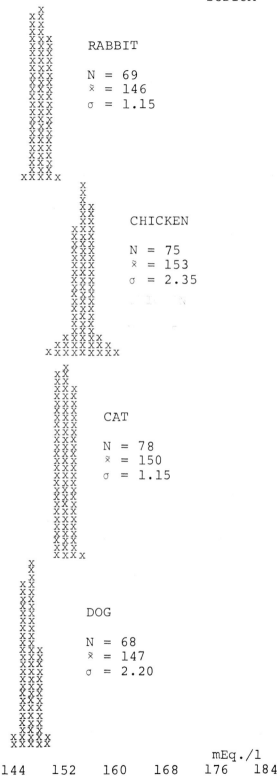

RABBIT

N = 69
x̄ = 146
σ = 1.15

CHICKEN

N = 75
x̄ = 153
σ = 2.35

CAT

N = 78
x̄ = 150
σ = 1.15

DOG

N = 68
x̄ = 147
σ = 2.20

mEq./l

120   128   136   144   152   160   168   176   184

Figure 11B. *Frequency distribution of sodium values in rabbits, chickens, cats, and dogs.*

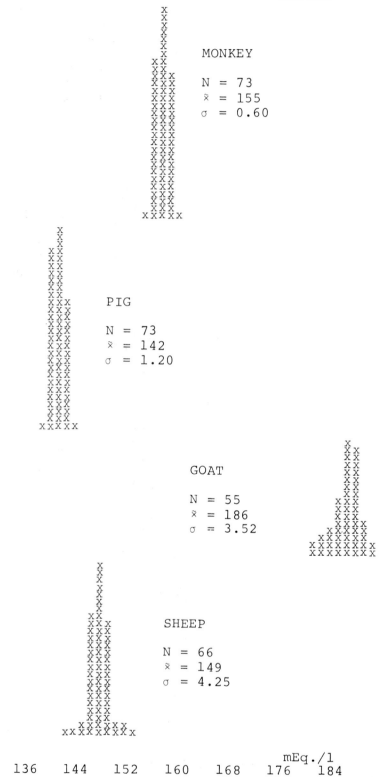

SODIUM

MONKEY

N = 73
x̄ = 155
σ = 0.60

PIG

N = 73
x̄ = 142
σ = 1.20

GOAT

N = 55
x̄ = 186
σ = 3.52

SHEEP

N = 66
x̄ = 149
σ = 4.25

mEq./l

120   128   136   144   152   160   168   176   184

Figure 11C. *Frequency distribution of sodium values in monkeys, pigs, goats, and sheep.*

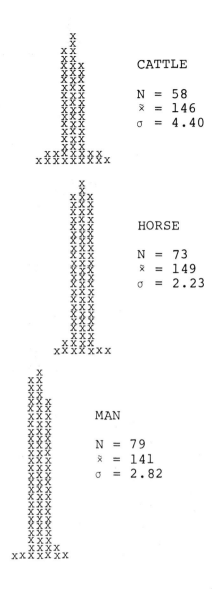

CATTLE

N = 58
x̄ = 146
σ = 4.40

HORSE

N = 73
x̄ = 149
σ = 2.23

MAN

N = 79
x̄ = 141
σ = 2.82

mEq./l
120   128   136   144   152   160   168   176   184

Figure 11D. *Frequency distribution of sodium values in cattle, horses, and man.*

295

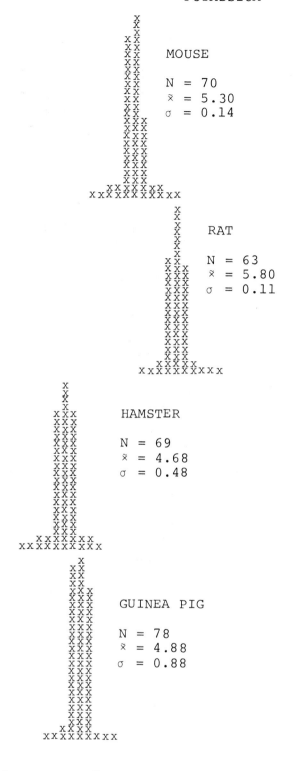

POTASSIUM

MOUSE

N = 70
x̄ = 5.30
σ = 0.14

RAT

N = 63
x̄ = 5.80
σ = 0.11

HAMSTER

N = 69
x̄ = 4.68
σ = 0.48

GUINEA PIG

N = 78
x̄ = 4.88
σ = 0.88

3   4   5   6 mEq./l

Figure 12A. *Frequency distribution of potassium values in mice, rats, hamsters, and guinea pigs.*

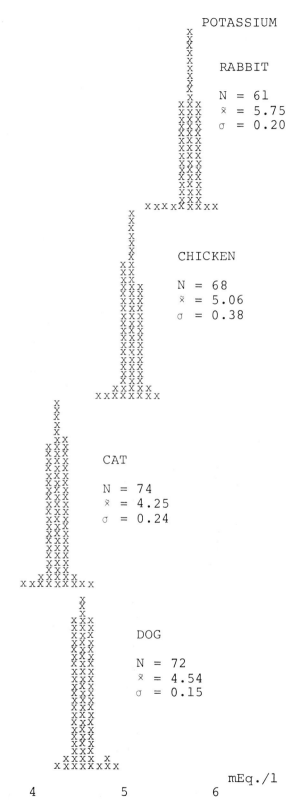

Figure 12B. *Frequency distribution of potassium values in rabbits, chickens, cats, and dogs.*

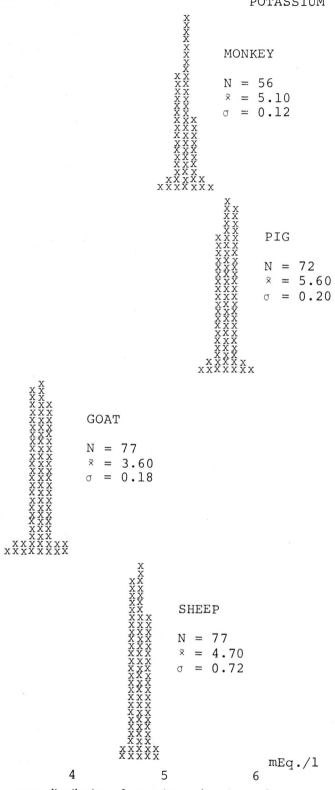

POTASSIUM

MONKEY

N = 56
x̄ = 5.10
σ = 0.12

PIG

N = 72
x̄ = 5.60
σ = 0.20

GOAT

N = 77
x̄ = 3.60
σ = 0.18

SHEEP

N = 77
x̄ = 4.70
σ = 0.72

mEq./1

3   4   5   6

Figure 12C. *Frequency distribution of potassium values in monkeys, pigs, goats, and sheep.*

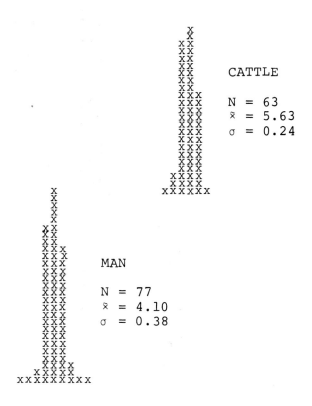

CATTLE

N = 63
x̄ = 5.63
σ = 0.24

MAN

N = 77
x̄ = 4.10
σ = 0.38

mEq./l

3   4   5   6

Figure 12D. *Frequency distribution of potassium values in cattle and man.*

299

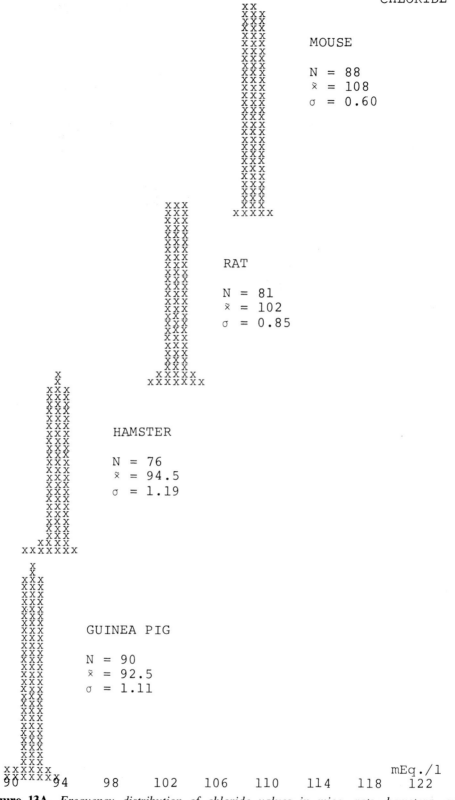

Figure 13A. *Frequency distribution of chloride values in mice, rats, hamsters, and guinea pigs.*

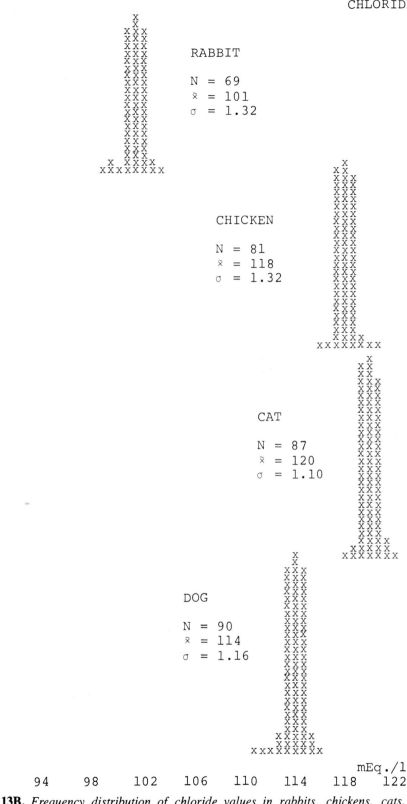

Figure 13B. *Frequency distribution of chloride values in rabbits, chickens, cats, and dogs.*

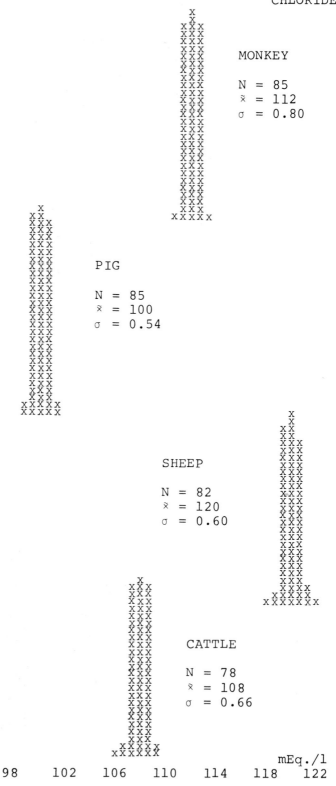

Figure 13C. *Frequency distribution of chloride values in monkeys, pigs, sheep, and cattle.*

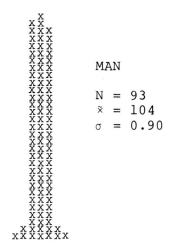

MAN

N = 93
x̄ = 104
σ = 0.90

mEq./l

90   94   98   102   106   110   114   118   122

Figure 13D. *Frequency distribution of chloride values in man.*

303

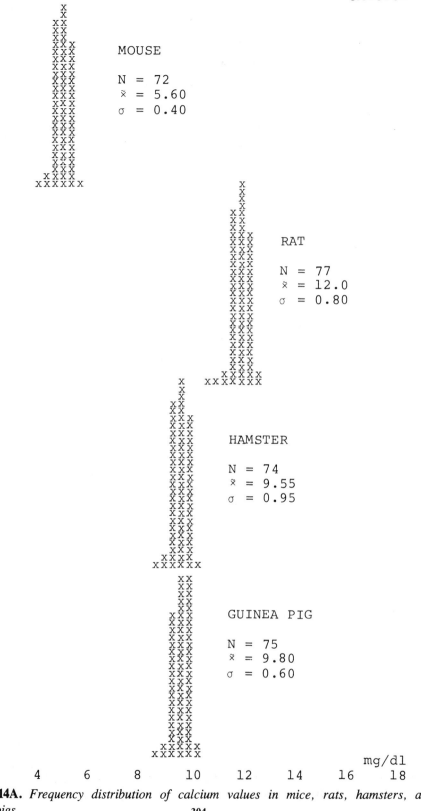

Figure 14A. *Frequency distribution of calcium values in mice, rats, hamsters, and guinea pigs.*

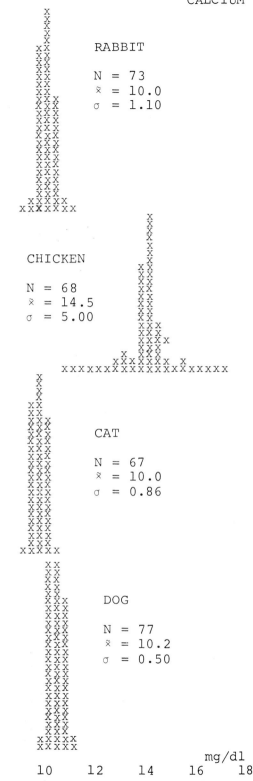

Figure 14B. *Frequency distribution of calcium values in rabbits, chickens, cats, and dogs.*

305

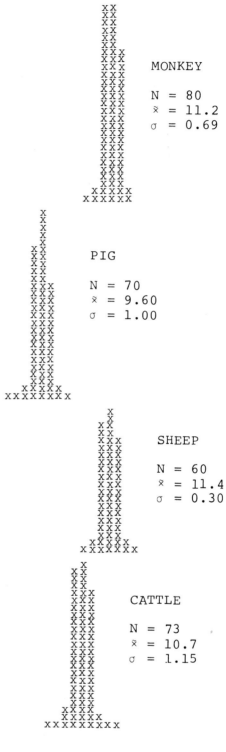

Figure 14C. Frequency distribution of calcium values in monkeys, pigs, sheep, and cattle.

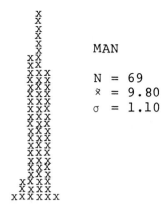

MAN

N = 69
x̄ = 9.80
σ = 1.10

mg/dl

2   4   6   8   10   12   14   16   18

Figure 14D. *Frequency distribution of calcium values in man.*

sent sera from normal laboratory personnel and hospital staff—20 men and 15 women between the ages of 20 to 55 years—weighing 50–90 kg. Although there are wide variations in human serum chemistries because of age, sex, race, food intake, and other biological variations, our data were within the range of normal values established in the Pepper Laboratory of the Hospital of the University of Pennsylvania and Dartmouth Medical Center. The main purpose of reporting the blood chemistries of humans in this section is to provide reference for comparison with values of experimentally useful animal species. Many publications, including research papers, monographs, and text and reference books in clinical pathology, discuss the importance of establishing normal values in human populations and the effects of many factors that may alter the values.

Frequency Distribution of Normal Values

A simple statistical concept has been applied in medicine to the normal range of blood chemical values. The concept is the Gaussian curve, which is often referred to as the "normal" curve because its values are found in samples from a group of normal people. The greatest density of values is somewhere near the middle of the range, and the density decreases steadily as the distance from the middle increases. If samples from a normal population actually conformed to this distribution, the mean ($\bar{X}$) ± 2 standard deviations (S.D.), which is derived easily from all the observed values, would represent the 95% limits of the normal range. The 95% of the normals could thus be calculated as $\bar{X}$ ± 2 S.D. if the values for these two parameters were truly those of the population, but actually the values are estimates from a relatively small sample. Therefore this method is applicable in experimental animal studies that use a

relatively small and select population in which sex, age, and nutritional and physiological conditions are clearly defined. However many statisticians claim that any defined population will not be Gaussian-distributed if the sample is too large. Pryce (1964), on the other hand, believes that the Gaussian curve should apply not only to repeated measurements of the same thing, but to normal values too, and that any deviation from a Gaussian distribution is the result of major sources of variation such as age, sex, and season. Further discussions of the validity of using frequency distribution curves to establish mean normal values are presented by Henry et al. (1974); Sunderman (1975); Caraway (1971); Amador et al. (1968); Harris and DeMets (1972); Neumann (1968); Thompson (1968); Henry (1971); Galen and Gambino (1976); Tietz (1976); and others.

The data on normal values presented in this book were derived from selected and well-defined animal species, and distribution curves were frequently plotted to determine the normal distribution of the values of samples from the populations (Figures 1–14). Comparative frequency distribution plots of bilirubin, cholesterol, creatinine, glucose, blood urea nitrogen (BUN), uric acid, total protein, alkaline phosphatase, AST, LDH, sodium, potassium, chloride, and calcium generally show a Gaussian population distribution, with some exceptions (e.g., lactic dehydrogenase in horses and other animal species).

Biochemical Components in Urines of Experimental Animals

There is a wide range in the physical and chemical composition of normal urine because of the variation in fluid and food intake, physical activity, and body temperature. Urine can be examined routinely because it is easily sampled from most experimental animals. Methods of collection, preservation, and processing urine from different animal species are described in Section II of this book.

Table 67 lists the comparative range of certain biochemical and physical parameters of normal experimental animals and humans. Detailed discussions and descriptions of these values and others can be found in the literature, some of which is cited in Table 67. In comparison with normal human urine values, the data show that the rat, rabbit, dog, and monkey had a higher output volume of urine per kilogram of body weight per day. The pH of urine was more acidic in cats, dogs, monkeys, and swine than in humans, and more alkaline in rats, rabbits, goats, sheep, cattle, and horses. Comparisons of the chemical components of urines in experimental animals and humans can be summarized as follows;

Calcium: Higher in rabbits and monkeys and lower in cats, dogs, goats, sheep, and cattle.

Magnesium: Higher in rabbits (highly variable), cattle, and monkeys; slightly higher in cats and dogs; lower in rats, goats, and sheep.

Inorganic Phosphorus: Higher in rats, rabbits, cats, and dogs, and lower in goats, sheep, cattle, and horses.

Sodium: Variably higher in dogs, much higher in goats, and lower in rabbits, sheep, and cattle.

Potassium: Higher in cats and dogs, and much higher in goats and monkeys.

Chloride: Higher in rabbits and goats, generally lower in other animals species, and very low in dogs.

Creatinine: Slightly higher in rabbits, dogs, swine, and monkeys; lower in sheep.

Urea Nitrogen: Higher in rats, rabbits, and cats; lower in sheep; and considerably lower in cattle.

Uric Acid: Higher in rats, rabbits, cats (variable), dogs, goats, and sheep; no significant difference between other animal species and humans.

Total Protein: Slightly higher in rats, goats, sheep, cattle, and monkeys; considerably higher in cats and dogs; lower in swine and horses.

Cerebrospinal Fluid

In man, examination of cerebrospinal fluid (CSF) has become essential for the diagnosis of diseases of the nervous system, such as meningitis, brain tumors, encephalitis, and infections. Lumbar puncture and examination of the aspirated fluid are ordinarily diagnostic tools, though at times the entire procedure itself may have therapeutic implications. In experimental animals this technique has not been utilized extensively.

Only a few reports on the examination of CSF in diseases of animals have appeared in the literature. The methods of sampling CSF from animals are described in Section II of this book. Table 68 presents biochemical and physical data on the CSF of experimental animals reported in the literature.

The CSF should be examined for general appearance, consistency, and tendency to clot. A cell count should be performed with an attempt to distinguish the types of cells present. In many cases determination of protein and sugar concentrations is desirable. Other tests are performed whenever there is clinical evidence that would suggest the presence of central nervous system disease. Occasionally such an examination can be of value as a prognostic method for the evaluation of the disease state or its response to treatment. When clinical conditions indicate, special tests are performed including examination of Gram-stained or acid-fast stained smears of the CSF sediment; culture for pyogenic bacteria, tubercle bacilli, or fungi; differential tests employing colloidal suspensions or protein partition; serological tests for syphilis; and miscellaneous chemical determinations such as bilirubin, urea, chlorides, bromides, and enzymes.

Table 67. Biochemical Compnents in Urines of Normal Experimental Animals and Humans

Component (mg/Kg body wt /day) or Property	Rat	Rabbit	Cat	Dog	Goat	Sheep
Volume (ml/Kg body wt /day)	150.-350.	20.0-350.	10.0-30.0	20.0-167.	7.0-40.0	10.0-40.0
Specific Gravity	1.040-1.076	1.003-1.036	1.020-1.045	1.015-1.050	1.015-1.062	1.015-1.045
pH	7.30-8.50	7.60-8.80	6.00-7.00	6.00-7.00	7.5-8.80	7.50-8.80
Calcium	3.00-9.00	12.1-19.0	0.20-0.45	1.00-3.00	1.00-3.40	1.00-3.00
Chloride	50.0-75.0	190.-300.	89.0-130.	5.00-15.0	186.-376.	----
Creatinine	24.0-40.0	20.0-80.0	12.0-30.0	15.0-80.0	10.0-22.0	5.80-14.5
Magnesium	0.20-1.90	0.65-4.20	1.50-3.20	1.70-3.00	0.15-1.80	0.10-1.50
Phosphorous, Inorganic	20.0-40.0	10.0-60.0	39.0-62.0	20.0-50.0	0.5-1.6	0.10-0.50
Potassium	50.0-60.0	40.0-55.0	55.0-120.	40.0-100.	250.-360.	300.-420.
Protein, Total	1.20-6.20	0.74-1.86	3.10-6.82	1.55-4.96	0.74-2.48	0.74-2.17
Sodium	90.4-110.	50.0-70.0	----	2.00-189.	140.-347.	0.80-2.00
Urea Nitrogen (g/Kg/day)	1.00-1.60	1.20-1.50	0.80-4.00	0.30-0.50	0.14-0.47	0.11-0.17
Uric Acid	8.00-12.0	4.00-6.00	0.20-13.0	3.1-6.0	2.00-5.00	2.00-4.00

Table 67 (con't.). Biochemical Components in Urines of Normal Experimental Animals and Humans

Component (mg/Kg body wt /day) or Property	Swine	Cattle	Horse	Monkey	Man
Volume (ml/Kg body wt /day)	5.00-30.0	17.0-45.0	3.0-18.0	70.0-80.0	8.60-28.6
Specific Gravity	1.010-1.050	1.025-1.045	1.020-1.050	1.015-1.065	1.002-1.040
pH	6.25-7.55	7.60-8.40	7.80-8.30	5.50-7.40	4.80-7.80
Calcium	----	0.10-3.60	----	10.0-20.0	0.60-8.30
Chloride	----	10.0-140.	81.0-120.	80.0-120.	40.0-180.
Creatinine	20.0-90.0	15.0-30.0	----	20.0-60.0	15.0-30.0
Magnesium	----	2.00-7.00	----	3.20-7.10	0.42-2.40
Phosphorous, Inorganic	----	0.01-6.20	0.05-2.00	9.00-20.6	10.0-15.0
Potassium	----	240.-320.	----	160.-245.	16.0-56.0
Protein, Total	0.33-1.49	0.25-2.99	0.62-0.99	0.87-2.48	0.81-1.86
Sodium	----	2.00-40.0	----	----	25.0-94.0
Urea Nitrogen (g/Kg/day)	0.28-0.58	0.05-0.06	0.20-0.80	0.20-0.70	0.20-0.50
Uric Acid	1.00-2.00	1.00-4.00	1.00-2.00	1.00-2.00	0.80-3.00

Source: Whelan (1925); Miller (1926); Morgulis and Spencer (1936); Morris and Ray (1937); Bernheim et al. (1945); Hawkins et al. (1946); Herrin (1947); Freidman (1948); Dinning and Day (1949); Ellinger and Abdel Kader (1949); Griffith (1949); Bass et al. (1950); Blosser and Smith (1950); Miller et al. (1951); Hansard et al. (1952); Rosenthal and Cravitz (1958); Worden et al. (1960); Benjamin (1961); Kesner and Muntwyler (1963); Thompson et al. (1966); Coles (1967); Jacobi and Fontaine (1967); Houpt (1970); Swenson et al. (1970); Flatt and Carpenter (1971); Kozma et al. (1974).

311

Table 68. Biochemical Components in Cerebrospinal Fluids (CSF) of Normal Experimental Animals and Humans

Components or Property	Chinchilla	Guinea Pig	Rabbit	Cat	Dog	Goat	Sheep
Specific Gravity	1.0035-1.0060	1.004-1.006	1.0035-1.0065	1.005-1.007	1.0060-1.0090	1.0030-1.0060	1.004-1.008
pH	7.40-7.80	7.35-7.55	7.40-7.85	7.40-7.60	7.35-7.39	7.25-7.35	7.30-7.40
No. of WBC/ml	0.00-5.00	0.00-5.0	0.00-4.00	0.00-3.00	1.00-8.00	1.00-5.00	0.00-15.0
Bicarbonate	30.5-38.6	35.5-45.5	41.2-48.5	20.6-29.8	20.8-31.8	18.5-26.5	24.6-36.0
Calcium (mg/dl)	4.00-6.00	5.75-6.25	5.10-5.80	5.20-6.00	5.48-5.72	8.40-9.60	4.90-6.57
Chloride (mEq/l)	134.-146.	117.-127.	169.-206.	125.-175.	122.-138.	181.-202.	212.-244.
Cholesterol (mg/dl)	0.00-0.20	0.00-0.10	0.00-0.30	0.00-0.40	0.00-0.50	----	----
Glucose (mg/dl)	111.-147.	60.-110.	55.0-90.0	55.0-115.	45.0-77.0	56.0-71.0	39.0-109.
Phosphorous, Inorganic (mg/dl)	----	1.80-2.90	2.10-2.40	0.60-1.60	2.82-3.47	----	2.00-4.00
Potassium (mEq/l)	2.60-3.12	3.25-4.75	2.64-3.36	4.40-7.35	2.80-3.90	2.90-3.60	3.05-3.64
Protein, Total (mg/dl)	22.5-30.6	16.0-24.0	15.0-19.0	17.0-25.0	11.0-55.0	7.00-24.0	8.00-70.0
Sodium (mEq/l)	160.-170.	145.-155.	129.-169.	148.-168.	143.-163.	136.-172.	140.-158.
Urea Nitrogen (mg/dl)	----	8.00-20.5	14.5-26.0	32.0-46.5	18.0-24.6	10.0-19.5	6.00-16.0
Uric Acid (mg/dl)	0.05-0.50	----	0.10-1.50	0.00-0.50	0.00-2.00	----	----

Table 68 (con't.). Biochemical Components in Cerebrospinal Fluids (CSF) of Normal Experimental Animals and Humans

Components or Property	Swine	Cattle	Horse	Monkey			Man
				Rhesus	Baboon	Chimpanzee	
Specific Gravity	1.004-1.008	1.005-1.008	1.004-1.008	1.004-1.005	1.004-1.013	1.004-1.005	1.0062-1.0082
pH	7.30-7.40	7.00-7.60	7.13-7.36	7.35-7.65	7.20-7.70	7.40-7.65	7.35-7.70
No. of WBC/ml	1.00-20.0	0.00-10.0	2.00-7.00	0.00-8.00	4.00-10.0	0.00-11.0	0.00-10.0
Bicarbonate	40.5-52.0	25.0-40.0	23.0-28.5	34.0-45.0	----	30.0-40.0	41.5-55.0
Calcium (mg/dl)	5.30-5.90	5.10-6.30	5.55-6.98	8.80-10.8	8.50-11.6	4.50-6.30	3.90-5.10
Chloride (mEq/l)	----	183.-204.	195.-224.	129.-137.	----	----	118.-127.
Cholesterol (mg/dl)	0.20-0.60	----	0.25-0.65	0.15-0.50	0.10-0.80	0.30-0.60	0.24-0.50
Glucose (mg/dl)	45.0-87.0	35.0-70.0	47.0-78.0	37.0-80.0	24.0-111.	34.7-95.0	·45.0-93.0
Phosphorous, Inorganic (mg/dl)	1.30-1.95	2.15-4.06	0.87-2.20	0.80-2.50	1.00-4.15	----	1.25-2.10
Potassium (mEq/l)	2.10-3.20	3.03-3.70	2.72-4.00	2.40-3.80	2.48-7.90	2.35-3.99	2.18-3.00
Protein, Total (mg/dl)	24.0-40.0	20.0-33.0	28.8-71.8	31.0-50.0	15.0-109.	20.0-67.4	12.0-43.0
Sodium (mEq/l)	198.-230.	138.-160.	270.-330	145.-158.	125.-154.	133.-207.	217.-235.
Urea Nitrogen (mg/dl)	12.0-20.0	8.00-15.0	23.0-31.0	20.0-28.0	26.0-32.5	42.0-60.5	10.5-18.6
Uric Acid mg/dl	0.40-2.50	----	0.08-2.00	0.10-1.00	----	----	0.50-2.80

Source: Carmichael and Jones (1939); Roeder and Rehm (1942); Ledoux (1943); Byers and Freidman (1949); Behrens (1953); Frankhauser (1953, 1963); Teunissen (1953); Sorensen et al. (1954); Criton et al. (1956); Davidson (1956); Criton and Exley (1957); Fedetov (1960); Yeary et al. (1960); Dixon and Bowie (1962); Pappenheimer et al. (1962); Freidman, et al. (1963); Ames et al. (1964); Held et al. (1964); Soliman (1965); Sokiman et al. (1965); Bito and Daveson (1966); Turbyfill et al. (1968, 1970); Medway et al. (1969); Derwelis et al. (1970); Himwich and Himwich (1970); Altman and Ditmer (1971, 1974); Butler and Wiley (1971); Jahn et al. (1974); Kozma et al. (1974).

In experimental medicine when an animal species is used for the study of a specific disease, changes in the CSF components of the experimental animals are compared to those in human CSF under similar clinical conditions. The normal CSF values of experimental animals compared with those of humans have the following characteristic features:

Bicarbonate: Lower in chinchillas, cats, dogs, goats, sheep, cattle, horses, and monkeys than in humans. No significant differences in other animal species.

Calcium: Higher in goats and monkeys than in humans.

Chloride: Higher in chinchillas, rabbits, cats, dogs, goats (very high), sheep (very high), horses (very high), and Rhesus monkeys than in humans.

Cholesterol: Lower in chinchillas, guinea pigs, cats (slightly), and sheep (slightly) than in human CSF.

Glucose: Higher in chinchillas, guinea pigs, cats (slightly), and sheep (slightly) than in humans.

Inorganic Phosphate: Higher in dogs, sheep, and cattle, and lower in cats and swine than in humans.

Potassium: Higher in guinea pigs, cats, and baboons than in humans. No significant difference between the K values of other animal species and those of humans has been reported.

Total Protein: Higher in dogs (slightly), sheep, horses, baboons, chimpanzees and Rhesus monkeys, and lower in guinea pigs, rabbits, cats, and goats.

Sodium: Higher in horses and lower in chinchillas, guinea pigs, rabbits, cats, dogs, goats, sheep, cattle, Rhesus monkeys, and baboons than in humans.

Urea Nitrogen: Higher in rabbits, cats (very high), dogs, horses, Rhesus monkeys, baboons, chimpanzees (very high), and lower in goats, sheep, and cattle (slightly) than in humans.

Uric Acid: Lower in chinchillas, cats, dogs (variable), horses (variable), and Rhesus monkeys than in humans. Other animal species have levels essentially similar to those reported for humans.

For further details on urine and other body fluid values in experimental animals and humans please refer to Free and Free (1976), Barsant and Finco (1979); Ahonen et al. (1978), Ruppaner et al. (1978); Munan et al. (1978), Kaplan and Timmons (1979); Ladenson (1980).

Interpretation of Reference Value Data

The useful interpretation of clinical laboratory results is dependent not only on knowledge of reference values that are obtained from "normal" animals or humans, but also on an appreciation of various influences that may alter test results. Individual laboratory tests are subject to a number of variables, and the physician and clinical investigator should be aware of this variability in his interpretations of any changes which he may note. The most common variations are due to sample collection, handling, preservation and storage conditions. These factors have been discussed in Section II of this book. Other factors which may cause misinterpretation of laboratory results are due to biological or technological variations. Effects of physiological and pathological conditions of experimental animal and methodological variations which influence the laboratory test results are emphasized in this Section. The most common variations in specimens received in clinical laboratory are those due to discoloration of serum resulting from hemolysis, lipemia, and hyperbilirubinemia, variable storage conditions of the specimens between the time the sample is obtained from the patient and the time the test is done in the laboratory; and the use of anticoagulants in blood specimens on which procedures intended for serum are to be used (Tasker, 1978; Lum and Gambino, 1974; Landerson et al., 1974; Hicks et al., 1976; Friedel and Mattehhimer, 1970; Laessig et al., 1976; Mia et al., 1976; Meites, 1977; Myers and Pierce, 1972; Street et al., 1968; Thurston, 1978). Variation in laboratory test results due to strain species or breed differences in experimental animals have been discussed in detail by Benirschke, 1978; Benjamin, 1978; Coles, 1980; Melby and Altman, 1976; Benjamin and McKelvie, 1978; Lumsden et al., 1979 and others. Individual variations in laboratory values of humans due to physiological, age, sex and nutritional factors have been well documented (Hayashi, 1979; Henry, 1979; Jones, 1980; Lundberg, 1980; Morrison et al., 1979; Siest, 1973; Speicher and Smith, 1980; Tallone-Lombardi et al., 1978; Munn et al., 1978; Sonnenwirth and Jarett, 1980; Frazer and Peake, 1980; Ash, 1980; Astrup, 1979; Cherian et al., 1979).

Effect of Methodological Variation on Clinical Laboratory Tests

Variation in methodology produce inconsistencies in the reporting of serum chemistries particularly of enzyme units, with the results comparable only when the same method is used. It is apparent that variation in test result can be highly significant due to different analytical methodology used for testing specimens from experimental animals and humans. In this day and age of advanced technology it is difficult for a laboratory scientist to keep up with the diversified test methodology using a vast number of automated and semiautomated procedures and reagent systems for assessing laboratory results of his experimental animals. Each type of automated or semiautomated analyzer uses its own unique method primarily suited to the mechanism of the instrument components. These instruments differ widely not only in performance, but also differ widely in test results (Tables 1 and 2). Another factor which causes variations in different instrument methods or within similar instrument systems is due to variations in reagents used by laboratories from different sources e.g. Technicon reagents, Fisher reagents or Worthington reagents for SMA type equipment. In an effort to correlate test data obtained with different methodology we compared results of normal dog sera analyzed with SMA(18/60, 12/60 or SMA II equipment), SMAC, ACA, KDA and other commonly used systems (Table 3). Based on these results and results from analysis of human sera, conversion factors were calculated. Table 4 represents data obtained during the last four years from analyses of human sera as well as experimental animal sera. Also included in these data are selected values from College of American Pathologist Comprehensive Survey Program. The experimental animals were laboratory conditioned and whenever possible a split sample was used for these analysis by different method. The manual methods used to calculate the factors are listed in Table 5. Although the conversion factors presented in this book are carefully derived, the purpose of using these factors is to provide the medical scientist with an aid in the interpretation of one set of data obtained from a particular type of methodology with another source using different methodology. It should serve as a guide with the current methodology being used in automated analyzer system. The medical scientist or clinician should be cautious in the application of these conversion factors to the diagnosis or therapeutic evaluation of patients or research subjects especially if a change in methodology or reagent is indicated by the instrument manufacturer.

Effect of Drugs on Clinical Laboratory Tests

Another important factor that may cause considerable variation in laboratory test results is medication of all types. Medication has both direct effects, such as elevation of thyroxine level owing to ingestion of thyroid hormone and indirect effects such as elevation of uric acid following the use of the diuretic ethacrynic acid. There are a number of publications on the subject containing some data on the effect of a few drugs on one method or simply list some interferences. Several investigators have published such a list (Elkin and Kabat, 1968; Lubran, 1969; Wirth and Thompson, 1965; Caraway, 1971; Christian, 1970; Sunderman, 1970). Van Peenen and Files (1969) studied the physiological variations produced by 42 drugs on the parameters of an SMA 12/60. They concluded that few drugs interfered significantly at a therapeutic level even when two compounds were administered simultaneously. O'Kell et al. (1971) reported the absence of interference by 13 antibiotics on the methodologies of an SMA12/60 when these compounds were

Table 1. Comparative Serum Chemistry Values of Normal Experimental Animals[1] and Normal Humans[2] Analyzed by SMAC System

CONSTITUENT	UNIT	Canine		Feline		Bovine		Equine (adult TB)		Humans 14 yrs		Humans 15-75yrs	
		x̄	S.D.	x̄	S.D.	x̄	S.D.	x̄	S.D.	x̄	S.D.	x̄	S.D.
Albumin	g/dl	4.6	0.4	3.3	0.3	3.7	0.3	4.2	0.4	4.3	0.3	4.3	0.3
Alkaline Phosphatase, 37°C	I.U./l	73	6.5	33	3.1	47	3.9	68	6.5	158	14	73	7.1
Bilirubin, Total	mg/dl	0.2	0.1	0.1	0.0	0.1	0.08	2.7	0.6	0.4	0.2	0.7	0.2
BUN/Creatinine		20	4.0	15	3.0	20	4.0	10	3.0	24	3.0	17	3.0
Calcium	mg/dl	11	0.7	10	0.4	9.2	0.5	12	0.5	10	0.5	9.4	0.4
Carbon Dioxide	mEq/l	22	1.8	20	1.7	92	2.1	31	2.5	27	2.8	27	2.0
Chloride	mEq/l	113	3.2	123	4.5	99	4.8	104	4.9	102	5.5	91	6.1
Cholesterol	mg/dl	220	13	136	12	190	15	98	10	132	6.9	240	16
Creatinine	mg/dl	0.9	0.1	1.4	0.2	1.0	0.1	1.7	0.2	0.8	0.1	1.1	0.2
CK, 37°C	mu/ml	82	10	45	5.6	32	5.2	25	4.6	77	9.8	66	8.5
Electrolyte Balance		11	2.0	11	2.5	18	3.0	4.5	2.5	11	2.0	12	2.0
Globulin	g/dl	1.7	0.2	3.9	0.4	3.8	0.4	1.7	0.2	3.4	0.4	3.0	0.3
A/G Ratio		2.6	0.2	0.8	0.1	1.0	0.1	3.4	0.2	1.3	0.4	1.4	0.4
Glucose	mg/dl	92	4.8	93	4.6	70	4.4	97	5.2	75	4.1	.05	4.8
Phosphorus	mg/dl	5.0	0.5	4.6	0.5	6.8	0.6	3.6	0.4	5.5	0.6	3.5	0.4

Table 1. (con't) Comparative Serum Chemistry Values of Normal Experimental Animals[1] and Normal Humans[2] Analyzed by SMAC System

CONSTITUENTS	UNIT	Canine		Feline		Bovine		Equine (adult TB)		Humans 14 yrs		Humans 15-75yrs	
		$\bar{X}$	S.D	$\bar{X}$	S.D.	$\bar{X}$	S.D.	$\bar{X}$	S.D.	$\bar{X}$	S.D.	$\bar{X}$	S.D.
LDH, 37°C	I.U./1	170	9.5	143	8.4	933	49	309	23	396	32	130	19
Potassium	mEq/1	4.6	0.1	4.3	0.1	4.9	0.1	3.2	0.1	4.3	0.1	4.3	0.1
Serum Iron	µg/dl	225	3.5	120	3.2	104	3.8	187	4.1	110	4.5	95	3.8
AST, 37°C	U/1	26	3.0	17	2.5	66	7.2	368	28	23	4.1	21	4.5
ALT, 37°C	U/1	79	6.5	58	6.0	39	4.1	34	3.8	23	3.1	23	3.4
Sodium	mEq/1	148	1.2	154	1.3	145	1.4	141	1.6	140	1.5	141	1.4
Total Protein	g/dl	6.3	0.2	7.2	0.2	7.3	0.2	6.1	0.3	7.6	0.3	7.3	0.2
Triglycerides	mg/dl	68	7.4	23	4.4	14	3.2	22	3.4	79	4.5	103	5.5
Urea nitrogen	mg/dl	13	0.4	26	0.9	37	1.6	18	1.1	20	1.0	18	1.1
Uric Acid	mg/dl	0.3	0.2	0.3	0.1	1.3	0.2	0.4	0.1	4.0	0.2	6.5	0.3

Source: [1]Values were established over a three-year period from conditioned laboratory animals, males and females, 148 mongorel dogs, 74 mongorel cats, 32 Holstein cows and 17 throroughbred horses

[2]Data from Southgate Medical Center and Laboratory, Cleveland, Ohio.

Table 2.   Reference Biochemical Values of Human Sera
Analyzed by DuPont Automatic Clinical Analyzer

Acid Phosphatase	0 - 0.8 IU/l
Albumin	3.8-4.8 g/dl
Alkaline Phosphatase	2.5-9.7 IU/l
Amylase	5.0-81 IU/dl Serum AML/Creat
(Urine)	1.3-4.2% Clearance Ratio
AML/Creat. Ratio in Urine	0.07-0.22 IU/mg (Males) 0.07-0.32 IU/ml (Females)
	5.0-38 IU/2 hrs AMY/2 hrs
Ammonia (Venous) Plasma	11-35 μmol/l
BUN	7-22 mg/dl
Calcium	8.7-10.2 mg/dl
Carbon Dioxide (Venous Plasma)	24-32 meq/l
Cholesterol	120-280 mg/dl general
	120-230 mg/dl 0-19 years
	120-240 mg/dl 20-29 years
	140-270 mg/dl 30-39 years
	150-310 mg/dl 40-49 years
	160-330 mg/dl 50-59 years
Chloride	98-108 meq/l
CK	50-180 IU/l (males)
	50-160 IU/l (females)
Creatinine	0.4-1.5 mg/dl
Glucose	70-110 mg/dl
AST	8-33 IU/l
ALT	3-36 IU/l
GGT	5-55 IU/l (female)
	15-85 IU/l (male)
HBDH	170-300 IU/l
Iron	42-135 μg/dl
	280-400 μg/dl (TIBC)

Table 2. (con't) Reference Biochemical Values of Human Sera
Analyzed by DuPont Automatic Clinical Analyzer

Lactic Acid	0.5-2.2 meq/l (Venous BCD pt. at rest)
	0.5-1.6 meq/l (arterial plasma)
LDH	100-190 IU/l
Lipase	4-24 IU/dl
Magnesium	1.8-2.4 mg/dl
Pseudocholinesterase (PCHE)	7-19 IU/ml
Phosphorus	2.5-4.9 mg/dl
Total Bilirubin	less than 1.5 mg/dl
Triglycerides	30-200 mg/dl (general)
	20-150 mg/dl 0-29 years
	20-160 mg/dl 30-39 years
	20-170 mg/dl 40-49 years
	20-200 mg/dl 50-59 years
Total Protein	6.4-8.2 g/dl
Uric Acid	3.8-7.1 mg/dl (male)
	2.6-5.6 mg/dl (female)
	2.0-5.5 mg/dl (children)

Source: DuPont De Nemour Company, Wilmington, Delaware

Table 3. Comparison of Serum Chemistry Values for Dogs Obtained with Manual and Automated Methods

Test	Unit	Manual Method	SMAC	SMA	ACA	Coulter	KDA
Albumin	g/dl	3.1-4.0	3.7-5.5	2.6-3.9	2.7-3.5	2.0-3.9	2.4-3.6
Alkaline Phosphatase, 37°C	I.U./l	7.1-25	16 -130	15 - 59	8.7- 30	6.0-106	19 - 77
Amylase, 37°C	I.U./l	105-295	-	-	63 -177	-	-
Bilirubin, Total	mg/dl	0.1-1.1	0.0-0.3	0.0-0.4	0.0-0.8	0.2-1.4	0.0-0.4
Calcium	mg/dl	9.3- 11	9.8- 12	9.5- 11	9.4- 11	9.0- 12	9.7- 11
Carbon Dioxide	mEq/l	17 - 27	17 - 27	19 - 29	18 - 30	-	-
Chloride	mEq/l	108-116	106-120	108-118	109-117	86 -123	106-116
Cholesterol	mg/dl	138-240	150-290	137-167	134-231	71 -441	116-142
Creatinine	mg/dl	1.1-1.9	0.6-1.1	1.1-1.6	1.0-1.8	0.4-2.1	1.1-1.5
CK, 37°C	I.U./l	30 -210	8.0-155	16 -108	43 -300	21 -112	-
Glucose	mg/dl	72 -120	65 -120	89 -122	77 -128	49 -116	93 -128
Phosphorus	mg/dl	3.5-4.8	2.5-7.5	3.4-5.2	5.3-7.3	2.2-5.4	3.5-5.4
LDH, 37°C	I.U./l	93 -120	75 -220	79 -143	90 -116	8.0- 97	62 -112
Lipase, 37°C	I.U./l	0.0-200 (Somogyi units/dl)			-		
Magnesium	mg/dl	1.5-2.6	-	-	1.8-3.1	-	-
Potassium	mEq/l	4.3-4.7	3.8-5.3	4.0-5.5	-	4.1-6.2	4.0-5.4
Protein, Total	g/dl	6.7-7.8	5.1-7.4	6.1-7.7	6.2-7.2	4.8-7.7	6.0-7.6
Serum, Iron	µg/dl	-	100-350	100-335	125-427	-	89 -299
AST, 37°C	I.U./l	33 - 75	8.0- 46	28 - 45	40 - 90	11 -116	37 - 60
ALT, 37°C	I.U./l	17 - 45	8.0- 65	22 - 47	20 - 56		18 - 39
Sodium	mEq/l	143-152	143-153	142-150	-	134-158	142-149
Triglyceride	mg/dl	70 -126	6.0-130	18 -145	66 -119	-	21 -168
Urea nitrogen	mg/dl	9.8- 22	8.0- 18	12 - 18	9.5- 21	7.0- 25	12 - 18
Uric Acid	mg/dl	0.2-0.7	0.0-0.5	0.2-0.8	0.2-0.8	0.0-0.0	0.2-0.6
A/G Ratio		0.9-1.6	1.6-6.3	-	-	-	-

Table 4. Conversion Factors for Normal Serum Chemistries with Commonly Used Manual and Automated Method

Test and Unit	Manual Method Value Mean	S.D.	Factor (Manual value/x)										
			ACA	SMA	SMAC	Roto Chem	Hy-cel	KDA	Coulter	Centri-fi-Chem	ABA	AGA	BMC
Albumin (g/dl)	4.20	0.10	0.98	0.94	1.05	0.98	1.02	0.98	0.95	0.86	1.06	0.98	–
Alkaline Phosphatase (37°C) (IU/l)	32.8	3.40	0.78	0.63	0.61	–	2.33	0.48	0.55	0.91	1.11	1.47	0.51
Calcium (mg/dl)	10.2	0.30	1.02	1.01	0.98	0.97	0.99	0.99	0.97	1.01	1.01	0.89	–
Carbon Dioxide (mEq/l)	26.0	1.00	0.91	0.96	0.90	–	–	–	–	–	–	–	–
Chloride (mEq/l)	99.5	1.10	0.99	0.99	1.01	–	0.97	1.01	0.96	–	–	0.97	–
(Enz)	169	3.60	1.10	1.07	1.05	1.03	0.72	1.05	0.94	1.02	1.32	3.08	1.02
Cholesterol (mg/dl) (L-B)	200	5.50	–	0.94	0.92	–	–	–	–	–	–	–	–
CK (37°C) (IU/l)	38.9	5.10	0.57	0.91	0.65	–	–	1.02	0.72	1.16	0.60	–	1.08
Creatinine mg/dl	1.41	0.08	1.26	1.77	1.64	1.79	1.76	1.61	1.25	1.22	–	1.10	–
Glucose mg/dl	102	4.90	1.02	0.96	0.99	0.96	0.96	0.97	0.96	1.03	1.10	0.98	0.95
AST (37°C) (IU/l)	53.7	5.40	1.36	3.15	2.80	1.27	1.81	2.18	–	0.90	1.17	1.46	1.58
ALT (37°C) (IU/l)	53.4	5.80	2.22	1.66	1.75	1.69	0.86	2.12	–	1.07	1.56	1.67	2.18
GGT (gamma-GT) (37°C) (IU/l)	21.2	2.40	0.77	1.16	1.54	2.71	–	–	–	–	–	1.60	1.08

Table 4. (con't) Conversion Factors for Normal Serum Chemistries with Commonly Used Manual and Automated Method

Test and Unit	Manual[1] Method Value Mean	S.D.	ACA 2	SMA 3	SMAC 4	Roto Chem 5	Hy-cel 6	KDA 7	Coulter 8	Cen-tri-fi-Chem 9	ABA 10	AGA 11	BMC 12
						Factor (Manual value/x)							
Inorganic Phosphorus (mg/dl)	3.9	0.20	0.94	1.05	1.18	1.04	0.97	1.04	–	1.11	0.85	0.98	–
Iron, total (µ/dl)	142	8.60	1.25	0.93	0.88	–	–	1.21	–	–	–	1.11	–
LDH (37°C) (IU/l)	189	8.70	1.16	1.06	1.19	1.82	0.60	1.33	1.32	1.10	0.64	0.90	1.87
Potassium (mEq/l)	5.0	0.10	–	1.01	0.98	–	1.00	1.01	1.03	–	–	–	–
Sodium (mEq/l)	141	1.20	–	0.98	0.99	–	1.00	0.99	0.98	–	–	–	–
Total Bilirubin (mg/dl)	1.6	0.10	3.01	1.97	3.27	2.33	2.23	2.97	1.18	1.51	1.11	2.38	–
Total Protein (g/dl)	7.1	0.30	1.05	1.04	1.09	1.07	1.01	1.03	1.04	0.98	1.03	0.91	–
Triglycerides (mg/dl)	144	4.80	1.24	1.27	1.52	1.34	0.86	1.23	1.05	1.59	0.96	1.13	1.17
Urea Nitrogen (mg/dl)	18.0	2.30	1.18	1.10	1.17	1.05	1.10	1.13	1.08	0.97	1.08	1.55	1.18
Uric Acid (mg/dl)	5.1	0.20	1.07	0.97	0.99	1.01	0.96	1.09	1.05	1.13	1.05	1.09	1.10

X = Automated Method Value

[1] See Table 5 for Methodology

Table 4. (con't) Conversion Factors for Normal Serum Chemistries with Commonly
Used Manual and Automated Method

[2]DuPont Automatic Clinical Analyzer (See Section V)

[3]Technicon's Sequential Multiple Analyzer. See Table 3 Section V for Methodology
SMA II and SMA 12/60 Data

[4]Technicon's Computer Controlled, High Speed, Biochemical Analyzer. See Table 3
Section V.

[5]American Instrument Co., Inc., Silver Spring, MD

[6]Hycel Super 17, Hycel, Inc. Houston, TX

[7]American Monitor Corporation, Indianapolis, IN

[8]Coulter Electronics, Inc., Hialeah, FL

[9]Union Carbide Corporation Diagnostics, Rye, NY

[10]Abbott Laboratories Diagnostic Division, Chicago, IL

[11]Autochemist, Metpath, Inc., Teterboro, NJ

[12]Bio Dynamics/bmc, Indianapolis, IN

Table 5. Manual Method Used For the Calculation of Conversion Factors

1. Albumin   BCG Dye Binding : Sclavo Diagnostic Reagents
2. Alkaline
   Phosphatase   PNP Worthington Reagent (AMP buffer)
                    (37°C, 405 nm)
3. Calcium   Fluorometric titration - Corning Reagents
4. Carbon Dioxide   Van Slyke - Manometric Method Harleco CO_2
                    Reagents and Apparatus
5. Chloride   Coulometry: Fiske Chlor-0-Counter
                    Instrument and Reagents
6. Cholesterol   a) Modified Abell-Kendall (Liebermann-
                    Burchard) CDC Methodology, 1976
                    b) Cholesterol Enzymatic Method: Pierce
                    Chemical Reagent
7. CK   Worthington Reagents (37°C, 340 nm)
                    (Røsalki, 1972)
8. Creatinine   Alkaline Picrate (Jaffe Reaction)
9. Glucose   Hexokinase Method (with protein-free
                    filtrate) (Federal Register 39,
                    No. 126, 24136, 1974)
10. AST (GOT)   Worthington Reagents (37°C, 340 nm)
                    (Henry et al, 1960)
11. ALT (GPT)   Worthington Reagents (37°C, 340 nm)
                    (Henry et al, 1960)
12. Gamma GTP   Worthington Reagents (37°C, 405 nm)
                    (Szasz, 1969)
13. Iron, total   Colorimetric (Magnesium Bathoplunantheroline
                    Sulfonate Reagent) AMC Reagents
14. LDH   Oxidation lactate by NAD (Wacker et al,
                    1956)
15. Phosphorus,
    Inorganic   Fiske and Subbarow (Molybdate Reagent and
                    Semidine)
16. Bilirubin,
    total   Jendrassik-Grof (Doumas, 1973)
17. Total Protein   Biuret-EDTA Reagent (SCLAVO Diagnostic
                    Reagents)
18. Triglycerides   Enzymatic Worthington Reagents and Kit
19. Urea Nitrogen   Diacetyl Reaction: Pfizer Diagnostics
                    "BUN-TEL" Reagents
20. Uric Acid   Direct Procedure using Fe^{+++}/TPTZ
                    Data Medical Associates Reagents
21. Sodium and
22. Potassium   Flame Photometry; Corning Flame Photometer

Table 6   Effects of Drugs on Clinical Chemistry Tests

Blood Test : Amylase

Increased

   Cholinergics (P); Ethanol (P); Opiates and Analogues (P); Steroids (P).

Decreased

   Citrate (M); Oxalate (M).

Blood Test : Bilirubin

Increased

   Anesthetics: Chloroform (P); Halothane (P).

   Anti-infectious Disease Agents: Cycloserine (P); Erythromycin Estolate (P); Griseofulvin (P); Isoniazid (P); Rifampin (P); Sulfonamides (P).

   Anti-inflammatory Agents: Acetominophen (P); Allopurinal (P); Colchicine (P); Ibufenac (P); Indomethacin (P); Phenylbutazone (P); Salicylates (P).

   Anti-neoplastic Agents: Methotrexate (P); Mithramycin (P); Puromycin (P).

   Drugs Used in Cardiovascular Disease: Methyldopa (P); Nicotinic Acid (P).

   Hormonal Derivatives: Anabolic/Androgenic Steroids - C-17 Alkylates Derivatives of Testosterone (P); Oral Contraceptives (P); Oral Hypoglycemics: Acetohexamide (P); Chlor-propamide (P); Tolazemide (P); Tolbutamide (P).

   Psychotropic and Anticonvulsant Agents: Chlordiazepoxide (P); Chlorpromazine (P); Iproniazid (P); Phenacemide (P); Phenytoin (P).

   Miscellaenous: Dextran (M); Gall Bladder Dyes (P); Novobiacin (P).

Table 6 (con't)

Blood Test : Bilirubin

Decreased

Ascorbic Acid (M); Barbiturates (P); Caffeine (M); Steroids (P); Theophylline (M).

Blood Test : Bicarbonate (Total CO2)

Increased

Aspirin (P); Prolonged Diuretic Therapy (P).

Decreased

Aspirin (P); Dimercaprol (P); Methanol (P); Triamterene (P).

Blood Test : Calcium

Increased

Anabolic/Androgenic Steroids (P); Calciferol (P); Dihydrotachysterol (P); Estrogens-Progestins (P); Thiazide Diuretics (P).

Decreased

Acetazolamide (P); Citrates (M); EDTA (M); Furosemide (P); Mithramycin (P); Phosphates (P); Steroids (P); Sulfates (M).

Blood Test : Chloride

Increased

Acetazolamide (P); Bromides (M); Chlorides (P); Oxyphenabutazone (P); Phenylbutazone (P).

327

Table 6   (con't)

Blood Test : Chloride

<u>Decreased</u>

   Diuretics: Ethacrynic Acid (P), Furosamide (P); Steroids (P).

Blood Test : Cholesterol, Total

<u>Increased</u>

   ACTH/Corticosteroids (P); Anabolic/Androgenic Hormones (P); Bile Salts (P); Chlorpromazine (P);
   Heparin (P).

<u>Decreased</u>

   Cholestyramine (P); Estrogens (P); Thyroid Hormones (P).

Blood Test : Creatinine

<u>Increased</u>

   Amphotericin B (P); Ascorbic Acid (M); Cephalosporins (M); Glucose (M); Keto Acids (M);
   Methyldopa (M); Triamterene (P). See Also Urea Nitrogen.

<u>Decreased</u>

   Anabolic/Androgenic Steroids (P).

Blood Test : Glucose

<u>Increased</u>

   Acetaminophen (M); ACTH/Corticosteroids (P); Aminosalicylic Acid (M); Ascorbic Acid (P);

Table 6   (con't)

Blood Test : Glucose

Increased

Dextran (M); Epinephrine (P); Ethacrynic Acid (M); Furosemide (P); Hydralazine (M); Iso-proterenol (M); Levodopa (M); Mercaptopurine (M); Methimazole (M); Methyldopa (M); Nalidixic Acid (M); Oxazepam (M); Phenytoin (P); Propylthiouracil (M); Thiazides (P).

Decreased

Ascorbic Acid (M); Methyldopa (M); Propranolol (P); Tetracycline (M).

Blood Test : Lactate Dehydrogenase

Increased

Anti-folate Compounds (P).

Decreased

Clofibrate (P); Oxalate (M); Theophylline (M).

Blood Test : Lipase

Increased

Cholinergics (P); Ethanol (P); Opiates and Analogues (P).

Decreased

Calcium Salts (M).

Table 6 (con't)

Blood Test : Phosphatase, Acid

<u>Increased</u>

Androgens in Female (P).

<u>Decreased</u>

Ethanol (M); Fluorides (M); Heparin (M).

Blood Test : Phosphatase, Alkaline

<u>Increased</u>

See Bilirubin - Drugs with (P) Effect.

<u>Decreased</u>

Fluorides (M); Oxalates (M); Theophylline (M).

Blood Test : Phosphorus, Inorganic

<u>Increased</u>

Calciferol (P); Methicillin (P); Tetracycline (P).

<u>Decreased</u>

Aluminum Hydroxide (P); Glucose Infusion (P); Insulin (P); Mithramycin (P).

Blood Test : Potassium

Table 6   (con't)

Blood Test : Potassium

Increased

    Heparin (P); Potassium Salts (P); Spironolactone (P).

Decreased

    ACTH/Corticosteroids (P); Amphotericin (P); Glucose Infusion (P); Insulin (P); Tetracyclines - Degraded (P).

Blood Test : Proteins, Total

Increased

    ACTH/Corticosteroids (P); Anabolic/Androgenic Steroids (P); Dextran (M).

Decreased

    Aspirin (M); Oral Contraceptives (P); Rifampin (P).

Blood Test : Transaminase, Aspartate (AST)

Increased

    See Bilirubin - Drugs with (P) Effect; Ascorbic Acid (M); Erythromycin (M); Isoniazid (M); Nitrofurantoin (M).

Decreased

    Fluorides (M); Progesterone (P); Trifluoperazine (P).

331

Table 6   (con't)

Blood Test : Sodium

Increased

Androgens (P); Calcium (M); Mannitol (P); Methyldopa (P); Phenylbutazone (P); Rauwolfia Preparations (P); Steroids (P).

Decreased

Ammonium Chloride (P); Diuretics (P); Heparin (P).

Blood Test : Uric Acid

Increased

ACTH/Corticosteroids (P); Ascorbic Acid (M); Busulfan (P); Glucose (M); Methyldopa (M); Purine Analogue Antimetabolites (P); Pyrazinamide (P); Quinethazone (P); Theophylline (M); Thiazides (P); Vincristine Sulfate (P).

Decreased

Acetylsalicylic Acid (P); Allopurinal (P); Chlorpromazine (P); Chlorprothixene (P); Oxyphenbutazone (P); Phenylbutazone (P).

Blood Test : Urea Nitrogen (BUN)

Increased

Acetaminophen (P); Alkaline Antacids (P); Ammonium Salts (M); Anabolic/Androgenic Steroids (P); Chloralhydrate (M); Fluorides (M); Furosamide (P); Gentamicin (P); Guanethidine (M); Neomycin (P).

332

Table 6 (con't)

Blood Test : Urea Nitrogen (BUN)

Decreased

Chloramphenicol (M); Phenothiazine (P); Streptomycin (M).

(M) = The drug exhibits primarily analytical interference. This may apply only to certain methods.

(P) = The drug effect is primarily pharmacologic, toxic, a hypersensitivity reaction, or results from the induced disease process.

Major References

Henry JB: Clinical Diagnosis and Management by Laboratory Methods, Sixteenth Edition, Philadelphia, W. B. Saunders Co., 1979.

Wright LA, Foster MG: Effect of some commonly prescribed drugs on certain chemistry tests. Clin Biochem 6:249-252, 1980.

Young DS, Pestaner LC, Gibberman V: Effects of drugs in clinical laboratory tests. Clin Chem 21:1D-432D, 1975.

Table 7   Effects of Pathological Conditions on Blood Chemistry Tests

Blood Test : Acid Phosphatase

Increased

Viral Hepatitis; Malignant Neoplasm of Bone, Breast, Prostate; Secondary
Malignant Neoplasm of Liver, Bone; Reticulum Cell Sarcoma; Multiple Myeloma;
Myelocytic Leukemia (Chronic); Myelofibrosis; Diabetes Mellitus; Hyperpara-
thyroidism; Gaucher's Disease; Niemann-Pick Disease; Hyperphosphatase;
Amyloidosis; Sickle Cell Anemias; Idiopathic Thrombocytopenic Purpura;
Friedreich's Ataxia; Acute Myocardial Infarction; Arterial Embolism and
Thrombosis; Mesenteric Artery Embolism; Pulmonary Embolism and Infarction;
Phlebitis and Thrombophlebitis; Influenza; Acute and Subacute Necrosis of
Liver; Laennec's or Alcoholic Cirrhosis; Extrahepatic Biliary Obstruction;
Acute Renal Failure; Chronic Renal Failure; Benign Prostatic Hypertrophy;
Prostatic Infarction; Osteitis Deformans; Osteogenesis Imperfecta; Osteo-
petrosis.

Decreased

Down's Syndrome; Malignant Neoplasm of Prostate; Niemann-Pick Disease;
Osteitis Deformans; Osteogenesis Imperfecta.

Blood Test : Alanine Aminotransferase (GPT)

Increased

Amebiasis; Disseminated Tuberculosis; Brucellosis; Tetanus; Septicemia;
Yellow Fever; Viral Hepatitis; Psittacosis; Infectious Mononucleosis; Cyto-
megalic Inclusion Disease; Malaria; Lymphogranuloma Venereum; Leptospirosis;
Actinomycosis; Histoplasmosis; Schistosomiasis; Hydatidosis; Trichinosis;

Table 7 (con't)

Blood Test : Alanine Aminotransferase (GPT)

Increased

Sarcoidosis; Malignant Neoplasm of Liver, Intrahepatic Bile Ducts, Pancreas; Secondary Malignant Neoplasm of Respiratory System; Secondary Malignant Neoplasm of Liver; Reticulum Cell Sarcoma; Lymphosarcoma; Hodgkin's Disease; Myelocytic Leukemia; Hyperthyroidism; Diabetic Acidosis; Forbes Disease; Anderson's Disease; Galactosemia; Hereditary Fructose Intolerance; A-Beta-lipoproteinemia; Gaucher's Disease; Niemann-Pick Disease; Porphyria Cutanea Tarda; Hepatolenticular Degeneration; Dubin-Johnson Syndrome; Hereditary (Congenital) Spherocytosis; Acquired Hemolytic Anemias (Autoimmune); Delirium Tremens; Alcoholism; Drug Dependence (Opium and Derivatives); Familial Progressive Spinal Muscular Atrophy; Progressive Muscular Dystrophy; Acute Myocardial Infarction; Acute Pericarditis; Congestive Heart Failure; Brain Infarction; Gangrene; Pulmonary Embolism and Infarction; Regional Enteritis or Ileitis; Ulcerative Colitis; Acute and Subacute Necrosis of Liver; Cirrhosis of Liver; Laennec's or Alcoholic Cirrhosis; Biliary Cirrhosis; Liver Abscess (Pyogenic); Chronic Active Hepatitis; Toxic Hepatitis; Reye's Syndrome; Hepatitis Failure; Acute Cholecystitis; Cholangitis; Extrahepatic Biliary Obstruction; Acute Pancreatitis; Chronic Pancreatitis; Chronic Renal Failure; Renal Infarction; Pre-Eclampsia; Eclampsia; Juvenile Rheumatoid Arthritis; Systemic Lupus Erythematosus; Dermatomyositis/Polymyositis; Shock; Hypothermia; Crus Injury (Trauma); Burn; Toxic Effects of Non-medicinal Metals, venom; Effects of X-Ray Irradiation; Heat Stroke.

Decreased

Exercise

335

Table 7 (con't)

Blood Test : Albumin

<u>Increased</u>

Leishmaniasis; Sarcoidosis; Hypothyroidism; Multiple Sclerosis; Laennec's or Alcoholic Cirrhosis; Dehydration; Heat Stroke.

<u>Decreased</u>

Gastroenteritis and Colitis; Pulmonary Tuberculosis; Leprosy; Septicemia; Whipple's Disease; Chickenpox; Herpes Zoster; Viral Hepatitis; Rocky Mountain Spotted Fever; Malaria; Leishmaniasis; Trypanosomiasis; Lymphogranuloma Venereum; Leptospirosis; Histoplasmosis; Schistosomiasis; Trichinosis; Ancylostomiasis (Hookworm Infestation); Strongyloidiasis; Sarcoidosis; Malignant Neoplasm of Esophagus, Stomach, Large Intestine, Liver, Pancreas, Bronchus and Lung, Breast, Corpus Uteri, Prostate, Bladder, Kidney; Secondary Malignant Neoplasm of Respiratory System; Secondary Malignant Neoplasm (Disseminated); Reticulum Cell Sarcoma; Lymphosarcoma; Hodgkin's Disease; Multiple Myeloma; Lymphocytic Leukemia; Myelocytic Leukemia; Monocytic Leukemia; Benign Neoplasm of Stomach; Hyperthyroidism; Diabetes Mellitus; Diabetic Acidosis; Zollinger-Ellison Syndrome; Carcinoid Syndrome; Vitamin C Deficiency; Protein Malnutrition; Celiac Sprue Disease; Malabsorption; Protein-Losing Enteropathy; Deficiency State (Unspecified); Lactosuria; Type II A and II B Hyperlipoproteinemia; Cystic Fibrosis; Hepatolenticular Degeneration; Analbuminemia; Waldenstrom's Macroglobulinemia; Heavy Chain Disease (Alpha and Gamma); Amyloidosis; Pernicious Anemia; Allergic Purpura; Multiple Sclerosis; Rheumatic Fever; Essential Benign Hypertension; Acute Myocardial Infarction; Bacterial Endocarditis; Constrictive Pericarditis; Chronic Renal Failure; Pregnancy; Pre-Eclampsia; Eclampsia; Pemphigus; Rheumatoid Arthritis; Systemic Lupus Erythematosus; Sjogren's Syndrome; Osteomyelitis; Hemolytic

Table 7 (con't)

Blood Test : Albumin

Decreased

Disease of Newborn (Erythroblastosis Fetalis); Burn; Stress.

Blood Test : Alkaline Phosphatase

Increased

Typhoid Fever; Amebiasis; Pulmonary Tuberculosis; Disseminated Tuberculosis; Brucellosis; Septicemia; Whipple's Disease; Herpes Zoster; Viral Hepatitis; Psittacosis; Infectious Mononucleosis; Cytomegalic Inclusion Disease; Malaria; Leptospirosis; Actinomycosis; Histoplasmosis; Blastomycosis; Schistosomiasis; Hydatidosis; Sarcoidosis; Malignant Neoplasm of Esophagus, Stomach, Large Intestine, Liver, Intrahepatic Bile Ducts, Pancreas, Bronchus and Lung, Bone, Breast, Ovary, Prostate; Secondary Malignant Neoplasm of Respiratory System, Liver, Bone; Reticulum Cell Sarcoma; Lymphosarcoma, Hodgkin's Disease; Multiple Myeloma; Lymphocytic Leukemia; Myelocytic Leukemia; Monocytic Leukemia; Myelofibrosis; Hyperthyroidism; Subacute and Hashimoto's Thyroiditis; Diabetes Mellitus; Hyperparathyroidism; Acromegaly; Adrenal Cortical Hyperfunction (Glucocorticoid Excess); Vitamin D Deficiency Rickets; Celiac Sprue Disease; Malabsorption, Cause Unspecified; Cytinosis; Gaucher's Disease; Niemann-Pick Disease; Cystic Fibrosis; Hemochromatosis; Hepatolenticular Degeneration; Osteomalacia; Hyperphosphatasia; Vitamin-D Resistant Rickets; Dubin-Johnson Syndrome; Proximal Renal Tubular Acidosis; Amyloidosis; Sickle Cell Anemias; Alcoholism; Friedreich's Ataxia; Acute Myocardial Infarction; Congestive Heart Failure; Mesenteric Artery Embolism; Gangrene; Cranial Arteritis and Related Conditions; Pulmonary Embolism and

337

Table 7 (con't)

Blood Test : Alkaline Phosphatase

<u>Increased</u>

Infarction; Phlebitis and Thrombophlebitis; Viral Pneumonia; Fistual of Stomach and Duodenum; Intestinal Obstruction; Regional Enteritis or Ileitis; Ulcerative Colitis; Acute and Subacute Necrosis of Liver; Cirrhosis of Liver; Laennec's or Alcoholic Cirrhosis; Biliary Cirrhosis; Liver Abscess (Pyogenic); Chronic Active Hepatitis; Toxic Hepatitis; Hepatic Failure; Acute Cholecystitis; Cholangitis Extrahepatic Biliary Obstruction; Pancreatitis; Pancreatic Cyst and Pseudocyst; Neprotic Syndrome; Chronic Renal Failure; Acute Pyelonephritis; Renal Dwarfism; Pregnancy; Pre-Eclampsia; Rheumatoid Arthritis; Rheumatoid (Ankylosing) Spondylitis; Osteomyelitis; Fracture of Bone; Osteoporosis; Osteitis Deformans; Osteogenesis Imperfecta; Polyostotic Fibrous Dysplasia; Toxic Effects of Non-Medicinal Metals; Postcholecystectomy Syndrome.

<u>Decreased</u>

Whipple's Disease; Hypothyroidism; Zollinger-Ellison Syndrome; Vitamin C Deficiency; Protein Malnutrition; Lactosuria; Hypophosphatasia; Hypervitaminosis; Pernicious Anemia; Chondrodystrophy; Milk-Alkali Syndrome.

Blood Test : Amylase

<u>Increased</u>

Viral Hepatitis; Mumps; Infectious Mononucleosis; Malignant Neoplasm of Pancreas, Bronchus and Lung, Ovary, Prostate; Diabetes Mellitus; Diabetic Acidosis; Malabsorption, Cause Unspecified; Type I and V Hyperlipoproteinemia;

Table 7   (con't)

Blood Test : Amylase

Increased

Acute Alcoholic Intoxication; Alcoholism; Drug Dependence (Opium and Derivatives); Encephalomyelitis; Dissecting Aortic Aneurysm; Mesenteric Artery Embolism; Diseases of the Salivary Gland; Peptic Ulcer, site unspecified; Acute Appendicitis; Intestinal Obstruction; Peritonitis; Cirrhosis of Liver; Acute Cholecystitis; Extrahepatic Biliary Obstruction; Spasm of Sphincter of Oddi; Pancreatitis; Pancreatic Cyst and Pseudocyst; Renal Failure; Benign Prostatic Hypertrophy; Prostatitis; Pregnancy; Shock; Hypothermia; Crush Injury (Trauma); Burn; Effects of X-Ray Irradiation; Postcholecystectomy Syndrome.

Decreased

Malignant Neoplasm of Pancreas; Hyperthyroidism; Diabetes Mellitus; Protein Malnutrition; Alcoholism; Congestive Heart Failure; Acute and Subacute Necrosis of Liver; Liver Abscess (Pyogenic); Toxic Hepatitis; Acute Cholecystitis; Acute Pancreatitis; Eclampsia; Burn.

Blood Test : Aspartate Aminotransferase (GOT)

Increased

Typhoid Fever; Amebiasis; Pulmonary Tuberculosis; Disseminated Tuberculosis; Brucellosis; Tetanus; Septicemia; Acute Poliomyelitis; Herpes Zoster; Yellow Fever; Viral Hepatitis; Psittacosis; Infectious Mononucleosis; Cytomegalic Inclusion Disease; Rocky Mountain Spotted Fever; Malaria; Lymphogranuloma Venereum; Leptospirosis; Histoplasmosis; Schistosomiasis; Hydatidosis;

Table 7 (con't)

Blood Test : Aspartate Aminotransferase (GOT)

Increased

Trichinosis; Sarcoidosis; Malignant Neoplasm of Esophagus, Stomach, Liver, Intrahepatic Bile Ducts, Pancreas, Bronchus and Lung, Bone, Breast, Prostate, Brain; Medullary Carcinoma of Thyroid; Secondary Malignant Neoplasm of Respiratory System, Liver; Reticulum Cell Sarcoma; Lymphosarcoma; Hodgkin's Disease; Multiple Myeloma; Lymphocytic Leukemia; Effects of Electric Current.

Decreased

Malaria; Pregnancy.

Blood Test : Bilirubin

Increased

Disseminated Tuberculosis; Chickenpox; Herpes Simplex; Measles; Yellow Fever; Viral Hepatitis; Infectious Mononucleosis; Cytomegalic Inclusion Disease; Malaria; Leishmaniasis; Relapsing Fever; Bartonellosis; Lymphogranuloma Venereum; Leptospirosis; Histoplasmosis; Aspergillosis; Schistosomiasis; Hydatidosis; Toxoplasmosis; Malignant Neoplasm of Liver, Intrahepatic Bile Ducts, Pancreas; Secondary Malignant Neoplasm of Liver; Reticulum Cell Sarcoma; Lymphosarcoma; Hodgkin's Disease; Histiocytosis X; Lymphocytic Leukemia; Erythroleukemia; Polycythemia Vera; Hyperthyroidism; Carcinoid Syndrome; Hereditary Fructose Intolerance; A-Betalipoproteinemia; Hemochroma-tosis; Hepatolenticular Degeneration; Dubin-Johnson Syndrome; Gilbert's Disease; Rotor's Syndrome; Crigler-Najjar Syndrome; Amyloidosis; Pernicious Anemia; Folic Acid Deficiency Anemia; Hereditary (Congenital) Spherocytosis;

Table 7 (con't)

Blood Test : Bilirubin

Increased

Hereditary Elliptocytosis; Anemias Due to Disorders of Glutathione Metabolism; Thalassemia Major; Hemoglobin C Disease; Sickle Cell Anemias; Acquired Hemolytic Anemias (Autoimmune); Paraxysmal Nocturnal Hemoglobinuria; Polycythemia, Secondary; Alcoholism; Congestive Heart Failure; Thrombotic Thrombocytopenic Purpura; Pulmonary Embolism and Infarction; Influenza; Bacterial Pneumonia; Mycoplasma Pneumoniae; Acute and Subacute Necrosis of Liver; Cirrhosis of Liver; Laennec's or Alcoholic Cirrhosis; Biliary Cirrhosis; Liver Abscess (Pyogenic); Chronic Active Hepatitis; Toxic Hepatitis; Hepatic Encephalopathy; Reye's Syndrome; Hepatic Failure; Acute Cholecystitis; Cholangitis; Extrahepatic Biliary Obstruction; Pancreatitis; Pancreatic Cyst and Psuedocyst; Eclampsia; Hemolytic Disease of Newborn (Erythroblastosis Fetalis); Jaundice Due to Hepatocellular Damage (newborn); Respiratory Distress Syndrome; Toxic Effects of Lead and Its Compounds (including Fumes); Toxic Effects of Non-Medicinal Metals; Effects of X-Ray Irradiation; Heat Stroke; Post-Cholecystectomy Syndrome.

Decreased

Iron Deficiency Anemia.

Blood Test : Calcium

Increased

Sarcoidosis; Malignant Neoplasm of Eosphagus, Liver, Pancreas, Bronchus and Lung, Bone, Breast, Bladder, Kidney; Secondary Malignant Neoplasm of Respira-

341

Table 7 (con't)

Blood Test : Calcium

Increased

tory System, Bone; Reticulum Cell Sarcoma; Lymphosarcoma; Hodgkin's Disease;
Multiple Myeloma; Lymphocytic Leukemia; Myelocytic Leukemia; Polycythemia
Vera; Pheochromocytoma; Hyperthyroidism; Hypothyroidism; Hyperparathyroidism;
Acromegaly; Anterior Pituitary Hypofunction; Adrenal Cortical Hypofunction;
Ovarian Hyperfunction; Adrenal Cortical Hyperfunction (Glucocorticoid
Excess); Type 11A and 11 B Hyperlipoproteinemia; Hypophosphatasia; Proximal
Renal Tubular Acidosis; Hypervitaminosis; Acute Alcoholic Intoxication;
Alcoholism; Epilepsy; Berylliosis; Peptic Ulcer, Site Unspecified; Acute
Pancreatitis; Renal Dwarfism; Rheumatoid Arthritis; Osteoporosis; Osteitis
Deformans; Milk-Alkali Syndrome; Toxic Effects of Non-Medicinal Metals.

Decreased

Leprosy; Whipple's Disease; LYmphocytic Leukemia (Acute): Diabetes Mellitus;
Pseudohypoparathyroidism; Hypoparathyroidism; Anterior Pituitary Hypofunction;
Ovarian Hypofunction; Zollinger-Ellison Syndrome; Vitamin-D Deficiency
Rickets; Protein Malnutrition; Celiac Sprue Disease; Malabsorption, Cause
Unspecified; Protein-Losing Enteropathy; Cystinosis; Lactosuria; Cystic
Fibrosis; Osteomalacia; Proximal Renal Tubular Acidosis; Distal Renal
Tubular Acidosis; Analbuminemia; Heavy Chain Disease (Alpha and Gamma);
Acute Alcoholic Intoxication; Alcoholism; Epilepsy; Regional Enteritis or
Ileitis; Ulcerative Colitis; Cirrhosis of Liver; Laennec's or Alcoholic
Cirrhosis; Acute Pancreatitis; Chronic Pancreatitis; Glomerulonephritis
(Focal and Rapidly Progressive); Neprotic Syndrome; Renal Failure; Kidney
Calculus; Renal Dwarfism; Urethritis; Pregnancy; Pemphigus; Osteoporosis;
Medullary Cystic Disease; Osteopetrosis; Hyperventilation Syndrome; Hypo-

Table 7 (con't)

Blood Test : Calcium

Decreased

magnesia; Metabolic Acidosis; Metabolic Alkalosis; Heat Stroke; Fat Embolism.

Blood Test : Chloride

Increased

Bacillary Dysentery; Cholera; Gastroenteritis and Colitis; Malignant Neoplasm of Stomach; Multiple Myeloma; Hyperparathyroidism; Diabetes Insipidus; Adrenal Cortical Hyperfunction (Mineralocorticoid Excess); Adrenal Cortical Hypofunction; Ovarian Hypofunction; Vitamin D Deficiency Rickets; Cystinosis; Galactosemia; Cystic Fibrosis; Proximal Renal Tubular Acidosis; Renovascular Hypertension; Fistula of Stomach and Duodenum; Intestinal Obstruction; Glomerulonephritis (Membranoproliferative); Acute Renal Failure; Chronic Pyelonephritis; Urethritis; Dehydration; Heat Stroke.

Decreased

Bacillary Dysentery; Gastroenteritis and Colitis; Tuberculosis Meningitis; Sarcoidosis; Diabetes Mellitus; Diabetes Acidosis; Adrenal Cortical Hyperfunction; Adrenal Cortical Hypofunction; Ovarian Hyperfunction; Adrenal Cortical Hyperfunction (Glucocorticoid Excess); Celiac Sprue Disease; Acute Intermittent Porphyria; Hepatolenticular Degeneration; Meningitis (Bacterial); Congestive Heart Failure; Gangrene; Pulmonary Emphysema; Chronic Obstructive Lung Disease; Peptic Ulcer, Site Unspecified; Pyloric Stenosis; Fistula of Stomach and Duodenum; Intestinal Obstruction; Ulcerative Colitis; Chronic Renal Failure; Eclampsia; Pemphigus; Diarrhea; Hyperhidrosis; Metabolis Alkalosis.

Table 7 (con't)

Blood Test : Cholesterol

Increased

Ancylostomiasis (Hookworm Infestation); Malignant Neoplasm of Intrahepatic Bile Ducts, Pancreas, Prostate; Multiple Myeloma; Myelocytic Leukemia; Hypothyroidism; Subacute Thyroiditis; Diabetes Mellitus; Anterior Pituitary Hypofunction; Ovarian Hypofunction; Adrenal Cortical Hyperfunction (Gluco-corticoid Excess); Von Gierke's Disease; Mcardle's Disease; Forbes Disease; Anderson's Disease; Type 11A, 11B and 1V Hyperlipoproteinemia; Type 1, 111, and V Hyperlipoproteinemia; Cystic Fibrosis; Gout; Analbuminemia; Amyloidosis; Friedreich's Ataxia; Malignant Hypertension; Essential Benign Hypertension; Renovascular Hypertension; Acute Myocardial Infarction; Chronic Ischemic Heart Disease; Arteriosclerosis; Polyarteritis-Nodosa; Acute and Subacute Necrosis of Liver; Laennec's or Alcoholic Cirrhosis; Biliary Cirrhosis; Toxic Hepatitis; Extrahepatic Biliary Obstruction; Chronic Pancreatitis; Acute Post-streptococcal Glomerulonephritis; Glomerulonephritis (minimal change); Glomerulonephritis (membranous); Nephrotic Syndrome; Chronic Renal Failure; Benign Prostatic Hypertrophy; Pregnancy; Systemic Lupus Erythematosus; Klinefelter's Syndrome; Toxic Effects of Non-Medicinal Metals; Stress.

Decreased

Pulmonary Tuberculosis; Leprosy; Septicemia; Whipple's Disease; Herpes Zoster; Yellow Fever; Viral Hepatitis; Infectious Mononucleosis; Leptospiro-sis; Malignant Neoplasm of Liver, Pancreas, Prostate; Secondary Malignant Neoplasm of Liver; Reticulum Cell Sarcoma; Hodgkin's Disease; Multiple Myeloma; Lymphocytic Leukemia; Myelocytic Leukemia; Hyperthyroidism; Hyper-parathyroidism; Ovarian Hyperfunction; Carcinoid Syndrome; Protein Malnutri-tion; Celiac Sprue Disease; Malabsorption, Cause Unspecified; Tangier

Table 7 (con't)

Blood Test : Cholesterol

Decreased

Disease; A-Betalipoproteinemia; Iron Deficiency Anemia; Folic Acid Deficiency Anemia; Hereditary (Congenital) Spherocytosis; Anemias Due to Disorders of Glutathione Metabolism; Sickle Cell Anemias; Hemophilia; Mental Retardation; Drug Dependence (Opium and Derivatives); Congestive Heart Failure; Brain Infarction; Mesenteric Artery Embolism; Viral Pneumonia; Bacterial Pneumonia; Chronic Obstructive Lung Disease; Intestinal Obstruction; Ulcerative Colitis; Acute and Subacute Necrosis of Liver; Cirrhosis of Liver; Chronic Active Hepatitis; Toxic Hepatitis; Hepatic Failure; Acute Pancreatitis; Rheumatoid Arthritis; Osteomyelitis.

Blood Test : Creatine Kinase

Increased

Typhoid Fever; Tetanus; Acute Poliomyelitis; Leptospirosis; Hydatidosis; Trichinosis; Malignant Neoplasm of Brain; Hypothyroidism; Subacute Thyroiditis; Diabetic Acidosis; Mcardle's Disease; Myoglobinuria; Hemophilia; Christmas Disease; Delirium Tremens; Acute Alcoholic Intoxication; Manic Depressive Disorder; Paranoid States and Other Psychoses; Alcoholism; Drug Dependence (Opium and Derivatives); Meningitis (Bacterial); Encephalomyelitis; Familial Progressive Spinal Muscular Atrophy; Progressive Muscular Dystrophy; Myotonia Atrophica; Myotonia Congenita; Epilepsy; Acute Myocardial Infarction; Acute Myocarditis; Cardiomyopathies; Cerebral Thrombosis; Brain Infarction; Cerebral Embolism; Arterial Embolism and Thrombosis; Pulmonary Embolism and Infarction; Influenza; Asthma (Intrinsic); Pulmonary Edema; Acute Pancreatitis; Pregnancy; Progressive Systemic Sclerosis; Systemic Lupus Erythematosus;

Table 7 (con't)

Blood Test : Creatine Kinase

Increased

Dermatomyositis/Polymyositis; Convulsions; Hypokalemia; Hypothermia; Crush
Injury (Trauma); Toxic Effects of Carbon Monoxide, Venom; Heat Stroke;
Effects of Electric Current.

Decreased

Hereditary (Congenital) Spherocytosis; Pregnancy.

Blood Test : Creatinine

Increased

Septicemia; Herpes Zoster; Psittacosis; Epidemic Typhus; Malaria; Lepto-
spirosis; Sarcoidosis; Malignant Neoplasm of Large Intestine, Corpus Uteri,
Prostate, Testis, Bladder; Reticulum Cell Sarcoma; Lymphosarcoma; Hodgkin's
Disease; Multiple Myeloma; Lymphocytic Leukemia; Myelocytic Leukemia; Mono-
cytic Leukemia; Hyperthyroidism; Diabetes Mellitus; Hyperparathyroidism;
Acromegaly; Adrenal Cortical Hypofunction; Cystinosis; Hepatolenticular
Degeneration; Gout; Amyloidosis; Acquired Hemolytic Anemias (Autoimmune);
Allergic Purpura; Meningitis (Bacterial); Myotonia Atrophica; Amyotrophic
Lateral Sclerosis; Essential Benign Hypertension; Acute Myocardial Infarc-
tion; Chronic Ischemic Heart Disease; Angina Pectoris; Bacterial Endocarditis;
Congestive Heart Failure; Arteriosclerosis; Arterial Embolism and Thrombosis;
Polyarteritis-Nodosa; Thrombotic Thrombocytopenic Purpura; Intestinal Obstruc-
tion; Hepatic Failure; Acute Pancreatitis; Acute Poststreptococcal Glomerulo-
nephritis; Glomerulonephritis (Rapidly Progressive and Membranoproliferative);

Table 7 (con't)

Blood Test : Creatinine

Increased

Nephrotic Syndrome; Acute Renal Failure; Chronic Renal Failure; Chronic Pyelonephritis; Acute Pyelonephritis; Hydronephrosis; Urethritis; Benign Prostatic Hypertrophy; Pre-Eclampsia; Progressive Systemic Sclerosis; Systemic Lupus Erythematosus; Shock; Vomiting; Diarrhea; Hyperhidrosis; Dehydration; Toxic Effects of Non-Medicinal Metals; Heat Stroke; Diet.

Decreased

Hepatolenticular Degeneration; Pregnancy; Pre-Eclampsia; Eclampsia.

Blood Test : Direct Bilirubin

Increased

Viral Hepatitis; Cytomegalic Inclusion Disease; Leptospirosis; Malignant Neoplasm of Pancreas; Hepatolenticular Degeneration; Dubin-Johnson Syndrome; Rotor's Syndrome; Amyloidosis; Acquired Hemolytic Anemias (Auto-immune); Acute and Subacute Necrosis of Liver; Cirrhosis of Liver; Laennec's or Alcoholic Cirrhosis; Biliary Cirrhosis; Chronic Active Hepatitis; Toxic Hepatitis; Extrahepatic Biliary Obstruction; Acute Pancreatitis; Pancreatic Cyst and Pseudocyst; Hemolytic Disease of Newborn (Erythroblastosis Fetalis); Jaundice Due to Hepatocellular Damage (Newborn).

Decreased

Crigler-Najjar Syndrome.

347

Table 7 (con't)

Blood Test : Globulins (Alpha 1)

Increased

Malignant Neoplasm of Bronchus and Lung; Secondary Malignant Neoplasm
(Disseminated); Hodgkin's Disease; Lymphocytic Leukemia; Myelocytic Leukemia;
Protein-Losing Enteropathy; Multiple Sclerosis; Rheumatic Fever; Polyarthri-
tis-Nodosa; Peptic Ulcer, Site Unspecified; Ulcerative Colitis; Acute Poststreptococcal Glomerulonephritis; Chronic Renal Failure; Chronic Pyelonephritis; Acute Pyelonephritis; Pregnancy; Pre-Eclampsia; Eclampsia; Pemphigus; Rheumatoid Arthritis; Systemic Lupus Erythematosus; Effects of X-Ray
Irradiation; Stress.

Decreased

Leprosy, Viral Hepatitis; Gout; Alpha 1-Antitrypsin Deficiency; Acute and
Subacute Necrosis of Liver; Nephrotic Syndrome.

Blood Test : Globulins (Alpha 2)

Increased

Viral Hepatitis; Lymphogranuloma Venereum; Leptospirosis; Trichinosis;
Sarcoidosis; Malignant Neoplasm of Large Intestine, Rectum, Liver, Kidney;
Secondary Malignant Neoplasm (Disseminated); Reticulum Cell Sarcoma; Lympho-
sarcoma; Hodgkin's Disease; Multiple Myeloma; Lymphocytic Leukemia (Acute);
Myelocytic Leukemia (Acute); Myelocytic Leukemia (Chronic); Diabetes Mellitus;
Hyperparathyroidism; Adrenal Cortical Hyperfunction (Glucocorticoid Excess);
Protein-Losing Enteropathy; Analbuminemia; Heavy Chain Disease (Alpha);
Amyloidosis; Meningitis (Bacterial); Multiple Sclerosis; Rheumatic Fever;
Chronic Ischemic Heart Disease; Polyarteritis-Nodosa; Cranial Arteritis and

Table 7 (con't)

Blood Test : Globulins (Alpha 2)

Increased

Related Conditions; Bacterial Pneumonia; Peptic Ulcer, Site Unspecified; Ulcerative Colitis; Acute and Subacute Necrosis of Liver; Laennec's or Alcoholic Cirrhosis; Biliary Cirrhosis; Extrahepatic Biliary Obstruction; Acute Poststreptococcal Glomerulonephritis; Nephrotic Syndrome; Chronic Renal Failure; Chronic Pyelonephritis; Acute Pyelonephritis; Pregnancy; Pre-Eclampsia; Eclampsia; Pemphigus; Reiter's Disease; Rheumatoid Arthritis; Systemic Lupus Erythematosus; Dermatomyositis/Polymyositis; Osteomyelitis; Effects of X-Ray Irradiation; Stress.

Decreased

Viral Hepatitis; Gout; Acute and Subacute Necrosis of Liver.

Blood Test : Globulins (Beta)

Increased

Chickenpox; Viral Hepatitis; Sarcoidosis; Hodgkin's Disease; Multiple Myeloma; Myelocytic Leukemia (Acute); Diabetes Mellitus; Type IIA, IIB, I, and III Hyperlipoproteinemia; Analbuminemia; Heavy Chain Disease (Alpha); Multiple Sclerosis; Polyarteritis-Nodosa; Biliary Cirrhosis; Extrahepatic Biliary Obstruction; Nephrotic Syndrome; Pregnancy; Rheumatoid Arthritis.

Decreased

Secondary Malignant Neoplasm (Disseminated); Reticulum Cell Sarcoma; Lympho-sarcoma; Hodgkin's Disease; Lymphocytic Leukemia; Monocytic Leukemia; Protein Malnutrition; Ulcerative Colitis; Acute and Subacute Necrosis of Liver.

Table 7 (con't)

Blood Test : Globulins (Gamma)

Increased

Typhoid Fever; Pulmonary Tuberculosis; Tuberculosis Meningitis; Disseminated Tuberculosis; Leprosy; Chickenpox; Viral Hepatitis; Psittacosis; Infectious Mononucleosis; Epidemic Typhus; Malaria; Leishmaniasis; Syphilis; Lymphogranuloma Venereum; Leptospirosis; Histoplasmosis; Schistosomiasis; Trichinosis; Strongyloidiasis; Larva Migrans (Visceralis); Toxoplasmosis; Sarcoidosis; Malignant Neoplasm of Bronchus and Lung; Hodgkin's Disease; Multiple Myeloma; Myelocytic Leukemia; Monocytic Leukemia; Benign Neoplasm of Cardiovascular Tissue; Hashimoto's Thyroiditis; Protein Malnutrition; Type IIA, IIB Hyperlipoproteinemia; Cystic Fibrosis; Analbuminemia; Waldenstrom's Macroglobulinemia; Amyloidosis; Agranulocytosis; Huntington's Chorea; Multiple Sclerosis; Rheumatic Fever; Bacterial Endocarditis; Congestive Heart Failure; Polyarteritis-Nodosa; Cranial Arteritis and Related Conditions; Pulseless Disease (Aortic Arch Syndrome); Bronchiectasis; Allergic Alveolitis; Berylliosis; Regional Enteritis or Ileitis; Acute and Subacute Necrosis of Liver; Laennec's or Alcoholic Cirrhosis; Biliary Cirrhosis; Chronic Active Hepatitis; Acute Streptococcal Glomerulonephritis; Pemphigus; Rheumatoid Arthritis; Rheumatoid (alkylosing) Spondylitis; Progressive Systemic Sclerosis; Systemic Lupus Erythematosus; Sjogren's Syndrome; Dermatomyositis/Polymyositis; Effects of X-Ray Irradiation; Reticulum Cell Sarcoma; Lymphosarcoma; Hodgkin's Disease; Multiple Myeloma; Lymphocytic Leukemia; Adrenal Cortical Hyperfunction (Glucocorticoid Excess); Protein Malnutrition; Protein-Losing Enteropathy; Agammaglobulinemia (Congenital Sex-linked); Immunodeficiency (Common Variable); Ataxia Telangiectasia; Heavy Chain Disease (Gamma); Amyloidosis; Myotonia Atrophica; Bronchiectasis; Nephrotic Syndrome; Pregnancy; Pre-Eclampsia; Eclampsia.

Table 7 (con't)

Blood Test : Globulins (Gamma)

Decreased

Yellow Fever; Reticulum Cell Sarcoma; Lymphosarcoma; Hodgkin's Disease;
Multiple Myeloma; Lymphocytic Leukemia; Adrenal Cortical Hyperfunction
(Glucocorticoid Excess); Protein Malnutrition; Protein-Losing Enteropathy;
Agammaglobulinemia (Congenital Sex-linked); Immunodeficiency (Common
Variable); Ataxia-Telangiectasia; Heavy Chain Disease (Gamma); Amyloidosis;
Myotonia Atrophica; Bronchiectasis; Nephrotic Syndrome; Pregnancy; Pre-
Eclampsia; Eclampsia.

Blood Test : Lactic Dehydrogenase

Increased

Typhoid Fever; Amebiasis; Tetanus; Septicemia; Acute Poliomyelitis; Herpes
Zoster; Viral Hepatitis; Psittacosis; Infectious Mononucleosis; Cytomegalic
Inclusion Disease; Lymphogranuloma Venereum; Actinomycosis; Schistosomiasis;
Hydatidosis; Trichinosis; Sarcoidosis; Malignant Neoplasm of Stomach, Small
Intestine, Large Intestine, Liver, Pancreas, Bronchus and Lung, Bone, Skin,
Breast, Cervix, Ovary, Prostate, Testis, Kidney; Secondary Malignant Neoplasm
of Respiratory System, Digestive System, Liver, Brain, Bone; Secondary Malig-
nant Neoplasm (Disseminated); Reticulum Cell Sarcoma; Lymphosarcoma; Hodgkin's
Disease; Lymphocytic Leukemia; Myelocytic Leukemia; Monocytic Leukemia;
Erythroleukemia; Myelofibrosis; Benign Neoplasm of Cardiovascular Tissue;
Hypothyroidism; Subacute Thyroiditis; Diabetes Mellitus; Diabetic Acidosis;
Mcardle's Disease; Gaucher's Disease; Hemochromatosis; Hepatolenticular
Degeneration; Hyperphosphatasia; Dubin-Johnson Syndrome; Pernicious Anemia;

Table 7 (con't)

Blood Test : Lactic Dehydrogenase

<u>Increased</u>

Folic Acid Deficiency Anemia; Hereditary (Congenital) Spherocytosis; Heredi-
tary Elliptocytosis; Anemias Due to Disorders of Glutathione Metabolism;
Hereditary Nonspherocytic Hemolytic Anemia; Thalassemia Major; Sickle Cell
Anemias; Acquired Hemolytic Anemias (Autoimmune); Paraxysmal Nocturnal
Hemoglobinuria; Aplastic Anemia; Delirium Tremens; Alcoholism; Drug Depen-
dence (Opium and Derivatives); Miningitis (Bacterial); Familial Progressive
Spinal Muscular Atrophy; Progressive Muscular Dystrophy; Myotonia Atrophica;
Myotonia Congenita; Amyotrophic Lateral Sclerosis; Rheumatic Fever; Acute
Myocardial Infarction; Acute Myocarditis; Cardiomyopathies; Congestive Heart
Failure; Cerebral Hemorrhage; Cerebral Thrombosis; Arterial Embolism and
Thrombosis; Mesenteric Artery Embolism; Gangrene; Pulmonary Embolism and
Infarction; Viral Pneumonia; Bacterial Pneumonia; Mycoplasma Pneumoniae;
Resolving Pneumonia; Silicosis; Pulmonary Alveolar Proteinosis; Intestinal
Obstruction; Regional Enteritis or Ileitis; Ulcerative Colitis; Acute and
Subacute Necrosis of Liver; Cirrhosis of Liver; Laennec's or Alcoholic
Cirrhosis; Toxic Hepatitis; Hepatic Failure; Extrahepatic Biliary Obstruction;
Acute Pancreatitis; Nephrotic Syndrome; Acute Renal Failure; Chronic Renal
Failure; Renal Infarction; Pregnancy; Ectopic Pregnancy; Pre-Eclampsia;
Dermatitis Herpetiformis; Rehumatoid Arthritis; Progressive Systemic Sclero-
sis; Dermatomyositis/Polymositis; Hemolytic Disease of Newborn (Erythroblas-
tosis Fetalis); Shock; Metabolic Acidosis; Hypothermia; Toxic Effects of
Venom; Effects of X-Ray Irradiation; Heat Stroke; Exercise; Effects of
Electric Current.

Table 7 (con't)

Blood Test : Lactic Dehydrogenase

<u>Decreased</u>

Effects of X-Ray Irradiation.

Blood Test : Magnesium

<u>Increased</u>

Lymphocytic Leukemia; Myelocytic Leukemia; Hypothyroidism; Diabetes Mellitus; Diabetic Acidosis; Adrenal Cortical Hypofunction; Acute and Subacute Necrosis of Liver; Acute Renal Failure; Chronic Renal Failure.

<u>Decreased</u>

Gastroenteritis and Colitis; Leprosy; Whipple's Disease; Viral Hepatitis; Secondary Malignant Neoplasm of Bone; Hyperthyroidism; Diabetes Mellitus; Diabetic Acidosis; Hyperparathyroidism; Hypoparathyroidism; Adrenal Cortical Hyperfunction (Mineralocorticoid Excess); Protein Malnutrition; Celiac Sprue Disease; Malabsorption, Cause Unspecified; Acute Intermittent Porphyria; Acute Alcoholic Intoxication; Alcoholism; Epilepsy; Chronic Ischemic Heart Disease; Regional Enteritis or Ileitis; Ulcerative Colitis; Cirrhosis of Liver; Laennec's or Alcoholic Cirrhosis; Chronic Active Hepatitis; Hepatic Failure; Acute Pancreatitis; Chronic Pancreatitis; Acute Poststreptococcal Glomerulonephritis; Nephrotic Syndrome; Chronic Renal Failure; Chronic Pyelonephritis; Acute Pyelonephritis; Hydronephrosis; Eclampsia; Osteoporosis; Burn.

Blood Test : Phosphorus

Table 7 (con't)

Blood Test : Phosphorus

Increased

Sarcoidosis; Secondary Malignant Neoplasm of Bone; Multiple Myeloma; Lympho-
cytic Leukemia; Myelocytic Leukemia; Diabetes Mellitus; Diabetic Acidosis;
Pseudohypoparathyroidism; Hypoparathyroidism; Acromegaly; Adrenal Cortical
Hypofunction; Ovarian Hyperfunction; Cystinosis; Hereditary Fructose Intoler-
ance; Hypervitaminosis; Hereditary (Congenital) Spherocytosis; Acquired
Hemolytic Anemias (Autoimmune); Acute Alcoholic Intoxication; Alcoholism;
Mesenteric Artery Embolism; Intestinal Obstruction; Acute and Subacute
Necrosis of Liver; Glomerulonephritis (Rapidly Progressive and Membrano-
proliferative); Nephrotic Syndrome; Renal Failure; Renal Dwarfism; Urethritis;
Eclampsia; Fracture of Bone; Osteitis Deformans; Milk-Alkali Syndrome;
Metabolis Acidosis; Heat Stroke; Exercise.

Decreased

Septicemia; Whipple's Disease; Malignant Melanoma of Skin, Prostate; Secon-
dary Malignant Neoplasm of Respiratory System; Multiple Myeloma; Benign
Neoplasm of Pancreas; Benign Neoplasm of Breast; Pheochromocytoma; Diabetes
Mellitus; Diabetic Acidosis; Hyperparathyroidism; Hypoparathyroidism; Anterior
Pituitary Hypofunction; Adrenal Cortical Hypofunction; Ovarian Hypofunction;
Adrenal Cortical Hyperfunction (Glucocorticoid Excess); Zollinger-Ellison
Syndrome; Vitamin D Deficiency Rickets; Protein Malnutrition; Celiac Sprue
Disease; Malabsorption, Cause Unspecified; Cystinosis; Von Gierke's Disease;
Hereditary Fructose Intolerance; Lactosuria; Hepatolenticular Degeneration;
Osteomalacia; Familial Periodic Paralysis; Vitamin-D Resistant Rickets;
Proximal Renal Tubular Acidosis; Distal Renal Tubular Acidosis; Gout; Hyper-
vitaminosis D; Thalassemia Major; Thalassemia Minor; Acute Alcoholic Intoxi-

Table 7 (con't)

Blood Test : Phosphorus

Decreased

cation; Alcoholism; Asthma (Intrinsic); Cirrhosis of Liver; Hepatic Failure;
Acute Pancreatitis; Pregnancy; Osteomyelitis; Osteopetrosis; Hyperventilation Syndrome; Vomiting; Burn.

Blood Test : Potassium

Increased

Arthropod-Borne Hemorrhagic Fever; Leptospirosis; Malignant Neoplasm of
Stomach; Myelocytic Leukemia; Polycythemia Vera; Diabetes Mellitus; Diabetic
Acidosis; Hyperparathyroidism; Acromegaly; Adrenal Cortical Hypofunction;
Carcinoid Syndrome; Cystinosis; Congestive Heart Failure; Pulmonary Emphysema;
Chronic Obstructive Lung Disease; Hepatic Encephalopathy; Renal Failure;
Urethritis; Pemphigus; Systemic Lupus Erythematosus; Respiratory Distress
Syndrome; Metabolic Acidosis; Crush Injury (Trauma); Heat Stroke.

Blood Test : Potassium

Decreased

Bacillary Dysentery; Cholera; Whipple's Disease; Malignant Neoplasm of Large
Intestine, Pancreas; Multiple Myeloma; Myelocytic Leukemia (Acute); Benign
Neoplasm of Rectum and Anus, Pancreas; Pheochromocytoma; Diabetes Mellitus;
Diabetic Acidosis; Hyperparathyroidism; Adrenal Cortical Hyperparathyroidism;
Adrenal Cortical Hyperfunction (Mineralocorticoid Excess); Adrenal Cortical
Hyperfunction (Glucocorticoid Excess); Zollinger-Ellison Syndrome; Protein
Malnutrition; Celiac Sprue Disease; Malabsorption, Cause Unspecified;
Cystinosis; Acute Intermittent Porphyria; Hepatolenticular Porphyria; Hepato-
lenticular Degeneration; Familial Periodic Paralysis; Proximal Renal Tubular

Table 7 (con't)

Blood Test : Potassium

Decreased

Acidosis; Distal Renal Tubular Acidosis; Pernicious Anemia; Folic Acid
Deficiency Anemia; Alcoholism; Malignant Hypertension; Essential Benign
Hypertension; Congestive Heart Failure; Asthma (Intrinsic); Chronic Obstruc-
tive Lung Disease; Peptic Ulcer, Site Unspecified; Pyloric Stenosis,
Acquired; Fistula of Stomach and Duodenum; Intestinal Obstruction; Regional
Enteritis or Ileitis; Ulcerative Colitis; Peritonitis; Cirrhosis of Liver;
Laennec's or Alcoholic Cirrhosis; Reye's Syndrome; Acute Pancreatitis;
Acute Renal Failure; Rental Failure (Chronic); Vomiting; Diarrhea; Metabolic
Alkalosis; Heat Stroke; Effects of Electric Current; Postgastrectomy
(Dumping) Syndrome.

Blood Test : Protein

Increased

Cholera; Pulmonary Tuberculosis; Leishmaniasis; Lymphogranuloma Venereum;
Sarcoidosis; Malignant Neoplasm of Bronchus and Lung; Secondary Malignant
Neoplasm of Respiratory System; Multiple Myeloma; Pheochromocytoma; Diabetic
Acidosis; Adrenal Cortical Hypofunction; Carcinoid Syndrome; Waldenstrom's
Macroglobulinemia; Huntington's Chorea; Rheumatic Fever; Cerebral Embolism;
Bacterial Pneumonia; Cirrhosis of Liver; Eclampsia; Rheumatoid Arthritis;
Respiratory Distress Syndrome; Dehydration; Exercise; Stress.

Decreased

Gastroenteritis and Colitis; Pulmonary Tuberculosis; Tuberculosis Meningitis;
Whipple's Disease; Epidemic Typhus; Rocky Mountain Spotted Fever; Malaria;

Table 7 (con't)

Blood Test : Protein

Decreased

Trichinosis; Malignant Neoplasm of Stomach, Small Intestine, Large Intestine, Rectum; Reticulum Cell Sarcoma; Lymphosarcoma; Hodgkin's Disease; Histiocytosis X; Lymphocytic Leukemia (Chronic); Hyperthyroidism; Adrenal Cortical Hyperfunction (Glucocorticoid Excess); Thiamine (B1) Deficiency; Protein Malnutrition; Celiac Sprue Disease; Hepatolenticular Degeneration; Immunodeficiency (Common Variable); Analbuminemia; Amyloidosis; Friedreich's Ataxia; Essential Benign Hypertension; Constrictive Pericarditis; Congestive Heart Failure; Mesenteric Artery Embolism; Peptic Ulcer, Site Unspecified; Gastritis; Fistula of Stomach and Duodenum; Regional Enteritis or Ileitis; Ulcerative Colitis; Cirrhosis of Liver; Toxic Hepatitis; Acute Cholecystitis; Acute Poststreptococcal Glomerulonephritis; Glomerulonephritis (Membranous); Chronic Renal Failure; Pregnancy; Pre-Eclampsia; Eclampsia; Pemphigus; Heat Stroke.

Blood Test : Sodium

Increased

Bacillary Dysentery; Cholera; Gastroenteritis and Colitis; Malignant Neoplasm of Brain; Secondary Malignant Neoplasm of Brain; Reticulum Cell Sarcoma; Lymphosarcoma; Lymphocytic Leukemia; Myelocytic Leukemia; Benign Neoplasm of Brain and CNS; Diabetic Acidosis; Acromegaly; Diabetes Insipidus; Adrenal Cortical Hyperfunction (Mineralocorticoid Excess); Ovarian Hypofunction; Adrenal Cortical Hyperfunction (Glucocorticoid Excess); Familial Periodic Paralysis; Meningitis (Bacterial); Encephalomyelitis; Congestive Heart Failure; Cerebral Hemorrhage; Cerebral Thrombosis; Cerebral Embolism; Intes-

Table 7 (con't)

Blood Test : Sodium

<u>Increased</u>

tinal Obstruction; Peritonitis; Hepatic Failure; Chronic Renal Failure;
Vomiting; Diarrhea; Dehydration; Diet (High Sodium).

<u>Decreased</u>

Bacillary Dysentery; Gastroenteritis and Colitis; Pulmonary Tuberculosis;
Tuberculosis Meningitis; Disseminated Tuberculosis; Epidemic Typhus;
Endemic Typhus; Brill's Disease; Mite-Borne Typhus; Rocky Mountain Spotted
Fever; Ricketsiae-pox; Malaria; Malignant Neoplasm of Bronchus and Lung;
Multiple Myeloma; Hypothyroidism; Subacute Thyroiditis; Diabetes Mellitus;
Diabetic Acidosis; Anterior Pituitary Hypofunction; Adrenal Cortical Hypo-
function; Ovarian Hyperfunction; Celiac Sprue Disease; Malabsorption, Cause
Unspecified; Type V Hyperlipoproteinemia; Cystic Fibrosis; Acute Inter-
mittent Porphyria; Proximal Renal Tubular Acidosis; Meningitis (Bacterial);
Essential Benign Hypertension; Acute Myocardial Infarction; Congestive
Heart Failure; Gangrene; Peptic Ulcer, Site Unspecified; Pyloric Stenosis,
Acquired; Fistula of Stomach and Duodenum; Regional Enteritis or Ileitis;
Ulcerative Colitis; Cirrhosis of Liver; Laennec's or Alcoholic Cirrhosis;
Biliary Cirrhosis; Reye's Syndrome; Hepatic Failure; Renal Failure; Urethri-
tis; Cystic Disease; Diarrhea; Hyperhidrosis; Hypothermia; Burn; Toxic
Effects of Non-Medicinal Metals.

Blood Test : Triglycerides

<u>Increased</u>

Viral Hepatitis; Hypothyroidism; Diabetes Mellitus; Von Gierke's Disease;

Table 7 (con't)

Blood Test : Triglycerides

Increased

Type IIA, IIB, IV, I, III and V Hyperlipoproteinemia; Tangier Disease; Gout;
Pernicious Anemia; Thalassemia Major; Alcoholism; Malignant Hypertension;
Essential Benign Hypertension; Renovascular Hypertension; Acute Myocardial
Infarction; Chronic Ischemic Heart Disease; Cerebral Thrombosis; Arterio-
sclerosis; Acute and Subacute Necrosis of Liver; Laennec's or Alcoholic
Cirrhosis; Biliary Cirrhosis; Extrahepatic Biliary Obstruction; Acute
Pancreatitis; Chronic Pancreatitis; Nephrotic Syndrome; Chronic Renal
Failure; Pregnancy; Fracture of Bone; Down's Syndrome; Respiratory Distress
Syndrome; Fat Embolism; Stress; Diet.

Decreased

Hyperthyroidism; Hyperparathyroidism; Protein Malnutrition; Deficiency State
(Unspecified); Lactosuria; A-Betalipoproteinemia; Hereditary (Congenital)
Spherocytosis; Brain Infarction; Chronic Obstructive Lung Disease; Exercise.

Blood Test : Urea Nitrogen

Increased

Cholera; Diphtheria; Septicemia; Herpes Zoster; Yellow Fever; Psittacosis;
Epidemic Typhus; Rocky Mountain Spotted Fever; Leptospirosis; Actinomycosis;
Aspergillosis; Malignant Neoplasm of Bladder; Lymphosarcoma; Multiple
Myeloma; Lymphocytic Leukemia; Myelocytic Leukemia; Monocytic Leukemia;
Hyperthyroidism; Hypothyroidism; Diabetes Mellitus; Diabetic Acidosis;
Hyperparathyroidism; Adrenal Cortical Hypofunction; Adrenal Cortical Hyper-
function (Glucocorticoid Excess); Cystinosis, Acute Intermittent Porphyria;

359

Table 7 (con't)

Blood Test : Urea Nitrogen

Increased

Hepatolenticular Degeneration; Gout; Waldenstrom's Macroglobulinemia; Heavy Chain Disease (Gamma); Amyloidosis; Acquired Hemolytic Anemias (Autoimmune); Allergic Purpura; Meningitis (Bacterial); Malignant Hypertension; Essential Benign Hypertension; Acute Myocardial Infarction; Bacterial Endocarditis; Congestive Heart Failure; Arteriosclerosis; Arterial Embolism and Thrombosis; Mesenteric Artery Embolism; Polyarteritis-Nodosa; Thrombotic Thrombocytopenic Purpura; Peptic Ulcer, Site Unspecified; Intestinal Obstruction; Peritonitis; Cirrhosis of Liver; Hepatic Failure; Acute Pancreatitis; Acute Poststrepto-coccal Glomerulonephritis; Glomerulonephritis (Rapidly Progressive and Mem-branous); Glomerulonephritis (Membranoproliferative); Nephrotic Syndrome; Acute Renal Failure; Chronic Renal Failure; Chronic Pyelonephritis; Acute Pyelonephritis; Hydronephrosis; Urethritis; Benign Prostatic Hypertrophy; Pre-Eclampsia; Rheumatoid Arthritis; Progressive Systemic Sclerosis; Systemic Lupus Erythematosus; Polycystic Kidney Disease; Medullary Cystic Disease; Respiratory Distress Syndrome; Shock; Vomiting; Diarrhea; Hyperhidrosis; Dehydration; Milk-Alkali Syndrome; Metabolic Alkalosis; Toxic Effects of Lead and Its Compounds (Including Fumes); Toxic Effects of X-Ray Irradiation; Heat Stroke; Exercise; Effects of Electric Current; Stress.

Decreased

Acromegaly; Protein Malnutrition; Celiac Sprue Disease; Cystic Fibrosis; Hepatolenticular Degeneration; Acute and Subacute Necrosis of Liver; Cirrhosis of Liver; Laennec's or Alcoholic Cirrhosis; Toxic Hepatitis; Hepatic Failure; Nephrotic Syndrome; Pregnancy; Pre-Eclampsia; Eclampsia; Exercise; Diet.

Table 7 (con't)

Blood Test : Uric Acid

Increased

Pulmonary Tuberculosis; Tuberculosis Meningitis; Septicemia; Viral Hepatitis; Leishmaniasis; Sarcoidosis; Malignant Neoplasm of Esophagus, Stomach, Large Intestine, Rectum, Bone, Pancreas, Breast, Corpus Uteri, Prostate, Testis, Bladder; Secondary Malignant Neoplasm of Respiratory System; Reticulum Cell Sarcoma; Lymphosarcoma; Hodgkin's Disease; Multiple Myeloma; Lymphocytic Leukemia; Myelocytic Leukemia; Monocytic Leukemia; Erythroleukemia; Polycythemia Vera; Myelofibrosis; Benign Neoplasm of Brain and CNS; Hypothyroidism; Diabetes Mellitus; Diabetic Acidosis; Hyperparathyroidism; Hypoparathyroidism; Acromegaly; Diabetic Insipidus; Protein Malnutrition; Celiac Sprue Disease; Xanthinuria; Alkaptonuria; Von Gierke's Disease; Mcardle's Disease; Forbes Disease; Anderson's Disease; Type IIA, IIB, IV, III and V Hyperlipoprotein-emia; Gout; Lesch-Nyhan Syndrome; Waldenstrom's Macroglobulinemia; Heavy Chain Disease (Gamma); Amyloidosis; Pernicious Anemia; Hereditary (Congenital) Spherocytosis; Anemias Due to Disorders of Glutathione Metabolism; Thalass-emia Major; Sickle Cell Anemias; Acquired Hemolytic Anemias (Autoimmune); Polycythemia, Secondary; Manic Depressive Disorder; Paranoid States and Other Psychoses; Depressive Neurosis; Alcoholism; Meningitis (Bacterial); Encepha-lomyelitis; Essential Benign Hypertension; Renovascular Hypertension; Acute Myocardial Infarction; Chronic Ischemic Heart Disease; Angina Pectoris; Congestive Heart Failure; Cerebral Hemorrhage; Cerebral Thrombosis; Brain Infarction; Cerebral Embolism; Transient Cerebral Ischemia; Arteriosclerosis; Resolving Pneumonia; Asthma (Intrinsic); Berylliosis; Regional Enteritis or Ileitis; Ulcerative Colitis; Cirrhosis of Liver; Laennec's or Alcoholic Cirrhosis; Glomerulonephritis (Focal); Nephrotic Syndrome; Renal Failure; Benign Prostatic Hypertrophy; Pregnancy Complicated by Intra-Uterine Death;

361

Table 7 (con't)

Blood Test : Uric Acid

Increased

Pre-Eclampsia; Eclampsia; Psoriasis; Rheumatoid Arthritis; Down's Syndrome; Respiratory Distress Syndrome; Convulsions; Toxic Effects of Lead and Its Compounds (including Fumes); Effects of X-Ray Irradiation; Heat Stroke; Exercise; Diet.

Decreased

Malignant Neoplasm of Bronchus and Lung; Hodgkin's Disease; Multiple Myeloma; Acromegaly; Celiac Sprue Disease; Hartnup Disease; Cystinosis; Xanthinuria; Hepatolenticular Degeneration; Familial Periodic Paralysis; Proximal Renal Tubular Acidosis; Pernicious Anemia; Folic Acid Deficiency Anemia; Cirrhosis of Liver; Laennec's or Alcoholic Cirrhosis; Biliary Cirrhosis; Pregnancy; Burn.

Table 8   Effects of Pathological Conditions on Hematology Tests

Blood Test : Basophils

Increased

Pulmonary Tuberculosis; Smallpox; Chickenpox; Ancylostomiasis (Hookworm Infestation); Hodgkin's Disease; Myelocytic Leukemia (Chronic); Polycythemia Vera; Myelofibrosis; Hypothyroidism; Diabetes Mellitus; Hereditary (Congenital) Spherocytosis; Acquired Hemolytic Anemias (Autoimmune); Influenza; Chronic Sinusitis; Ulcerative Colitis; Nephrotic Syndrome; Pregnancy; Urticaria.

Decreased

Myelofibrosis; Hyperthyroidism; Pregnancy; Urticaria.

Blood Test : Eosinophils

Increased

Amebiasis; Scarlet Fever; Malaria; Coccidioidomycosis; Aspergillosis; Schistosomiasis; Hydatidosis; Diphyllobothriasis (Intestinal); Trichinosis; Ancylostomiasis (Hookworm Infestation); Ascariasis; Strongyloidiasis; Trichuriasis; Larva Migrans (Visceralis); Toxoplasmosis; Sarcoidosis; Malignant Neoplasm of Bone, Skin, Ovary, Testis, Brain; Secondary Malignant Neoplasm (Disseminated); Hodgkin's Disease; Myelocytic Leukemia; Polycythemia Vera; Myelofibrosis; Simple Goiter; Hyperthyroidism; Anterior Pituitary Hypofunction; Adrenal Cortical Hypofunction; Protein-Losing Enteropathy; Waldenstrom's Macroglobulinemia; Heavy Chain Disease (Gamma); Pernicious Anemia; Allergic Purpura; Idiopathic Thrombocytopenic Purpura; Agranulocytosis; Drug Dependence (Opium and Derivatives); Otitis Media; Acute Myocarditis; Loeffler's Endocarditis; Cardiomyopathies; Cerebral Hemorrhage; Polyarteritis-Nodosa; Thrombotic

Table 8 (con't)

Blood Test : Eosinophils

Increased

Thrombocytopenic Purpura; Phlebitis and Thrombophlebitis; Viral Pneumoniae;
Bacterial Pneumoniae;Mycoplasma Pneumoniae; Chronic Bronchitis; Asthma
(Intrinsic); Allergic Alveolitis; Chronic Obstructive Lung Disease; Hay
Fever; Pulmonary Collapse; Idiopathic Pulmonary Hemosiderosis; Gastritis;
Regional Enteritis or Ileitis; Ulcerative Colitis; Acute Poststreptococcal
Glomerulonephritis; Nephrotic Syndrome; Acute Renal Failure; Prostatitis;
Atopic Eczema; Dermatitis Herpetiformis; Pemphigus; Erythema Multiforme;
Psoriasis; Urticaria; Rheumatoid Arthritis; Progressive Systemic Sclerosis;
Sjogren's Syndrome; Dermatomyositis/Polymyositis; Toxic Effects of Non-
Medicinal Metals; Toxic Effects of Venom; Effects of X-Ray Irradiation.

Decreased

Typhoid Fever; Epidemic Typhus; Malaria; Diabetic Acidosis; Adrenal Cortical
Hyperfunction (Glucocorticoid Excess); Familial Periodic Paralysis; Pemphigus; Shock; Stress.

Blood Test : Erythrocyte Sedimentation Rate

Increased

Amebiasis; Pulmonary Tuberculosis; Tuberculosis Meningitis; Disseminated
Tuberculosis; Tularemia; Brucellosis; Leprosy; Diptheria; Tetanus; Septicemia;
Whipple's Disease; Herpes Simplex; Viral Hepatitis; Mumps; Psittacosis; Cyto-
megalic Inclusion Disease; Malaria; Leishmaniasis; Relapsing Fever; Syphilis;
Lymphogranuloma Venereum; Leptospirosis; Actinomycosis; Coccidioidomycosis;
Blastomycosis; Schistosomiasis; Trichinosis; Sarcoidosis; Malignant Neoplasm

Table 8   (con't)

Blood Test : Erythrocyte Sedimentation Rate

Increased

of Stomach, Large Intestine, Liver, Breast, Kidney: Secondary Malignant Neo-
plasm of Bone; Secondary Malignant Neoplasm (Disseminated); Reticulum Cell
Sarcoma; Lymphosarcoma; Hodgkin's Disease; Histiocytosis X; Multiple Myeloma;
Benign Neoplasm of Cardiovascular Tissue; Hyperthyroidism; Hypothyroidism;
Subacute Thyroiditis; Hashimoto's Thyroiditis; Hyperparathyroidism; Cystino-
sis; Hepatolenticular Degeneration; Gout; Analbuminemia; Cryoglobulinemia
(Essential Mixed); Waldenstrom's Macroglobulinemia; Amyloidosis; Sickle Cell
Anemias; Allergic Purpura; Agranulocytosis; Meningitis (Bacterial); Intra-
cranial Abscess; Guillain-Barre Syndrome; Otitis Media; Rheumatic Fever; Acute
Myocardial Infarction; Bacterial Endocarditis; Cardiomyopathies; Cerebral
Hemorrhage; Dissecting Aortic Aneurysm; Arterial Embolism and Thrombosis;
Polyarteritis-Nodosa; Cranial Arteritis and Related Conditions; Pulseless
Disease (Aortic Arch Syndrome); Pulmonary Embolism and Infarction; Influenza;
Mycoplasma Pneumoniae; Chronic Bronchitis; Allergic Alveolitis; Abscess of
Lung; Silicosis; Anthracosis; Asbestosis; Berylliosis; Peptic Ulcer, Site
Unspecified; Acute Appendicitis; Diverticular Disease of Intestine; Diverticu-
litis of Colon; Regional Enteritis or Ileitis; Ulcerative Colitis; Acute and
Subacute Necrosis of Liver; Cirrhosis of Liver; Laennec's or Alcoholic
Cirrhosis; Acute Cholecystitis; Acute Pancreatitis; Acute Poststreptococcal
Glomerulonephritis; Glomerulonephritis (Focal); Chronic Renal Failure; Preg-
nancy; Ectopic Pregnancy; Pemphigus; Acute Arthritis (Pyogenic); Reiter's
Disease; Rheumatoid Arthritis; Juvenile Rheumatoid Arthritis; Rheumatoid
(Alkylosing) Spondylitis; Osteoarthritis; Progressive Systemic Sclerosis;
Systemic Lupus Erythematosis; Sjogren's Syndrome; Dermatomyositis/Polymyosi-
tis; Osteomyelitis; Infantile Cortical Hyperostosis; Hemolytic Disease of

Table 8 (con't)

Blood Test : Erythrocyte Sedimentation Rate

Increased

Newborn (Erythroblastosis Fetalis); Toxic Effects of Lead and Its Compounds (including Fumes); Toxic Effects of Non-Medicinal Metals; Postcardiotomy Syndrome.

Decreased

Typhoid Fever; Infectious Mononucleosis; Trichinosis; Polycythemia Vera; A-Betalipoproteinemia; Cryoglobulinemia (Essential Mixed); Sickle Cell Anemias; Congenital Afibrinogenemia; Factor V Deficiency; Rheumatic Fever; Congestive Heart Failure; Glomerulonephritis (Membranoproliferative); Progressive Systemic Sclerosis.

Blood Test : Erythrocytes

Increased

Cholera, Malignant Neoplasm of Liver, Kidney, Brain; Multiple Myeloma; Polycythemia Vera; Pheochromocytoma; Adrenal Cortical Hyperfunction (Glucocorticoid Excess); Cystic Fibrosis; Iron Deficiency Anemia; Thalassemia Minor; Hemoglobin C Disease; Polycythemia, Secondary; Cardiomyopathies; Cor Pulmonale; Arteriosclerosis; Pulmonary Emphysema; Chronic Obstructive Lung Disease; Silicosis; Anthracosis; Asbestosis; Berylliosis; Hydroneprosis; Shock.

Decreased

Iron Deficiency Anemia; Folic Acid Deficiency Anemia; Vitamin B6 Deficiency Anemia; Hereditary (Congenital) Spherocytosis; Thalassemia Major; Thalassemia

Table 8 (con't)

Blood Test : Erythrocytes

Decreased

Minor; Sickle Cell Anemias; Paraxysmal Nocturnal Hemoglobinuria; Aplastic Anemia; Sideroblastic Anemia; Allergic Purpura; Hypersplenism; Cirrhosis of Liver; Disseminated Tuberculosis; Malaria; Ancylostomiasis (Hookworm Infestation); Trichurias; Toxoplasmosis; Sarcoidosis; Secondary Malignant Neoplasm of Bone; Hodgkin's Disease; Multiple Myeloma; Lymphocytic Leukemia; Leukemic Reticuloendotheliosis; Myelofibrosis; Malabsorption, Cause Unspecified; Gaucher's Disease; Niemann-Pick Disease; Chronic Renal Failure; Pregnancy; Osteopetrosis; Hemolytic Disease of the Newborn (Erythroblastosis Fetalis); Shock; Toxic Effects of Lead and Its Compounds; Toxic Effects of Non-Medicinal Metals; Effects of X-Ray Irradiation.

Blood Test : Hematocrit

Increased

Cholera; Gastroenteritis and Colitis; Arthropod-Borne Hemorrhagic Fever; Infectious Mononucleosis; Myelocytic Leukemia (Chronic); Polycythemia Vera; Benign Neoplasm of Brain and CNS; Pheochromocytoma; Diabetic Acidosis; Adrenal Cortical Hypofunction; Adrenal Cortical Hypofunction (Glucocorticoid Excess); Polycythemia, Secondary; Acute Myocardial Infarction; Transient Cerebral Ischemia; Mesenteric Artery Embolism; Pulmonary Emphysema; Asthma (Intrinsic); Chronic Obstructive Lung Disease; Chronic Sinusitis; Silicosis; Anthracosis; Berylliosis; Fistula of Stomach and Duodenum; Intestinal Obstruction; Peritonitis; Acute Pancreatitis; Pre-Eclampsia; Eclampsia; Dermatomyositis/Polymyositis; Diarrhea; Dehydration; Exercise; Effects of Electric Current.

Table 8 (con't)

Blood Test : Hematocrit

Decreased

Typhoid Fever; Amebiasis; Pulmonary Tuberculosis; Disseminated Tuberculosis;
Leprosy; Diphtheria; Septicemia; Whipple's Disease; Chickenpox; Herpes Zoster;
Herpes Simplex; Measles; Tick-Borne Fever; Viral Hepatitis; Infectious Mono-
nucleosis; Cytomegalic Inclusion Disease; Epidemic Typhus; Rocky Mountain
Spotted Fever; Malaria; Leishmaniasis; Trypanosomiasis; Relapsing Fever;
Bartonellosis; Syphilis; Lymphogranuloma Venereum; Leptospirosis; Actino-
mycosis; Histoplasmosis; Balstomycosis; Aspergillosis; Diphyllobothriasis
(Intestinal); Trichinosis; Ancylostomiasis (Hookworm Infestation); Strongy-
loidiasis; Trichuriasis; Toxoplasmosis; Malignant Neoplasm of Stomach;
Malignant Neoplasm of Sma-l Intestine; Malignant Neoplasm of Large Intestine;
Malignant Neoplasm of Rectum; Malignant Neoplasm of Liver; Malignant Neoplasm
of Pancreas; Malignant Neoplasm of Bone; Malignant Neoplasm of Corpus Uteri;
Malignant Neoplasm of Prostate; Malignant Neoplasm of Kidney; Secondary
Malignant Neoplasm of Respiratory System; Secondary Malignant Neoplasm of
Bone; Reticulum Cell Sarcoma; Lymphosarcoma; Hodgkin's Disease; Histiocyto-
sis X; Multiple Myeloma; Lymphocytic Leukemia (Acute); Lymphocytic Leukemia
(Chronic); Myelocytic Leukemia (Acute); Myelocytic Leukemia (Chronic); Mono-
cytic Leukemia; Leukemic Reticuloendotheliosis; Erythroleukemia; Myelofibrosis;
Benign Neoplasm of Stomach; Benign Neoplasm of Cardiovascular Tissue; Hyper-
thyroidism; Hyperparathyroidism; Anterior Pituitary Hypofunction; Adrenal
Cortical Hyperfunction (Mineralocorticoid Excess); Adrenal Cortical Hypofunction; Testicular Hypofunction; Adrenal Cortical Hyperfunction (Glucocor-
ticoid Excess); Thiamine (Bl) Deficiency; Vitamin C Deficiency; Protein
Malnutrition; Celiac Sprue Disease; Malabsorption, Cause Unspecified; Protein-
Losing Enteropathy; Cystinosis; Von Gierke's Disease; A-Betalipoproteinemia;

368

Table 8 (con't)

Blood Test : Hematocrit

<u>Decreased</u>

Gaucher's Disease; Niemann-Pick Disease; Cystic Fibrosis; Congenital Erythropoietic Porphyria (Gunther's Disease); Erythropoietic Protoporphyria; Erythropoietic Coproporphyria; Acute Intermittent Porphyria; Porphyria Cutanea Tarda; Hereditary Coproporphyria; Hepatolenticular Degeneration; Lesch-Nyhan Syndrome; Agammaglobulinemia (Congeital Sex-linked); Immunodeficiency (Common Variable); Waldenstrom's Macroglobulinemia; Heavy Chain Disease (Gamma); Dysgammaglobulinemia (Selective Immunoglobulin Deficiency); Amyloidosis; Iron Deficiency Anemia; Vitamin B6 Deficiency Anemia; Hereditary (Congenital) Spherocytosis; Hereditary Elliptocytosis; Hereditary Nonspherocytic Hemolytic Anemia; Thalassemis Major; Thalassemia Minor; Hemoglobin C Disease; Hemoglobin E Disease; Hemoglobin H Disease; Sickle Cell Anemias; Acquired Hemolytic Anemias (Autoimmune); Paroxysmal Nocturnal Hemoglobinuria; Aplastic Anemia; Congenital Aplastic Anemia; Sideroblastic Anemia; Hemophilia; Allergic Purpura; Idiopathic Thrombocytopenic Purpura; Agranulocytosis; Hypersplenism; Depressive Neurosis; Meningitis (Bacterial); Myasthenia Gravis; Rheumatic Fever; Malignant Hypertension; Acute Myocardial Infarction; Bacterial Endocarditis; Cardiomyopathies; Congestive Heart Failure; Polyarteritis-Nodosa; Crnaial Arteritis and Related Conditions; Thrombotic Thrombocytopenic Purpura; Influenza; Viral Pneumonia; Bacterial Pneumonia; Mycoplasma Pneumoniae; Resolving Pneumonia; Bronchiectasis; Chronic Obstructive Lung Disease; Abscess of Lung; Silicosis; Anthracosis; Asbestosis; Berylliosis; Peptic Ulcer, Site Unspecified; Gastritis; Fistula of Stomach and Duodenum; Hernia Diaphragmatic; Diverticular Disease of Intestine; Regional Enteritis or Iliietis; Ulcerative Colitis; Acute and Subacute Necrosis of Liver; Cirrhosis of Liver; Laennec's or Alcoholic Cirrhosis; Liver Abscess (Pyogenic); Chronic Active

Table 8 (con't)

Blood Test : Hematocrit

Decreased

Hepatitis; Acute Pancreatitis; Acute Poststreptococcal Glomerulonephritis; Glomerulonephritis (Rapidly Progressive); Glomerulonephritis (Membrano-proliferative); Acute Renal Failure; Chronic Renal Failure; Chronic Pyelo-nephritis; Hydronephrosis; Benign Prostatic Hypertrophy; Pregnancy; Pemphigus; Reiter's Disease; Rheumatoid Arthritis; Juvenile Rheumatoid Arthritis; Rheumatoid (Ankylosing) Spondylitis; Progressive Systemic Sclero-sis; Systemic Lupus Erythematosus; Sjogen's Syndrome; Osteomyelitis; Poly-cystic Kidney Disease; Medullary Cystic Disease; Osteopetrosis; Hemolytic Disease of Newborn (Erythroblastosis Fetalis); Toxic Effects of Lead and Its Compounds (including Fumes); Toxic Effects of Non-Medicinal Metals; Toxic Effects of Venom; Effects of X-Ray Irradiation; Postgastrectomy (Dumping) Syndrome.

Blood Test : Hemoglobin

Increased

Cholera; Gastroenteritis and Colitis; Infectious Mononucleosis; Myelocytic Leukemia (Chronic); Benign Neoplasm of Brain and CNS; Pheochromocytoma; Benign Neoplasm of Cardiovascular Tissue; Adrenal Cortical Hyperfunction (Glucocorticoid Excess); Vitamin B6 Deficiency Anemia; Polycythemia, Secon-dary; Transient Cerebral Isochemia; Pulmonary Emphysema; Asthma (Intrinsic); Chronic Obstructive Lung Disease; Chronic Sinusitis; Silicosis; Anthracosis; Asbestosis; Berylliosis; Intestinal Obstruction; Peritonitis; Acute Pancrea-titis; Diarrhea; Dehydration; Heat Stroke; Exercise.

Table 8 (con't)

Blood Test : Hemoglobin

Decreased

Typhoid Fever; Amebiasis; Pulmonary Tuberculosis; Leprosy; Diphtheria; Septicemia; Whipple's Disease; Herpes Zoster; Tick-Borne Fever; Viral Hepatitis; Infectious Mononucleosis; Cytomegalic Inclusion Disease; Epidemic Typhus; Rocky Mountain Spotted Fever; Malaria; Leishmaniasis; Trypanosomiasis; Relapsing Fever; Syphilis; Lymphogranuloma Venereum; Leptospirosis; Actinomycosis; Histoplasmosis; Blastomycosis; Aspergillosis; Diphyllobothriasis (Intestinal); Trichinosis; Ancylostomiasis (Hookworm Infestation); Strongyloidiasis; Trichuriasis; Toxoplasmosis; Malignant Neoplasm of Esophagus; Malignant Neoplasm of Stomach; Malignant Neoplasm of Small Intestine; Malignant Neoplasm of Large Intestine; Malignant Neoplasm of Rectum; Malignant Neoplasm of Liver; Malignant Neoplasm of Pancreas; Malignant Neoplasm of Bone; Malignant Neoplasm of Corpus Uteri; Malignant Neoplasm of Prostate; Malignant Neoplasm of Bladder; Malignant Neoplasm of Kidney; Secondary Malignant Neoplasm of Respiratory System; Secondary Malignant Neoplasm of Bone; Reticulum Cell Sarcoma; Lymphosarcoma; Hodgkin's Disease; Histiocytosis X; Multiple Myeloma; Lymphocytic Leukemia (Acute); Lymphocytic Leukemia (Chronic); Myelocytic Leukemia (Acute); Myelocytic Leuekmia (Chronic); Monocytic Leukemia; Leukemic Reticuloendotheliosis; Erythroleukemia; Myelofibrosis; Benign Neoplasm of Stomach; Benign Neoplasm of Cardiovascular Tissue; Hyperthyroidism; Hypothyroidism; Hyperparathyroidism; Anterior Pituitary Hypofunction; Adrenal Cortical Hypofunction; Testicular Hypofunction; Adrenal Cortical Hyperfunction (Glucocorticoid Excess); Thiamine (B1) Deficiency; Vitamin C Deficiency; Protein Malnutrition; Celiac Sprue Disease; Malabsorption, Cause Unspecified; Protein-Losing Enteropathy; Cystinosis; Von Gierke's Disease; A-Betalipoproteinemia; Gaucher's Disease; Niemann-Pick Disease;

Table 8   (con't)

Blood Test : Hemoglobin

Decreased

Cystic Fibrosis; Congenital Erythropoietic Porphyria (Gunther's Disease);
Erythropoietic Protoporphyria; Erythropoietic Coproporphyria; Acute Inter-
mittent Porphyria; Porphyria Cutanea Tarda; Hereditary Coproporphyria;
Hepatolenticular Degeneration; Lesch-Nyhan Syndrome; Agammaglobulinemia
(Congenital Sex0linked); Immunodeficiency (Common Variable); Waldenstrom's
Macroglobulinemia; Heavy Chain Disease (Gamma); Dysgammaglobulinemia (Selec-
tive Immunoglobulin Deficiency); Amyloidosis; Iron Deficiency Anemia; Plummer-
Vinson Syndrome; Pernicious Anemia; Folic Acid Deficiency Anemia; Hereditary
(Congenital) Spherocytosis; Hereditary Elliptocytosis; Anemias Due to Dis-
orders of Glutathione Metabolism; Hereditary Nonspherocytic Hemolytic Anemia;
Thalassemia Major; Thalassemia Minor; Hemoglobin C Disease; Hemoglobin E
Disease; Hemoglobin H Disease; Sickle Cell Anemias; Acquired Hemolytic Anemias
(Autoimmune); Paroxysmal Nocturnal Hemoglobinuria; Aplastic Anemia; Congenital
Aplastic Anemia; Sideroblastic Anemia; Hemophilia; Allergic Purpura; Idio-
pathic Thrombocytopenic Purpura; Agranulocytosis; Hypersplenism; Depressive
Neurosis; Meningitis (Bacterial); Myasthenia Gravnis; Rheumatic Fever; Malig-
nant Hypertension; Bacterial Endocarditis; Cardiomyopathies; Polyarteritis-
Nodosa; Cranial Arteritis and Related Conditions; Thrombotic Thrombocytopenic
Purpura; Viral Pneumonia; Bacterial Pneumonia; Mycoplasma Pneumoniae;
Resolving Pneumonia; Bronchiectasis; Chronic Obstructive Lung Disease; Abscess
of Lung; Silicosis; Anthracosis; Asbestosis; Berylliosis; Peptic Ulcer, Site
Unspecified; Gastritis; Fistula of Stomach and Duodenum; Hernia Deaphragmatic;
Diverticular Disease of Intestine; Regional Enteritis or Ileitis; Ulcerative
Colitis; Acute and Subacute Necrosis of Liver; Cirrhosis of Liver; Laennec's
or Alcoholic Cirrhosis; Liver Abscess (Pyogenic); Chronic Active Hepatitis;

Table 8 (con't)

Blood Test : Hemoglobin

Decreased

Acute Pancreatitis; Acute Poststreptococcal Glomerulonephritis; Glomerulo-
nephritis (Rapidly Progressive); Glomerulonephritis (Membranoproliferative);
Acute Renal Failure; Chronic Renal Failure; Chronic Pyelonephritis; Hydro-
Nephrosis; Benign Prostatic Hypertrophy; Pregnancy; Pemphigus; Reiter's
Disease; Rheumatoid Arthritis; Juvenile Rheumatoid Arthritis; Rheumatoid
(Ankylosing) Spondylitis; Progressive Systemic Lupus Erythematosus;
Sjorgren's Syndrome; Dermatomyositis/Polymyositis; Osteomyelitis; Fracture
of Bone; Polycystic Kidney Disease; Medullary Cystic Disease; Osteopetrosis;
Hemolytic Disease of Newborn (Erythroblastosis Fetalis); Toxic Effects of
Lead and Its Compounds (including Fumes); Toxic Effects of Non-Medicinal
Metals; Toxic Effects of Venom; Effects of X-Ray Irradiation; Fat Embolism;
Postgastrectomy (Dumping) Syndrome.

Blood Test : Leukocytes

Increased

Typhoid Fever; Bacillary Dysentery; Amebiasis; Cholera; Pulmonary Tuberculo-
sis; Disseminated Tuberculosis; Tularemia; Brucellosis; Leprosy; Diphtheris;
Whooping Cough; Meningococcal Meningitis; Tetanus Septicemia; Whipple's
Disease; Acute Poliomyelitis; Smallpox; Measles; Western Equine Encephalitis;
Eastern Equine Encephalitis; Rabies; Mumps; Psittacosis; Infectious Mono-
nucleosis; Cytomegalic Inclusion Disease; Epidemic Typhus; Rocky Mountain
Spotted Fever; Relapsing Fever; Bartonellosis; Syphilis; Lymphogranuloma
Venereum; Leptospirosis; Moniliasis; Actinomycosis; Coccidioidomycosis; Histo-

Table 8 (con't)

Blood Test : Leukocytes

Increased

plasmosis; Blastomycosis; Schistosomiasis; Trichinosis; Ancylostomiasis (Hookworm Infestation); Strongyloidiasis; Trichuriasis; Larva Migrans; Toxoplasmosis; Malignant Neoplasm of Esophagus; Malignant Neoplasm of Stomach, Large Intestine, Rectum, Liver, Pancreas, Kidney; Secondary Malignant Neoplasm of Respiratory System, Brain; Reticulum Cell Sarcoma; Lymphosarcoma; Hodgkin's Disease; Histiocytosis X; Lymphocytic Leukemia (Acute and Chronic); Myelocytic Leukemia (Acute and Chronic); Monocytic Leukemia; Leukemic Reticuloendotheliosis; Erythroleukemia; Polycythemia Vera; Myelofibrosis; Benign Neoplasm of Cardiovascular Tissue; Subacute Thyroiditis; Diabetic Acidosis; Adrenal Cortical Hyperfunction (Glucocorticoid Excess); Carcinoid Syndrome; Cystic Fibrosis; Acute Intermittent Porphyria; Porphyria Variegata; Hereditary Coproporphyria; Familial Periodic Paralysis; Gout; Agammaglobulinemia (Congenital Sex-linked); Waldenstrom's Macroglobulinemia; Amyloidosis; Hereditary (Congenital) Spherocytosis; Hereditary Nonspherocytic Hemolytic Anemia; Thalassemia Major; Sickle Cell Anemias; Acquired Hemolytic Anemias (Autoimmune); A-lergic Purpura; Idiopathic Thrombocytopenic Purpura; Polycythemia, Secondary; Meningitis (Bacterial); Intracranial Abscess; Polyneuritis; Guillain-Barre Syndrome; Otitis Externa; Otitis Media; Rheumatic Fever; Acute Myocardial Infarction; Acute Pericarditis; Bacterial Endocarditis; Loeffler's Endocarditis; Cerebral Hemorrhage; Dissecting Aortic Aneurysm; Arterial Embolism and Thrombosis; Mesenteric Artery Embolism; Gangrene; Polyarteritis-Nodosa; Cranial Arteritis and Related Conditions; Thrombotic Thrombocytopenic Purpura; Pulseless Disease (Aortic Arch Syndrome); Pulmonary Embolism and Infarction; Phlebitis and Thrombophlebitis; Acute Tonsilitis; Influenza; Viral Pneumonia; Bacterial Pneumonia; Mycoplasma

Table 8   (con't)

Blood Test : Leukocytes

Increased

Pneumoniae; Resolving Pneumonia; Chronic Bronchitis; Asthma (Intrinsic);
Bronchiectasis; Allergic Alveolitis; Chronic Obstructive Lung Disease;
Emphysema; Abscess of Lung; Silicosis; Anthracosis; Asbestosis; Berylliosis;
Pulmonary Collapse; Pneumomediastinum; Peptic Ulcer, Site Unspecified;
Gastritis; Fistula of Stomach and Duodenum; Acute Appendicitis; Intestinal
Obstruction; Diverticulitis of Colon; Regional Enteritis or Ileitis; Ulcera-
tive Colitis; Peritonitis; Acute and Subacute Necrosis of Liver; Laennec's
or Alcoholic Cirrhosis; Liver Abscess (Pyogenic); Acute Cholecystisis;
Cholangitis; Extrahepatic Biliary Obstruction; Acute Pancreatitis; Chronic
Pancreatitis; Acute Poststreptococcal Glomerulonephritis; Nephrotic Syndrome;
Acute Renal Failure; Chronic Renal Failure; Acute Pyelonephritis; Hydronephro-
sis; Renal Infarction; Pregnancy; Ectopic Pregnancy; Eclampsia; Pemphigus;
Acute Arthritis (Pyogenic); Reiter's Disease; Rheumatoid Arthritis; Juvenile
Rheumatoid Arthritis; Dermatomyositis/Polymyositis; Osteomyelitis; Infantile
Cortical Hyperostosis; Hemolytic Disease of Newborn (Erythroblastosis Fetalis);
Shock; Metabolis Acidosis; Crush Injury (Trauma); Burn; Toxic Effects of Non-
Medicinal Metals, Carbon Monoxide, Venom; Effects of X-Ray Irradiation; Heat
Stroke; Exercise; Effects of Electric Current; Fat Embolism; Postcholecystec-
tomy Syndrome.

Decreased

Typhoid Fever; Bacillary Dysentery; Disseminated Tuberculosis; Tularemia;
Brucellosis; Diphtheria; Septicemia; Smallpox; Chickenpox; Measles; Rubella;
Yellow Fever; Western Equine Encephalitis; Eastern Equine Encephalitis;
Anthropod-Borne Hemorrhagic Fever; Phlebotomus Fever; Tick-Borne Fever; Viral

375

Table 8 (con't)

Blood Test : Leukocytes

Decreased

Hepatitis; Mumps; Psittacosis; Infectious Mononucleosis; Lymphocytic Chorio-
meningitis; Epidemic Typhus; Endemic Typhus; Mite-Borne Typhus; Rocky
Mountain Spotted Fever; Rickettsialpox; Malaria; Leishmaniasis; Relapsing
Fever; Bartonellosis; Lymphogranuloma Venereum; Leptospirosis; Histoplasmosis;
Toxoplasmosis; Sarcoidosis; Secondary Malignant Neoplasm of Bone; Reticulum
Cell Sarcoma; Lymphosarcoma; Hodgkin's Disease; Histiocytosis X; Multiple
Myeloma; Lymphocytic Leukemia; Myelocytic Leukemia (Acute): Monocytic
Leukemia; Leukemia Reticuloendotheliosis; Myelofibrosis; Hyperthyroidism;
Hyperparathyroidism; Anterior Pituitary Hypofunction; Adrenal Cortical
Hypofunction; Protein Malnutrition; Celiac Sprue Disease; Gaucher's Disease;
Niemann-Pick Disease; Hepatolenticular Degeneration; Agammaglobulinemia
(Congenital Sex-linked); Waldenstrom's Macroglobulinemia; Heavy Chain Disease;
Amyloidosis; Iron Deficiency Anemia; Pernicious Anemia; Folic Acid Deficiency
Anemia; Vitamin B6 Deficiency Anemia; Hereditary (Congenital) Spherocytosis;
Thalassemia Major; Hemoglobin C Disease; Acquired Hemolytic Anemias (Auto-
immune); Paroxysmal Nocturnal Hemoglobinuria; Aplastic Anemia; Congenital
Aplastic Anemia; Sideroblastic Anemia; Agranulocytosis; Hypersplenism;
Encephalomyelitis; Bacterial Endocarditis; Influenza; Bacterial Pneumonia;
Acute and Subacute Necrosis of Liver; Cirrhosis of Liver; Laennec's or
Alcoholic Cirrhosis; Chronic Active Hepatitis; Rheumatoid Arthritis; Felty's
Syndrome; Systemic Lupus Erythematosus; Sjogren's Syndrome; Osteopetrosis;
Shock; Toxic Effects of Lead and Its Compounds (including Fumes); Toxic
Effects of Non-medicinal Metals; Effects of X-Ray Irradiation.

Blood Test : Lymphocytes

Table 8 (con't)

Blood Test : Lymphocytes

Increased

Typhoid Fever; Pulmonary Tuberculosis; Tuberculosis Meningitis; Brucellosis; Whooping Cough; Septicemia; Smallpox; Rubella; Tick-Borne Fever; Viral Hepatitis; Mumps; Infectious Mononucleosis; Lymphocytic Choriomeningitis; Cytomegalic Inclusion Disease; Rickettsialpox; Malaria; Trypanosomiasis; Syphilia; Lymphogranuloma Venereum; Histoplasmosis; Toxoplasmosis; Malignant Neoplasm of Breast, Cervix; Lymphosarcoma; Hodgkin's Disease; Multiple Myeloma; Lymphocytic Leukemia (Chronic); Hyperthyroidism; Anterior Pituitary Hypofunction; Adrenal Cortical Hypofunction; Waldenstrom's Macroglobulinemia; Hereditary (Congenital) Spherocytosis; Thalassemia Major; Idiopathic Thrombo-cytopenic Purpura; Influenza; Viral Pneumonia; Mycoplasma Pneumoniae; Acute and Subacute Necrosis of Liver; Osteopetrosis; Effects of X-Ray Irradiation; Heat Stroke.

Decreased

Gastroenteritis and Colitis; Tuberculosis Meningitis; Tetanus; Septicemia; Whipple's Disease; Acute Poliomyelitis; Chickenpox; Measles; Yellow Fever; Viral Hepatitis; Toxoplasmosis; Sarcoidosis; Malignant Neoplasm of Esophagus, Stomach, Skin, Brain; Secondary Malignant Neoplasm of Respiratory System; Hodgkin's Disease; Myelocytic Leukemia; Diabetes Mellitus; Diabetic Acidosis; Adrenal Cortical Hyperfunction (Glucocorticoid Excess); Zollinger-Ellison Syndrome; Celiac Sprue Disease; Lactosuria; Agammaglobulinemia (Congenital Sex-linked); Immunodeficiency (Common Variable); Ataxia-Telangiectasia; Aplastic Anemia; Agranulocytosis; Multiple Sclerosis; Myasthenia Gravis; Guillain-Barre Syndrome; Constrictive Pericarditis; Influenza; Viral Pneumonia; Acute Appendicitis; Acute and Subacute Necrosis of Liver;

377

Table 8 (con't)

Blood Test : Lymphocytes

Decreased

Nephrotic Syndrome; Acute Renal Failure; Chronic Renal Failure; Systemic Lupus Erythematosus; Osteopetrosis; Effects of X-Ray Irradiation.

Blood Test : MCHC

Increased

Hereditary (Congenital) Spherocytosis; Congenital Aplastic Anemia.

Decreased

Whipple's Disease; Histoplasmosis; Ancylostomiasis (Hookworm Infestation); Strongyloidiasis; Trichuriasis; Malignant Neoplasm of Stomach, Small Intestine, Large Intestine, Rectum, Bladder, Kidney; Reticulum Cell Sarcoma; Lymphosarcoma; Polycythemia Vera; Benign Neoplasm of Stomach; Hyperthyroidism; Vitamin C Deficiency; Celiac Sprue Disease; Erythropoietic Protoporphyria; Iron Deficiency Anemia; Plummer-Vinson Syndrome; Vitamin B6 Deficiency Anemia; Thalassemia Minor; Sideroblastic Anemia; Hemophilia; Allergic Purpura; Polycythemia, Secondary; Polyarteritis-Nodosa; Peptic Ulcer, Site Unspecified; Gastritis; Diverticular Disease of Intestine; Diverticulitis of Colon; Regional Enteritis or Ileitis; Ulcerative Colitis; Chronic Renal Failure; Progressive Systemic Sclerosis; Toxic Effects of Lead and Its Compounds (including Fumes); Postgastrectomy (Dumping) Syndrome.

Blood Test : MCH

Table 8 (con't)

Blood Test : MCH

Increased

Diphyllobothriasis (Intestinal); Myelocytic Leukemia; Hypothyroidism; Pernicious Anemia; Folic Acid Deficiency Anemia; Hereditary (Congenital) Spherocytosis; Congenital Aplastic Anemia; Sideroblastic Anemia; Myasthenia Gravis; Hemolytic Disease of Newborn (Erythroblastosis Fetalis).

Decreased

Whipple's Disease; Histoplasmosis; Ancylostomiasis (Hookworm Infestation); Strongyloidiasis; Trichuriasis; Malignant Neoplasm of Stomach, Small Intestine, Large Intestine, Rectum, Bladder, Kidney; Reticulum Cell Sarcoma; Lymphosarcoma; Polycythemia Vera; Benign Neoplasm of Stomach; Hyperthyroidism; Hypothyroidism; Vitamin C Deficiency; Celiac Sprue Disease; Malabsorption, Cause Unspecified; Congenital Erythropoietic Porphyria (Gunther's Disease); Erythropoietic Protoporphyria; Erythropoietic Coproporphyria; Acute Intermittent Porphyria; Porphyria Variegata; Porphyria Cutanea Tarda; Hereditary Coproporphyria; Iron Deficiency Anemia; Plummer-Vinson Syndrome; Vitamin B6 Deficiency Anemia; Thalassemia Major; Thalassemia Minor; Sideroblastic Anemia; Hemophilia; Allergic Purpura; Polycythemia, Secondary; Polyarteritis-Nodosa; Peptic Ulcer, Site Unspecified; Gastritis; Hernia Diaphragmatic; Diverticular Disease of Intestine; Diverticulitis of Colon; Regional Enteritis or Ileitis; Ulcerative Colitis; Rheumatoid Arthritis; Progressive Systemic Sclerosis; Toxic Effects of Lead and Its Compounds (including Fumes); Postgastrectomy (Dumping) Syndrome.

Blood Test : MCV

Table 8 (con't)

Blood Test : MCV

Increased

Whipple's Disease; Diphyllobothriasis (Intestinal); Malignant Neoplasm of Stomach; Multiple Myeloma; Myelocytic Leukemia; Erythroleukemia; Hypothyroidism; Vitamin C Deficiency; Celiac Sprue Disease; Malabsorption, Cause Unspecified; Deficiency State (Unspecified); Immunodeficiency (Common Variable); Pernicious Anemia; Folic Acid Deficiency Anemia; Hereditary (Congenital) Spherocytosis; Hereditary Nonspherocytic Hemolytic Anemia; Sickle Cell Anemias; Paroxysmal Nocturnal Hemoglobinuria; Aplastic Anemia; Congenital Aplastic Anemia; Sideroblastic Anemia; Idiopathic Thrombocytopenic Purpura; Myasthenia Gravis; Congestive Heart Failure; Gastritis; Regional Enteritis or Ileitis; Cirrhosis of Liver; Laennec's or Alcoholic Cirrhosis; Chronic Renal Failure; Pregnancy; Rheumatoid Arthritis; Down's Syndrome; Hemolytic Disease of Newborn (Erythroblastosis Fetalis); Toxic Effects of Lead and Its Compounds (including Fumes).

Blood Test : Monocytes

Increased

Pulmonary Tuberculosis; Tuberculosis Meningitis; Disseminated Tuberculosis; Brucellosis; Septicemia; Herpes Zoster; Viral Hepatitis; Mumps; Infectious Mononucleosis; Rocky Mountain Spotted Fever; Malaria; Leishmaniasis; Trypanosomiasis; Syphilis; Lymphogranuloma Venereum; Toxoplasmosis; Sarcoidosis; Malignant Neoplasm of Esophagus, Stomach, Large Intestine, Rectum, Liver, Pancreas, Bone, Corpus Uteri, Prostate, Bladder, Brain; Secondary Malignant Neoplasm of Respiratory System; Reticulum Cell Sarcoma; Lymphosarcoma; Hodgkin's Disease; Histiocytosis X; Multiple Myeloma; Lymphocytic

Table 8 (con't)

Blood Test : Monocytes

Increased

Leukemia (Acute); Myelocytic Leukemia (Acute); Myelocytic Leukemia (Chronic);
Monocytic Leukemia; Polycythemia Vera; Myelofibrosis; Hyperthyroidism; Celiac
Sprue Disease; Gaucher's Disease; Waldenstrom's Macroglobulinemia; Vitamin B6
Deficiency Anemia; Thalassemia Major; Sickle Cell Anemias; Acquired Hemolytic
Anemias (Autoimmune); Sideroblastic Anemia; Idiopathic Thrombocytopenic
Purpura; Agranulocytosis; Bacterial Endocarditis; Polyarteritis-Nodosa;
Cranial Arteritis and Related Conditions; Regional Enteritis or Ileitis;
Ulcerative Colitis; Benign Prostatic Hypertrophy; Dermatitis Herpetiformis;
Urticaria; Rheumatoid Arthritis; Osteoarthritis; Systemic Lupus Erythematosus;
Osteomyelitis; Heat Stroke.

Decreased

Lymphocytic Leukemia (Acute); Lymphocytic Leukemia (Chronic); Leukemic
Reticuloendotheliosis; Aplastic Anemia.

Blood Test : Neutrophils

Increased

Bacillary Dysentery; Amebiasis; Gastroenteritis and Colitis; Disseminated
Tuberculosis; Diphtheria; Meningococcal Meningitis; Tetanus; Septicemia;
Acute Poliomyelitis; Infectious Mononucleosis; Leptospirosis; Moniliasis;
Trichinosis; Malignant Neoplasm of Stomach, Liver, Pancreas, Bronchus and
Lung, Bladder; Secondary Malignant Neoplasm of Respiratory System; Hodgkin's
Disease; Lymphocytic Leukemia (Chronic); Myelocytic Leukemia (Acute);

Table 8   (con't)

Blood Test   :   Neutrophils

Increased

Myelocytic Leukemia (Chronic); Polycythemia Vera; Diabetic Acidosis; Adrenal Cortical Hyperfunction (Glucocorticoid Excess); Familial Periodic Paralysis; Gout; Sickle Cell Anemias; Acquired Hemolytic Anemias (Autoimmune); Allergic Purpura; Idiopathic Thrombocytopenic Purpura; Meningitis (Bacterial); Intracranial Abscess; Polyeuritis; Guillain-Barre Syndrome; Otitis Media; Rheumatic Fever; Acute Myocardial Infarction; Bacterial Endocarditis; Cerebral Hemorrhage; Arterial Embolism and Thrombosis; Gangrene; Polyarteritis-Nodosa; Cranial Arteritis and Related Conditions; Thrombotic Thrombocytopenic Purpura; Pulmonary Embolism and Infarction; Phlebitis and Thrombophlebitis; Influenza; Viral Pneumonia; Bacterial Pneumonia; Mycoplasma Pneumoniae; Resolving Pneumonia; Chronic Bronchitis; Asthma (Intrinsic); Allergic Alveolitis; Chronic Obstructive Lung Disease; Abscess of Lung; Gastritis; Acute Appendicitis; Intestinal Obstruction; Diverticulitis of Colon; Peritonitis; Acute and Subacute Necrosis of Liver; Liver Abscess (Pyogenic); Acute Cholecystitis; Cholangitis; Acute Pancreatitis; Chronic Pancreatitis; Acute Poststreptococcal Glomerulonephritis; Nephrotic Syndrome; Acute and Chronic Renal Failure; Acute Pyelonephritis; Benign Prostatic Hypertrophy; Prostatitis; Pregnancy; Eclampsia; Dermatitis Herpetiformis; Urticaria; Acute Arthritis; Reiter's Disease; Rheumatoid Arthritis; Osteomyelitis; Hemolytic Disease of Newborn (Erythroblastosis Fetalis); Convulsions; Vomiting; Crush Injury (Trauma); Burn; Toxic Effects of Venom; Effects of X-Ray Irradiation; Exercise; Effects of Electric Current.

Decreased

Typhoid Fever; Bacillary Dysentery; Pulmonary Tuberculosis; Brucellosis;

Table 8 (con't)

Blood Test : Neutrophils

Decreased

Septicemia; Smallpox; Chickenpox; Measles; Rubella; Yellow Fever; Arthropod-
Borne Hemorrhagic Fever; Phlebotomus Fever; Tick-Borne Fever; Psittacosis;
Epidemic Typhus; Endemic Typhus; Rocky Mountain Spotted Fever; Malaria;
Leishmaniasis; Secondary Malignant Neoplasm of Bone; Reticulum Cell Sarcoma;
Lymphosarcoma; Multiple Myeloma; Lymphocytic Leukemia (Acute); Lymphocytic
Leukemia (Chronic); Monocytic Leukemia; Leukemic Reticuloendotheliosis;
Hyperthyroidism; Adrenal Cortical Hypofunction; Waldenstrom's Macroglobulin-
emia; Dysgammaglobulinemia (Selective Immunoglobulin Deficiency); Iron
Deficiency Anemia; Folic Acid Deficiency Anemia; Vitamin B6 Deficiency
Anemia; Acquired Hemolytic Anemias (Autoimmune); Paroxysmal Nocturnal Hemo-
blobinuria; Aplastic Anemia; Congenital Aplastic Anemia; Sideroblastic
Anemia; Agranulocytosis; Hypersplenism; Otitis Media; Acute Myocarditis;
Influenza; Viral Pneumonia; Acute and Subacute Necrosis of Liver; Felty's
Syndrome; Systemic Lupus Erythematosus; Effects of X-Ray Irradiation.

Blood Test : Platelets

Increased

Pulmonary Tuberculosis; Malignant Neoplasm of Stomach, Large Intestine, Liver,
Bronchus and Lung, Testis; Secondary Malignant Neoplasm of Respiratory System,
Digestive System, Liver, Brain, Bone; Hodgkin's Disease; Lymphocytic Leukemia
(Acute); Myelocytic Leukemia (Acute); Myelocytic Leukemia (Chronic); Poly-
cythemia Vera; Myelofibrosis; Diabetes Mellitus; Carcinoid Syndrome; Celiac
Sprue Disease; Amyloidosis; Iron Deficiency Anemia; Vitamin B6 Deficiency

383

Table 8 (con't)

Blood Test : Platelets

Increased

Anemia; Hereditary (Congenital) Spherocytosis; Hereditary Nonspherocytosis Hemolytic Anemia; Thalassemia Major; Sicile Cell Anemias; Si-eroblastic Anemia; Hemophilia; Polycythemia, Secondary; Polyarteritis-Nodosa; Pulmonary Embolism and Infarction; Regional Enteritis or Ileitis; Ulcerative Colitis; Cirrhosis of Liver; Chronic Pancreatitis; Nephrotic Syndrome; Chronic Renal Failure; Rheumatoid Arthritis; Crush Injury (Trauma); Burn; Exercise.

Decreased

Disseminated Tuberculosis; Septicemia; Smallpox; Herpes Simplex; Rubella; Arthropod-Borne Hemorrhagic Fever; Tick-Borne Fever; Viral Hepatitis; Cyto-megalic Inclusion Disease; Endemic Typhus; Rocky Mountain Spotted Fever; Malaria; Bartonellosis; Leptospirosis; Histoplasmosis; Sarcoidosis; Secondary Malignant Neoplasm of Respiratory System, Digestive System, Brain, Bone; Secondary Malignant Neoplasm (Disseminated); Reticulum Cell Sarcoma; Lympho-sarcoma; Hodgkin's Disease; Histiocytosis X; Multiple Myeloma; Lymphocytic Leukemia (Acute); Lymphocytic Leukemia (Chronic); Myelocytic Leukemia (Acute); Myelocytic Leukemia (Chronic); Monocytic Leukemia; Leukemic Reticuloendotheli-osis; Erythroleukemia; Myelofibrosis; Benign Neoplasm of Cardiovascular Tissue; Gaucher's Disease; Niemann-Pick Disease; Hepatolenticular Degeneration; Immunodeficiency (Common Variable); Waldenstrom's Macroglobulinemia; Heavy Chain Disease (Gamma); Iron Deficiency Anemia; Pernicious Anemia; Folic Acid Deficiency Anemia; Vitamin B6 Deficiency Anemia; Hereditary (Congenital) Spherocytosis; Thalassemia Major; Hemoglobin C Disease; Sickle Cell Anemias; Acquired Hemolytic Anemias (Autoimmune); Paroxysmal Nocturnal Hemoglobinuria; Aplastic Anemia; Sideroblastic Anemia; Von Willebrand's Disease; Congenital

Table 8 (con't)

Blood Test : Platelets

Decreased

Afibrinogenesia; Disseminated Intravascular Coagulopathy; Allergic Purpura;
Idiopathic Thrombocytopenic Purpura; Thrombasthenia (Glanzmann's); Agranulo-
cytosis; Hypersplenism; Alcoholism; Bacterial Endocarditis; Thrombotic
Thrombocytopenic Purpura; Influenza; Mycoplasm Pneumoniae; Acute and Subacute
Necrosis of Liver; Cirrhosis of Liver; Laennec's or Alcoholic Cirrhosis;
Liver Abscess (Pyogenic); Chronic Active Hepatitis; Hepatic Failure; Cholan-
gitis; Glomerulonephritis (Rapidly Progressive); Acute Renal Failure; Hydati-
diform Mole; Pre-Eclampsia; Eclampsia; Progressive Systemic Sclerosis;
Systemic Lupus Erythematosus; Fracture of Bone; Osteopetrosis; Hemolytic
Disease of Newborn (Erythroblastosis Fetalis); Jaundice Due to Hepatolenticu-
lar Damage (Newborn); Burn; Toxic Effects of Lead and Its Compounds (including Fumes); Toxic Effects of Non-medicinal Metals; Toxic Effects of Venom;
Effects of X-Ray Irradiation; Heat Stroke; Fat Embolism.

Table 9.   Effect of Stress on Serum Chemistries and
           Hematology Values

Increased	Decreased
Cholesterol	Albumin
Cortisol	Eosinophils
Corticotropin	
Epinephrine	
Factor VIII	
Fat	
Alpha - 1 Globulin	
Alpha - 2 Globulin	
Glucocorticoids	
Glucose	
Growth Hormone	
Nonesterified fatty acids	
Protein	
Triglycerides	
Urea Nitrogen	

added at a therapeutic concentration to a normal pool of human serum. Young *et al.* (1972) published the first version of a computerized review of the literature covering physiological, methodological and other interferences (such as due to bilirubin, hemolysis, lipemia, etc.) by drugs. Young *et al.* (1975) presented an up-dated version of their computerized review of the literature. Young and Pauck (1976) and Pauck *et al.* (1978) tested eighty-four different drugs on a SMAC Analyzer at concentrations reported to be lethal or toxic. Vinet and Letllier (1977) compared the effects on SMAC of ten drugs known to interfere using six different control sera and a fresh human pool of serum. A study dealing with the effect of 36 drugs on 21 procedures including those of an SMA 12/60 and an SMA 6/60 was presented by Wright (1976). Letllier and Gagne (1980) described a new way of studying drug interferences on a great number of methods using a non-lyophilized pool of human serum. Other studies dealing with drug interference on laboratory tests have recently been reported by Beselt *et al.* (1975), Jones (1980), Pauck *et al.* (1978); Vinet and Letllier (1977); Winek (1976); Dooley (1979); Wright and Foster (1980). Effects of Drugs on clinical biochemical and hematological test values are summarized in Table 6.

Effect of Pathological Conditions on Clinical Laboratory Tests

The results of laboratory tests are first of all interpreted as "normal" or "abnormal" based upon the limits of normal values established in the laboratory. However, undiagnosed pathological conditions such as parasitic infestations or chronic bacterial or viral infections in laboratory animals and humans may affect test results. Borderline cases and chronic infection in the selected healthy population may have significant effects on establishing normal range. Effects of pathological conditions on clinical bio-

chemical and hematological tests are summarized in Tables 7 and 8. Although these lists are not all inclusive, they may be useful in interpretations of laboratory data. Most of the data presented in these Tables are derived from the works of Friedman *et al.* (1980). Effect of stress on laboratory tests is summarized in Table 9. Variations of plasmatic or seric enzymes due to exercise have been frequently studied in humans and animals. It has been reported that AST (GOT) generally increases in humans, trained or not. In some cases (ALT (GPT) decreases sometimes as much as 50% of the initial value (Schlang (*et al.,* 1961). Lactic dehydrogenase too exhibits variations sometimes contradictory, but there is little significance in the decrease observed. Creatine kinase increases sometimes considerably following rigorous exercise. Other effects of stress including exercise have been discussed in literature (Siest, 1973, Payne *et al.,* 1976; Henry, 1979; Jones, 1980; Benirschke *et al.,* 1978; Friedman *et al.,* 1980; Benjamin, 1978).

Hematology Tests

The reliability and precision of the analytical methods as well as presentation and evaluation of results in view of possible physiological, nutritional and pathological variation are prerequisites of meaningful interpretations of laboratory data by the clinician and experimental scientist. The data presented in this book are primarily for use in experimental biology rather than diagnostic work. Reference values reported in this work were obtained from experimental animals which were properly defined and documented by method, sample handling, their pathophysiologic and nutritional status. Such definition provides the medical investigator with guidelines which may be useful in interpretation of reference value data in his own experimental design and work.

References

Section 1

1. Albritton AB: Standard Values in Blood. Philadelphia, W.B. Saunders, 1952

2. Altman PL, Dittmer DS: Blood and Other Body Fluids. Washington D.C., Federation of American Societies for Experimental Biology, 1971

3. Altman PL, Dittmer DS: Biology Data Book. Washington D.C., Federation of American Societies for Experimental Biology, 1974

4. Altman PL, Katz DD: Human Health and Disease, Federation of American Societies for Experimental Biology, Bethesda, MD, 1977

5. Amador E, Hsi PB, Massod MF: An evaluation of the average of normals and related methods of quality control. Am J Clin Pathol 50: 369, 1968

6. Anido G, Van Kampin EJ, Rosalki SB, Rubin M: Quality Control in Clinical Chemistry, Transactions of the VIth International Symposium Geneva, 1975; Walter de Gruyter, Berlin, New York, 1975

7. Anonymous: A Review of Variations in the Concentrations of Metabolites in the Blood of Beef and Dairy Cattle Associated with Physiology, Nutrition and Disease, with Particular Reference to the Interpretation of Metabolic Profiles. World Rev. Nutr. Diet 35: 172, 1980

8. Archer PG: The Predictive Value of Clinical Laboratory Test Results. Am J Clin Path 69, 1978

9. Bailey RM: Clinical Laboratories and the Practice of Medicine. Berkley, McCutcheon Publishing Corporation, 1979

10. Benirschke K, Garner FM, Jones TC (Editors): Pathology of Laboratory Animals Vol I and II. New York, Springer-Verlag, 1978

11. Burns KF, DeLannor CEJ: Compendium of normal blood values of laboratory animals with indication of variations. Toxic Appl Pharmacol 8: 429, 1966

12. Caraway WT: Accuracy in Clinical Chemistry. Clin Chem 17: 63, 1971

13. Cembrowski GS, Westgard JO, Conover WJ, Toren EC: Statistical Analysis of Method Comparison Data. Testing Normality. Am J Clin Pathol 72: 21, 1979

14. Cherian AG, Hill GJ: Percentile Estimates of Reference Values for Fourteen Clinical Constituents in Sera of Children and Adolescents. Am J Clin Pathol 69: 24, 1978

15. Coffman JR: A Perspective in Clinical Chemistry in the Horse. Vet Med Small Anim Clin 65: 1091, 1970

16. Coles EH: Veterinary Clinical Pathology, Second Edition, Philadelphia, W.B. Saunders, 1973, 1980 (Revised)

17. Corley, GS: Regional Quality Control for Medical Laboratories. Lab Med 1: 22, 1970

18. Ducan JR, Prasse KW: Veterinary Laboratory Medicine. Ames, Iowa, Iowa State University Press, 1977

19. Dybkaer R: Concepts and Nomenclature in Theory of Reference Values. Scand J Clin Lab Invest 29: 191, 1972

20. Fox RR, Laird CW, Blau EM, Schultz HS, Mitchell BP: Biochemical Parameters of Clinical Significance in Rabbits. I. Strain Variation. II. Diurnal Variation. J Hereditary 61: 261, 1970

21. Galen RS, Gambino SR: Beyond Normality. The Predictive Value and Efficiency of Medical Diagnosis. New York, John Wiley and Sons, 1975

22. Gras R, Wiest G, Wiling P, Williams GZ, Whitehead TP: Provisional Recommendation on the Theory of Reference Values. The Concept of Reference Values. Clin Chem 25: 1506, 1979

23. Hafez ESE: Reproduction and Breeding

Techniques for Laboratory Animals. Philadelphia, Lea and Febiger, 1970

24. Hafez ESE: Comparative Reproduction of Non-human Primates. Springfield, Illinois, Charles C Thomas, 1971

25. Hawkey CM: Comparative Mammalian Haematology, London, Heinemann, 1975

26. Henry JB, Todd, Sanford, Davidsohn: Clinical Diagnosis and Management by Laboratory Methods. Philadelphia, London, Toronto, W.B. Saunders Company, 1979

27. Henry RJ: Use of the Control Chart in Clinical Chemistry. Clin Chem 5: 309, 1959

28. Henry RJ, Reed AH: Normal Values and the Use of Laboratory Results for the Detection of Disease. Clin Chem 20: 343, 1974

29. Hoffman RG, Waid ME: The Number Plus Method of Quality Control of Laboratory Accuracy. Am J Clin Pathol 40: 263, 1963

30. Hoffman RG, Waid ME: The "Average of Normals" Method of Quality Control. Am J Clin Pathol 43: 134, 1965

31. Holland RR: Decision Tables: Their Use for the Presentation of Clinical Algorithms. J Am Med Assoc 233: 455, 1975

32. Holt J, Lumsden JH, Mullen K: On Transforming Biological Data to Gaussian Form. Canad J Comp Med 44:43, 1980

33. Jennings FW, Mulligan A: Levels of Some Chemical Constituents in Normal Horse Serum. J Comp Pathol Therap 62: 286, 1953

34. Jungner G, Jungner I: Multiple Laboratory Screening. Edited by Benson ES, Strandjord PE. New York, Academic Press, 1970

35. Kanenko JJ, Cornelius CE, eds: Clinical Biochemistry of Domestic Animals. Volumes I,II. 2nd Edition. New York, Academic Press, 1970

36. Koller S: Problems in Defining Normal Values, Standardization, Documentation and Normal Values in Hematology. Bibl Haematol (Basel) 21: 125, 1965

37. Laird CW, Fox RR, Schultz HS, Mitchell BP, Blau EM: Strain Variations in Rabbits; Biochemical Indicators of Thyroid Functions. Life Sci 9: 203, 1970

38. Laird CW, Fox RR: The Effect of Age on Cholesterol and PBI Levels in II/III$_c$ Hybrid Rabbits. Life Sci 9: 1243, 1970

39. Laird CW: Representative Values for Animal and Veterinary Populations and their Clinical Significances. Houston, Texas, Hycel Inc., 1972

40. Laird CW: Clinical Pathology, Blood Chemistry. Handbook of Laboratory Animal Science. Vol II. Edited by Melby EC Jr. and Altman AH. Cleveland, Ohio, CRC Press, pp. 347–436, 1974

41. Levey S, Jennings ER: Use of Control Charts in the Clinical Laboratory. Am J Clin Pathol 20: 1059, 1950

42. Lumsden JH, Mullen K: On Establishing Reference Values. Canad J Comp Med 42: 293, 1978

43. Lusted LB: Introduction to Medical Decision Making. Springfield, Illinois, Charles C Thomas, 271, 1968

44. Martin HF, Gudzinowica BJ, Fanger H: Normal Values in Clinical Chemistry. A Guide to Statistical Analysis of Laboratory Data. New York, Marcel Dekker, Inc. 1975

45. Martinek RG, Becktel JM, Forster GF, Morrissey RA: Comparison of Methods for Conducting Laboratory Proficiency Surveys. Health Lab Sci 5: 239, 1968

46. McNeil BJ, Keller E, Adelstein SJ: Primer of Certain Elements of Medical Decision Making. New England J Med 293: 212, 1975

47. McPherson K, Healy MJ, Flynn FV, Piper KA, Garcia, Webb P: The Effect of Age, Sex and Other Factors on Blood Chemistry in Health, Clin Chim Acta 84: 373, 1978

48. Medway W, Prier JE, Wilkinson JS (eds): A Textbook of Veterinary Clinical Pathology. Baltimore, Williams and Wilkins, 1969

49. Moss ML, Horton CA, White JC: Clinical Biochemistry. Annual Review of Biochemistry 40: 573, 1971

50. Neumann GJ: The Determination of Normal Ranges from Routine Laboratory Data. Clin Chem 14: 979, 1968

51. Peplow AM, Peplow PV, Hafez ESE: Parameters of Reproduction, in Handbook of Laboratory Animal Science, Vol I, Melby EC, Altman NH(eds), Cleveland, CRC Press, Inc., 1974

52. Phillis JW: Veterinary Physiology, Philadelphia, W.B. Saunders Company, 1976

53. Roussel JD, *et al*.: Effects of Seasonal Climatic Changes on the Productive Traits, Blood Glucose, Body Temperature and Respiration Rate of Lactating Dairy Cows (abstract). J. Dairy Science 54: 458, 1971

54. Spector WS: Handbook of Biological Data. Philadelphia, W.B. Saunders Company, 1961

55. Sunderman FW Jr: Current Concepts of

"Normal Values," "Reference Values," and "Discrimination Values" in Clinical Chemistry. Clin Chem 21: 1873, 1975

56. Swenson MJ: Dukes Physiology of Domestic Animals, Eighth Edition. Ithaca, Cornell University Press, 1970

57. Tsay JY, Chen IW, Maxon HR, Heminger L: A Statistical Method for Determining Normal Ranges from Laboratory Data Including Values Below Minimum Detectable Value. Clin Chem 12: 2011, 1979

58. Waid ME, Hoffman RG: The Quality Control of Laboratory Precision. Am J Clin Pathol 25: 585, 1955

59. Werner M, Marsh L: Normal Values. Theoretical and Practical Aspects. CRC Crit Rev Clin Lab Sci 81–100, 1975

60. Whitehead TP: Advances in Quality Control In. Advances in Clin Chem Vol 19 (O Bodansky and AL Latner eds), pp. 175–205, 1977

61. Wilson JMG: Current Trends and Problems in Health Screening. Am J Clin Pathol 26: 555, 1973

62. Wirth WA, Thompson RL: The Effects of Various Conditions and Substances on the Results of Laboratory Procedures. Am J Clin Pathol 43: 579, 1965

63. Zieve L: Misinterpretation and Abuse of Laboratory Tests by Clinicians. Ann NY Acad Sci 134: 563, 1966

Section II

1. Amador E, Massod F, Franey RF: Reliability of Glutamic-oxaloacetic transaminase methods. Am J Clin Pathol 47: 419, 1967

2. Archer RF: Hematological Technic for Use on Animals, Oxford, Blackwell Scientific Publication, 1965

3. Bentinck-Smith J: A Textbook of Veterinary Cinical Pathology. Edited by Medway W, Prier JE, Wilkinson JS. Baltimore, The Williams and Wilkins Company, 1969

4. Berchelmann ML, Kalter SS and Britton HA: Comparison of Hematologic Values from Peripheral Blood of the Ear and Venous Blood of Infant Baboons (Papio cynocephalus). Lab Anim Sci 23: 48, 1973

5. Berchelmann ML, Kalter SS: Hematology. Primates in Medicine, Kalter SS (editor), Karger, Basel, 8: 51, 1973

6. Bone JF: Crystallization patterns in vaginal and cervical mucus smears as related to bovine ovarian activity and pregnancy. Am J Vet Res 15: 542, 1954

7. Bustad LK: Housing and Handling of Miniature Swine, Symposium on Swine in Biomedical Research. 3–8, 1965

8. Bustad LK, McClellan RO: Swine in Biomedical Research. Seattle, Frayn Printing Co., 1966

9. Coles EH: Veterinary Clinical Pathology. Philadelphia, WB Saunders, 1967; 1980 (Revised)

10. Foote RH: Estimation of Bull Sperm by packed cell volume. J Dairy Sci 41: 1109, 1958

11. Frankhouser R: Der Liquor Cerebrospinalis in der Veterinamedizin, Zbl Vet Med 1: 136, 1953

12. Frank JJ, Bermes EW, Bikel MJ, Watkins BF: Effect of in vitro Hemolysis on Chemical Values for Serum. Clin Chem 24: 1966, 1978

13. Fuchs F: Symposium on Amniotic Fluids. Clin Obstet Gynec 9: 425, 1966

14. Hoffman RA, Robinson PF and Magalhae, H: The Golden Hamster. Its Biology and Use in Medical Research. Ames, Iowa, The Iowa State University Press, 1968

15. Huhn GG, Osweiler GD, Switzer WP: Application of the Orbital Sinus Bleeding Technique to Swine. Lab Anim Care, 19: 403, 1969

16. Jacobs P, Andreaensseus L: A simple method for repeated Blood Sampling in Small Animals. J Lab Clin Med 76: 1013, 1970

17. Kaplan HM, Timmons EH: The Rabbit: A Model for the Principles of Mammalian Physiology and Surgery. New York, Academic Press, 1979

18. Laessig RH, Indriksons AA, Hassemer DJ, Paskey TA, Schwartz TH: Changes in Serum Chemical Values as a Result of Prolonged Contact with the Clot. Am J Clin Pathol 6: 598, 1976

19. Ladenson JN: Non Analytical Sources of Variation in Clinical Chemistry Results, In: Gradwohl's Clinical Laboratory Methods and Diagnosis. AC Sonnenwirth and L Jarett (editors), St Louis, The CV Mosby Company, 1980 Chap 9, pp. 149–192

20. Lang CM: The Laboratory Care and Clinical Management of Saimiri (Squirrel Monkeys).

The Squirrel Monkeys LA Rosenblum and RW Cooper (editors), New York, Acad Press, 405, 1968

21. Mackellar JC: Collection of Blood Samples and Smears for Diagnosis. Vet Rec 86: 302, 1976

22. Mann T: Reproduction in Domestic Animals. Edited by Coles EH, Cupps PT. New York, Academic Press, 2: 1959

23. Mann T: The Artificial Insemination of Farm Animals. Edited by Percy EJ. New Brunswick, New Jersey, Rutgers University Press, 1961

24. McGrath JT: Neurological Examination of the Dog, 2nd Edition. Philadelphia, Lea and Febiger, 1960

25. Ono T, Kitaguchi K, Takehora M, Shieba M, Hayami K: Serum Constituents Analysis. Effect of Duration and Temperature of Storage of Clotted Blood. Clin Chem 27: 35, 1981

26. Rosenkrantz H, Langille J, Mason MM: The Chemical Analysis of Normal Canine Fluid. Am J Vet Res 22: 1057, 1961

27. Schalm OW: Veterinary Hematology. Philadelphia, Lea and Febiger, 1961; 1975 (Revised)

28. Schermer S: The Blood Morphology of Laboratory Animals. Philadelphia, FA Davis Company, 1967

29. Sonnenwirth AC, Jarett L: Gradwohl's Clinical Laboratory Methods and Diagnosis. Volumes I and II. St Louis, Toronto, London, The CV Mosby Company, 1980

30. Tumbleson ME, Donemert AR, Middleton CC: Techniques for Handling Miniature Swine for Laboratory Procedures. Lab Anim Care 18: 584, 1968

31. Vagtborg H (editor): The Baboon in Medical Research. Vol II, Austin and London, University of Texas Press, 1967

32. Valerio DA, Miller RL, Innes JRM, Courtney RD, Palotta AJ, Guttmacher RM: *Macaca Mulatta,* Management of Laboratory Breeding Colony, New York, Academic Press, 1969

33. Wagner JE, Manning PJ: The Biology of the Guinea Pig. New York, Academic Press, 1976

34. Weisbroth SH, Flatt RE, Krans AL: The Biology of the Laboratory Rabbit. New York, London, Academic Press, 1974

Section III

1. Balogh K, Cohen RB: Histochemical Demonstration of Diaphorases and Dehydrogenases in Normal Human Leukocytes and Platelets. Blood 17: 491, 1961

2. Benzel JE, Egan JJ, Hart DJ, Christopher EA: Evaluation of an Automated Differential Leukocyte counting system II. Normal Cell Identification. Am J Clin Pathol 62: 530, 1974

3. Blood DC, Henderson JA, Radostits N: Veterinary Medicine, 5th Edition, Philadelphia, Lea and Febiger, 1979

4. Bointon DF, Farquhar MG: Origin of Granules in Polymorphnuclear Leukocytes. J Cell Biol 28: 277, 1966

5. Brecher G, Schneiderman M, Williams GZ: Evaluation of an Electronic Blood Cell Counter. Am J Clin Pathol 26: 1439, 1956

6. Bull BS, Schneiderman, MA, Brecker G: Platelet Counts with the Coulter Counter. Am J Clin Pathol 44: 678, 1965

7. Dacie JV, Lewis SM: Practical Hematology. 5th Edition, Edinburgh, Churchill Livingstone, 1975

8. Dutcher TF, Benzel JE, Egan JJ, Hart DJ, Christopher EA.: Evaluation of an Automated Differential Leukocyte Counting System I. Instrument Description and Reproducibility Studies. Am J Clin Pathol 62: 525, 1974

9. Egan JJ, Benzel JE, Hart DJ, Christopher EA: Evaluation of an Automated Differential Leukocyte Counting System III. Detection of Abnormal Cells. Am J Clin Pathol 62: 537, 1974

10. Hawkey CM: Comparative Mammalian Haematology—London, Heinemann, 1975

11. Kaplow LS: Cytochemistry of Leukocyte Alkaline Phosphatase. Use of Complex Naphthol Phosphate in Azo Dyes Coupling Technique. Am J Clin Pathol 39: 439, 1963

12. Kaplow LS, Burstone MS: Cytochemical Demonstration of Acid Phosphatase in Hematopoietic Cells in Health and in Various Hematological Disorders using Azo Dye Techniques. J Histochem Cytochem 12: 805, 1964

13. Kaplow LS: Simplified Myeloperoxidone Stain Using Benzidine Dihyrochloride. Blood 26: 215, 1965

14. Koepke JA: Tips in Technology in Medical Laboratory Observer, 13, 17, 1981

15. Kurnick NB: Pyronin Y in the Methyl Green Pyronin Histological Stain. Stain Technol 30: 213, 1955

16. Lappin TR, Lamont A, Nelson MB: An Evaluation of the Auto-Analyzer SMA4 J. Clin Pathol 22: 11, 1969

17. McManns JFA: Histological Demonstration of Mucin After Periodic Acid. Nature 158: 202, 1946

18. Megla GK: The LARC Automatic White Blood Cell Analyzer Acta Cytol 17: 3, 1973

19. Nelson MG: Multichannel Continuous Flow Analysis on SMA-4/7A J. Clin Pathol 22: 20, 1969

20. Ohringer P: PCV Centrifuge on Automatic Microhematocrit. Adv Automation Anal 2: 407, 1970

21. Payne BJ, Lewis HB, Muchison TE, Hart EA: Hematology of Laboratory Animal, In, Melby EC, Jr, Altman NH (editors), Handbook of Laboratory Animal Science, Vol III, 383, Cleveland, CRC Press Inc, 1976

22. Pierre RV, O'Sullivan MB: Evaluation of the Hemalog D Automated Differential Leukocyte Counter. Mayo Clinic Proc 49: 70, 1974

23. Richar WJ, Breakell ES: Evaluation of an Electronic Particle Counter for the Counting of White Blood Cells. Am J Clin Pathol 31: 384, 1959

24. Roasles CL et al.: Histochemical Demonstration of Leukocyte Acid Phosphatase in Health and Disease. Brit J Hematol 12: 172, 1966

25. Saunders AM, Scott F: Hematologic Automation by Use of Continuous Flow Systems. J Histochem Cytochem 22: 707, 1974

26. Simmons A, Elbert G: Hemalog-D and Manual Differential Leukocyte Counts: A Laboratory Comparison of Results Obtained with Blood of Hospitalized Patients. Am J Clin Pathol 64: 512, 1975

27. Sipe CR, Cronkite EP: Studies on the Application of the Coulter Electronic Counter in Enumeration of Platelets. Ann NY Acad Sci 99: 262, 1962

28. Williams WJ, Bentler E, Erslev AJ, Rundles RW: Hematology, 2nd Edition McGraw-Hill Book Company, New York, 1977

29. Wintrobe MM: Clinical Hematology. Philadelphia, Lea and Febiger, 1967

30. Wintrobe MM, Lee GR, Boggs DR, Bithell TC, Athens JW, Foerster WR: Clinical Hematology. 7th Edition, Philadelphia, Lea and Febiger, 1974

Section IV

1. Albritton AB: Standard Values in Blood, Philadelphia, W.B. Saunders, 1952

2. Albritton AB: Standard Values in Blood, Philadelphia, W.B. Saunders, 1958

3. Albritton AB: Standard Values in Blood, Philadelphia, W.B. Saunders, 1961

4. Altman PL, Dittmer P: Biology Data Book. Washington, D.C., Fed Am Soc Exp Biol, 1971

5. Altman PL, Dittmer P: Biology Data Book. Washington, D.C., Fed AM Soc Exp Biol, 1974

6. Altman PL, Katz DD: Human Health and Disease. Bethesda, MD, Fed Am Soc Exp Biol, 1977

7. Anderson AC, Gee W: Normal Blood Values in the Beagle, Vet Med 53: 135, 1958

8. Anderson DR: Normal Values for Clinical Blood Chemistry Tests for Macaca Mulatta Monkey. Am J Vet Res 27: 1484, 1966

9. Anderson JM, Benirschke K: The armadillo, Dasypus Novemcinctus, in Experimental Biology. Lab Anim Care 16: 202, 1966

10. Anderson ET, Lewis JP, Passovoy M, Trobaugh FE Jr: Marmosets as Laboratory Animals II. The Hematology of Laboratory Kept Marmosets. Lab Anim Care 17: 30, 1967

11. Anderson ET, Schalm OW: Haematology in the Beagles as an Experimental Dog. Edited by Anderson AC, Ames Iowa State University Press, 1970

12. Andrews W: Comparative Hematology. New York, Grune and Stratton, 1965

13. Atland PD, Kirkland BC: Red Cell Life Span in the Turtle and Toad. Amer J Physiol, 203: 1188, 1962

14. Atwal OS, McFarland LZ, Wilson WO: Hematology of Coturnix from Birth to Maturity. Poultry Sci 43: 1392, 1969

15. Benirschke K, Garner FM, Jones TC(Eds): Pathology of Laboratory Animals. Vols I and II. New York, Springer-Verlag, 1978

16. Benjamin MM: Outline of Veterinary Pathology, Ames, Iowa, The Iowa State University Press, 1978

17. Bergman RAM: The Erythrocytes of Snakes. Folia Haematol 75: 92, 1957

18. Brown HE, Dougherty TF: The Diurnal

Variation of Blood Leukocytes in Normal and Adrenalectomized Mice. Endocrinol 58: 365, 1956

19. Blair CB Jr: The Case and Maintenance of the Nine-banded Armadillo, *Dasypus Novemcinctus Mexicanus,* in a Laboratory Environment. Bull Ga Acad Sci 28: 113, 1970

20. Burns KF, DeLannoy CE Jr: Compendium of Normal Blood Values of Laboratory Animals with Indication of Variations. Toxic Appl Pharmacol 8: 429, 1966

21. Burns KF, *et al.*: Compendium of Normal Blood Values for Baboons, Chimpanzees, and Marmosets. Am J Clin Path 48: 484, 1967

22. Burns KF, Timmons EH, Poiley SM: Serum Chemistry and Hematological Values for Axenic (Germfree) and Environmentally Associated Inbred Rats. Lab Anim Science 21: 415, 1971

23. Busby SRM: Methods in Toxicology. Edited by Porget GE, Oxford, Blackwell, 1970, pp 338–71

24. Casella RL: The Peripheral Hemogram in the Chinchilla. Mod Vet Prac 44: 51, 1963

25. Chapman Al: Leukocyte Values of Normal and Adrenalectomized Male and Female C57BL/Cn Mice. Lab Animal Care 18: 616, 1968

26. Coles EH: Veterinary Clinical Pathology. Philadelphia, London, Toronto, W.B. Saunders Company, 1980

27. Dieterich RA: Hematologic Values for Five Northern Microtines. Lab Anim Sci 22: 390, 1972

28. Didisheim D, Hattori K, Lewis JH: Hematologic and Coagulation Studies in Various Animal Species. J Lab Clin Med 53: 866, 1959

29. Dillon GW, Gilomski CA: The Mongolian Gerbil: Quantitative and Qualitative Aspects of Cellular Blood Picture. Lab Anim 9: 283, 1975

30. Dontenwill W, *et al.*: Biochemical and Haematological Investigations in Syrian Golden Hamster After Cigarette Smoke Inhalation. Lab Anim 8: 217, 1974

31. Dumas J: Les Animaux de Laboratoire: Anatomie, Particularites Physiologiques, Hematologie, Maladies Naturelles, Experimentation. Paris, Editions Medicales Flammarion, 1953

32. Earl FL, Melveger BE, Wilson RL: The Hemogram and Bone Marrow Profile of Normal Neonatal and Weaning Beagle Dogs. Lab Anim Sci 23: 690, 1973

33. Ernst RA, Ringer RK: The Effect of DDT Zeitran and Zytron on the Packed Cell Volume, Total Erythrocytic Count and Mean Corpuscular Volume of Japanese Quail. Poultry Sci 47: 639, 1968

34. Ewing GO, Schalm OW, Smith RS: Hematologic Values of Normal Basenji Dogs. J Am Vet Med Assoc 161: 1661, 1972

35. Farris EJ, Griffith JQ Jr: The Rat in Laboratory Investigation, Second Edition, New York, Hafner Publishing Co, 1963

36. Fetters MD: Hematology of the Black-tailed Jack Rabbit *(Lepus Californicus)* Lab Anim Sci 22: 546, 1972

37. Finch CE, Foster JR: Hematologic and Serum Electrolyte Values of the C57BL/6J Male Mouse in Maturity and Senescence. Lab Anim Sci 23: 339, 1973

38. Garcia FG, Hunt RD: The Hematogram of the Squirrel Monkey. Lab Anim Care 16: 50, 1966

39. Giorno R, Clifford J, Beverly S, Rossing RS: Analysis by Different Statistical Technics and Variation with Age and Sex. J Clin Pathol 74: 765, 1980

40. Godwin KO, Fraizer FJ, Ibbotson RN: Haematological Observations on Healthy (SPF) Rats. Brit J Exp Pathol 45: 514, 1964

41. Grunsell CS: The Marrow Cells of Normal Sheep. J Comp Path Therap 65: 8, 1955

42. Hall A III, *et al.*: Mystromys Albicaudatus (the African White-tailed Rat) as a Laboratory Species. Lab Anim Care 17: 180, 1967

43. Hambleton P, Harris-Smith PW, Baskerville A, Bailey NE, Pavey KJ: Normal Values for Some Whole Blood and Serum Components of Grivet Money *(Cercopithecus aethiops).* Lab Anim 13: 87, 1969

44. Hamrick PE, McRee DI, Zinkl JG, Thaxton P, Parkhurst CR: Hematology of Neonatal Japanese Quail. Lab Anim Sci 25: 495, 1975

45. Hanser HJ: Atlas of Comparative Primate Hematology, New York, Academic Press, 1970

46. Hardy J, Cotchin E, Roe FJC (Eds): Hematology of Rats and Mice. Philadelphia, F.A. Davis Co., 1967

47. Harrison SD Jr: Burdeshaw JA, Crosby RG, Cusic AM, Denine EP: Hematology and Clinical Chemistry Reference Values. Cancer Res 38: 2636, 1978

48. Heady JT, Rogers JE: Turtle Blood Cell Morphology. Iowa Acad Sci Proc 69: 587, 1962

49. Henry JB: Todd, Sanford, Davidsohn's Clinical Diagnosis and Management by Laboratory Method. Philadelphia, W.B. Saunders Company, 1979

50. Holmann HH: The Blood Picture of the Cow. Brit Vet J 3: 440, 1955

51. Holmann HH: Changes Associated with Age in the Blood Picture of Calves and Heifers. Brit Vet J 112: 91, 1956

52. Jain NC: A Staining Technique to Demonstrate Erythrocyte Refractile Bodies in Dog Blood. Brit Vet J 125: 437, 1969

53. Jones RT: Normal Values for Some Biochemical Constituents in Rabbits. Lab Anim Sci 9: 143, 1975

54. Kennedy GH: Cytology of the Blood of Normal Mink and Raccoon II. The Numbers of the Blood Elements in Normal Mink. Canad J Res 12: 484, 1935

55. King TO, Gargus JL: Normal Blood Values of the Adult Female Monkey *(Macaca Mulatta)* Lab Anim Care 17: 391, 1967

56. Kleinburger C: Die Blutmorphologie der Labrotoriumstiere. Leipzig, Barth, 1927

57. Kozma CK, Weisbroth SH, Stratman SL, Conejeros M: Normal Biological Values for Long Evans Rats: Part II. Lab Anim Care 19: 697, 1969

58. Kraft VH: The Morphological Blood Picture of the *Chinchilla Villigera*. Blut 6: 386, 1959

59. Krise GM: Hematology of the Normal Monkey. Ann NY Acad Sci 85: 803, 1960

60. Krise GM Jr, Wald N: Normal Blood Picture of the *Macaca Mulatta* Monkey. J Appl Physiol 12: 482, 1958

61. Kunze H: Die Erythropoese Bei Einer Erblicken Anamie Rontgenmutierti Mause. Folia Hematol 72: 392, 1954

62. Lewis JH: Comparative Hematology: Studies on Goats. Am J Vet Res 37: 601, 1976

63. Lewis JH, Doyle AP: Coagulation, Protein and Cellular Studies on Armadillo Blood. Comp Biochem Physiol 26: 61, 1969

64. Lewis JK, Phillips LL, Hann C: Coagulation and Hematological Studies in Primitive Australian Mammals. Comp Biochem Physiol 25: 1129, 1968

65. Loeb WF, Bannerman RM, Rininger BF, Johnson AJ: Hematological Disorders in, Pathology of Laboratory Animals. Vol I,

Benirschke K, Garner FM, Jones TC (Eds), New York, Springer-Verlag, 1978 pp. 890–1050

66. Lucas AM, Jamroz C: Atlas of Avian Hematology: Agriculture Monograph 25. Washington, D.C., United States Dept. of Agriculture, 1961

67. Lumsden JH, Mullen K, Rowe R: Hematology and Biochemistry Reference Values for Female Holstein Cattle. Can J Comp Med 44: 24, 1980

68. Lumsden JH, Rowe R, Mullen K: Hematology and Biochemistry Reference Values for the Light Horse. Can J Comp Med 44: 32, 1980

69. Lumsden JH, Mullen K, McSherry BJ: Canine Hematology and Biochemistry Reference Values. Canad J Comp Med H3: 125, 1979

70. Mammeric M, Lorenz RJ, Straub OC, *et al.*: Bovine Hematology IV Comparative Breed Studies on the Erythrocyte Parameters of 16 European Cattle Breeds as Determined in the Common Reference Laboratory. Zentralbl Veterinaermed (B) 25: 484, 1978

71. Manning PJ, Lehner NDM, Feldner MA, *et al.*: Selected Hematologic, Serum Chemical and Arterial Blood Gas Characteristics of Squirrel Monkeys *(Saimiri Sciureus)*. Lab Anim Care 19: 831, 1969

72. Mays A Jr: Baseline Hematological and Blood Biochemical Parameters of the Mongolian Gerbil *(Meriones Ungriculatus)*. Lab Anim Care 19: 838, 1969

73. Mays A Jr, Loew FM: Hemograms of Laboratory Confined Opossums *(Didelphis Virginiana)*. J Am Vet Med Assoc 153: 800, 1968

74. Melby EC Jr, Altman NH: Handbook of Laboratory Animal Science. Cleveland, CRC Press, Inc. 1976

75. McClure HM, Guillons NB, Keeling ME: Clinical Pathology Data for the Chimpanzee and Other Anthropoid Apes. The Chimpanzee, Bourne GH (Ed), Vol VI, Baltimore, University Park Press, 1973

76. Melville GS Jr, Whitcomb WH, Martinez RS: Hematology of the *Macaca Mulatta* Monkey. Lab Anim Care 17: 189, 1967

77. Michaelson SM, Scheer K, Gilt S: The Blood of the Normal Beagle. J Am Vet Med Assoc 148: 532, 1966

78. Middleton CC, Rosal J: Weights and Mea-

surements of Normal Squirrel Monkeys (*Saimiri Sciureus*). Lab Anim Sci 22: 583, 1972

79. Moore W: Hemogram of the Chinese Hamster. Amer J Vet Res 27: 608, 1966

80. New AE: The Squirrel Monkey. Edited by Rosenblum LA, Cooper RW, New York, Academic Press, 1968

81. Newberne PM: A Preliminary Report on the Blood Picture of the South American Chinchilla. J Am Vet Med Assoc 122: 221, 1953

82. Nirmalan GP, Robinson GA: Hematology of the Japanese Quail (*Coturnix Coturnix Japanica*). Br Poultry Sci 12: 475, 1971

83. Osborne JC, Meredith JH: Hematological Values of the Normal Weaning Piglet. Cornell Vet 61: 13, 1971

84. Oser F, Lang RE, Login EE, *et al.:* Blood Values in Stumptailed Macaques (*M. Arctoides*) under Laboratory Conditions. Lab Anim Care 20: 462, 1970

85. Pelger S: Nuclear Anomaly of Leucocytes, in Fol Ham, Schweiz Med Wsch, Undritz E(Ed)., 67: 249, 1943

86. Pick JR, Eubank JW: A Clinicopathologic Study of Heterogenous Dog Populations in North Carolina. Lab Anim Care 15: 11, 1965

87. Porter JA Jr: Hematology of the Night Monkey, *Aotus Trivirgatus*. Lab Anim Care 19: 470, 1969

88. Porter JA Jr: Hematologic Values of the Panamanian Marmoset (*Saguinus Geoffroyi*). Am J Vet Res 31: 379, 1970

89. Porter JA Jr: Hematologic Values of the Black Spider Monkey (*Ateles Fusciceps*), Red Spider Monkey (*A. Geoffroyi*), White Monkey (*Cebus Caepicinus*) and Black Howler Monkey (*Alouatta Villosa*). Lab Anim Sci 21: 426, 1971

90. Porter JA Jr, Canaday WR Jr: Hematologic Values in Mongrel and Greyhound Dogs Being Screened for Research Use. J Am Vet Med Assoc 159: 1603, 1971

91. Potkay S, Zinn RD: Effects of Collection Interval, Body Weight and Season on the Hemograms of Canine Blood Donors. Lab Animal Care 19: 192, 1969

92. Rahaman H, Srihari K, Krishnamoorthy RV: Comparative Hematology, Hemochemistry and Electrocardiography of the Slender Loris and Bonnet Monkey. Lab Anim 9: 69, 1975

93. Robinson FR, Ziegler RF: Clinical Laboratory Data Derived from 102 *M. Mulatta*. Lab Anim Care 18: 50, 1968

94. Rollins JB, Hobbs CH, Spertzel RO, McCounell S: Hematologic Studies of the Rhesus Monkey (*Macaca Mulatta*). Lab Anim Care 20, 1970

95. Ross JG, Christie G, Halliday WG, Jones RM: Haematological and Blood Chemistry "Comparison Values For Clinical Pathology in Poultry." Vet Rec 102: 29, 1978

96. Rothstein R, Hunsaker D II: Baseline Hematology and Blood Chemistry of The South American Wooly Opossum *Galuromys Derbianus*. Lab Anim Sci 22: 227, 1972

97. Ruhren R: Normal Values for Hemoglobin Concentration and Cellular Elements in the Blood of Mongolian Gerbils. Lab Anim Care 15: 313, 1965

98. Russel ES, Neufeld EI, Higgins CT: Comparison of Normal Blood Picture of Young Adults from 18 Inbred Strains of Mice. Proc Soc Expt Biol Med 78: 761, 1951

99. Sato T, Oda K, Kubo M: Hematological and Biochemical Values of Thoroughbred Foals in the First 6 months of Life. Cornell Vet 69: 3, 1978

100. Schalm OW: Veterinary Hematology. Philadelphia, Lea and Febiger, 1965

101. Schalm OW, Jain NC: Veterinary Hematology. Philadelphia, Lea and Febiger, 1974

102. Schermer S: The Blood Morphology of Laboratory Animals. Philadelphia, F.A. Davis Co., 1967

103. Secord DC, Russell JC: A Clinical Laboratory Study of Conditioned Mongrel Dogs and Labrador Retrievers. Lab Anim Sci 23: 567, 1973

104. Shellenberger TE, *et al.:* Erythrocyte and Leukocyte Evaluations of *Coturnix* Quail. Poultry Sci 44: 1332, 1965

105. Sherwood BF, Rowlands DT Jr, Haekel DB, Lemay JC: The Opossum, *Didelphis Virginiana*, as a Laboratory Animal. Lab Anim Care 19: 494, 1969

106. Sonnenwirth AC, Jarett L: Gradwohl's Clinical Laboratory Methods and Diagnosis, Vol I and II. St. Louis, Toronto, London, The CV Mosby Company, 1980

107. Spiller GA, Spaur CL, Amen RJ: Selected Blood Values From *Macaca Nemestrina* Fed Semi-Purified Fiber-free Liquid Diets. Lab Anim Sci 25: 341, 1975

108. Stanley RR, Cramer MB: Hematologic Values of the Monkey (*Macaca Mulatta*) Am J Vet Res 29: 1041, 1968

109. Street RP Jr, Highman B: Blood Chemistry Values in Normal *Mystromys Albicaudatus* and Osborne-Mendel Rats. Lab Anim Sci 21: 394, 1971

110. Strike TA: Hemogram and Bone Marrow Differential of the Chinchilla. Lab Anim Care 20: 33, 1970

111. Timmons EH, Marques PA: Blood Chemical and Hematological Studies in the Lab Confined Unaesthetised Opossum. *Didelphis Virginiana*. Lab Anim Care 19: 342, 1969

112. Vogin EE, Oser F: Comparative Blood Values in Several Species of Non-Human Primates. Labnim Sci 21: 937, 1971

113. Vondruska JF: Certain Hematologic and Blood Chemical Values in Adult Stumptailed Macaques (*Macaca Arctoides*). Lab Anim Care 20: 97, 1970

114. Weisbroth SH, Flatt RE, Kraus AL (Eds): Biology of the Laboratory Rabbit. New York, Academic Press, 1974, p. 552

115. Wellde BT, Johnson AJ, Williams JS, Langbehn HR, Sadun EH: Hematologic, Biochemical and Parasitologic Parameters of the Night Monkey (*Aotus Trivirgatus*). Lab Anim Sci 21: 575, 1971

116. Whitney RA Jr, Johnson DJ, Cole WC: Laboratory Primate Handbook, New York, Academic Press Inc., 1973

117. Wolf HG, Moshe S, Klein AK, *et al.*: Hematologic Values For Laboratory Reared *Marmosa Mitis*. Lab Anim Sci 21: 249, 1971

118. Youatt WG, Fay LD, Howe DL, *et al.*: Hematologic Data on Some Small Mammals. Blood 18: 758, 1961

Section V

1. Abel LL, Levy BB, Brady BB, Kirwall FE: Standard Methods of Clinical Chemistry, Vol 2, Edited by Seligson D, New York, Academic Press, 1958

2. Amador E, Dorfman LE, Walker WEC: Serum Lactic Dehydrogenase Activity. An Analytical Assessment of Current Assays. Clin Chem 9: 391, 1963

3. Amador E, Urban J: Simplified Serum Phosphorus Analysis by Continuous Flow Ultraviolet Spectrophotometry. Clin Chem 18: 601, 1972

4. Astrup P: The Need for Standardization of Quantities and Units in Clinical Chemistry. Scand J Clin Lab Invest 25: 125, 1970

5. Babson AL, Greeley SJ, Coleman CM, Phillips GE: Phenolphthalein Monophosphate as a Substitute for Serum Alkaline Phosphatase. Clin Chem 12: 482, 1966

6. Babson AL, Reed PA, Phillips GE: The Importance of the Substrate in Assays of Acid Phosphatase in Serum. Am J Clin Path 32: 167, 1959

7. Baron DN: SI Units in Pathology: The Next Stage. Am J Clin Path 26: 729, 1973

8. Baron DN, Broughton PMG, Cohn M, Lansley TS, Lewis SM, Shinton NK: The Use of SI Units in Reporting Results Obtained in Hospital Laboratories. Am Clin-Biochem 11: 194, 1974

9. Bauer JD, Ackermann PG, Toro G: Bray's Clinical Laboratory Methods. St. Louis, Mosby, 1975

10. Benirschke K, Garner FM, Jones TC (editors): Pathology of Laboratory Animals. Volumes I and II, New York, Springer-Verlag, 1978

11. Benjamin MM: Outline of Veterinary Clinical Pathology, Ames, Iowa, The Iowa State University, 1978

12. Bergmeyer HU, Bernt E: Glutamate-Oxaloacetate Transaminase and Glutamate-Pyruvate Transaminase *In* Methods of Enzymatic Analysis. Bermeyer HV (editor). New York, Academic Press, Inc., 1965, pp. 837, 846

13. Bergmeyer HU: Methods of Enzymatic Analysis. 2nd Edition, New York, Academic Press, Inc, 1974

14. Berry JW, Chappell DG, Barnes RD: Sodium and Potassium Measurement by Flame Photometry. Ind Eng Chem Anal 18: 19, 1946, *In* Manual of Operation of Flame Photometer. Corning, New York, Corning Med Instrument Co.

15. Bessey OA, Lowry OH, Brock MJ: A Method for the Rapid Determination of Alkaline Phosphatase with Five Cubic Millimeters of Serum. J. Biol Chem 164: 321, 1946

16. Bouda J: Determination of Iron with Bathophenanthroline without Deproteinization. Clin Chem Acta 21, 159, 1968

17. Bowers GN, McComb C: A Continuous

Spectrophotometric Method for Measuring the Activity of Serum Alkaline Phosphatase. Clin Chem 12: 70, 1966

18. Bucalo G, David H: Quantitative Determination of Serum Triglycerides by the Use of Enzymes. Clin Chem 19: 476, 1973

19. Caraway WT: A Stable Starch Substrate for the Determination of Amylase in Serum and Other Body Fluids. Am J Clin Path 32: 97, 1959

20. Chasson AL, Grady HJ: Stanley MA Determination of Creatinine by Means of Automatic Chemical Analysis. Am J Clin Path 35: 83, 1961

21. Coles EH: Veterinary Clinical Pathology. Philadelphia, W.B. Saunders, 1967, 1980 (Revised)

22. Cornelius CE, Kaneko JJ: Clinical Biochemistry of Domestic Animals. New York, Academic Press, 1971

23. Cotlove E; Standard Methods of Clinical Chemistry, Vol 3, Edited by Seligson D, New York, Academic Press, 1961

24. Connerty HV, Lau HSC, Briggs AR: Spectrophotometric Determination of Magnesium by Use of Methylthymol Blue. Clin Chem 17: 661, 1971

25. Crocker C: Rapid Determination of Urea Nitrogen in Serum or Plasma Without Deproteinization. Am J Med Technol 33: 361, 1967

26. Doumas BT: Standards for Total Serum Protein Assays—A Collaborative Study. Clin Chem 21: 1159, 1975

27. Doumas BT, Perry BW, Sasse EA, Stroumfjord JV: Standardization in Bilirubin Assays: Evaluation of Selected Methods and Stability of Bilirubin Solutions. Clin Chem 19: 984, 1973

28. Doumas BT, Watson WA, Biggs HG: Albumin Standards and the Measurement of Serum Albumin with Bromocresolgreen. Clin Chim ACTA, 31: 87, 1971

30. Dubowski KM: An O-toluidine Method for Body Fluid Glucose Determination. Clin Chem 8: 215, 1962

31. Dybkaer R: Standard Methods of Clinical Chemistry, Vol 6, Edited by Seligson, D. New York, Academic Press, Inc., 1970, p 223

32. Dybkaer R, Jorgensen K: Quantities and Units in Clinical Chemistry Including Recommendation 1966 of the Commission of Clinical Chemistry of the International Union of Pure and Applied Chemistry and of International Federation of Clinical Chemistry: Munksgaard, Copenhagen, 1967. Baltimore, Williams and Wilkins, 1967

33. Ferris CD: Guide to Medical Laboratory Instruments. Boston, Little Brown and Company, 1980

34. Feteris WA: A Serum Glucose Method without Protein Precipitation. Am J Med Technol 31: 17, 1965

35. Frankel S, Reitman S, Sonnenwirth AC: Gradwohl's Clinical Laboratory Methods and Diagnosis, 2nd Edition. St. Louis, C.V. Mosby, 1970

36. Gal EM, Roth E: Spectrophotometric Methods for Determination of Cholinesterase Activity. Clin Chem ACTA 2: 316, 1957

37. Gambino RS: Modified Jendrassik and Grof Method of Bilirubin Determination. Std Methods in Clin Chem 5: 55, 1965

38. Gambino SR, Schreiber R: Determination of Total and Conjugated Bilirubin. Automation in Analytical Chemistry Technicon Symposia, White Plains, New York, Mediad Inc., 1964

39. Gay RJ, McComb RB, Bowers GN: Optimum Reaction Conditions for the Human Lactate Dehydrogenase Isoenzymes as They Affect Total Lactate Dehydrogenase Activity. Clin Chem 14: 740, 1968

41. Gitelman HJ: An Improved Automated Procedure for the Determination of Calcium in Biological Specimens. Anal Biochem 18: 521, 1967

42. Gochman N, Schmitz JM: Application of a New Peroxide Indicator Reaction to Specific Automated Determination of Glucose with Glucose Oxidase. Clin Chem 18: 943, 1972

43. Goldenberg H, Fernandez A: Simplified Method for the Estimation of Inorganic Phosphorus in Body Fluids. Clin Chem 12: 871, 1966

44. Goodale RH, Widmann FK: Clinical Laborarory Interpretation of Laboratory Tests, 6th Edition. Philadelphia, F.A. Davis, 1979

40. Giovaniello TJ, Di Beneditto G, Palmer DW, Peters T, Jr: Fully and Semiautomated Methods for the Determination of Serum Iron and Total Iron Binding Capacity. J Lab Clin Med 71: 874, 1968

45. Goodwin JF: Quantification of Serum Inor-

ganic Phosphorus, Phosphatase and Urinary Phosphate without Preliminary Treatment. Clin Chem 16: 776, 1970

46. Hamilton RH: A Direct Photometric Method for Chloride in Biological Fluids, Employing Mercuric Thiocyanate and Perchloric Acid. Clin Chem 12: 1, 1966

47. Henry RH, Sobel C, Berkman S: On the Determination of Pancreatitis Lipase in Serum. Clin Chem 3: 77, 1957

48. Henry RJ, Chismori N, Golub OJ, Berkman S: Revised Spectrophotometric Methods for the Determination of Glutamic-oxaloacetic Transaminase and Glutamic Pyruvic Transaminase and Lactic Acid Dehydrogenase. Am J Clin Pathol 34: 381, 1960

49. Henry RJ, Canon DC, Winkelman JW: Clinical Chemistry Principles and Techniques. Harper and Row, Publishers, Hagerstown, Maryland, 1974, 1979 (Revised)

50. Henry RJ, Chiamori N: Study of Saccharogenic Method for the Determination of Serum and Urine Amylase. Clin Chem 6: 434, 1960

51. Henley KS, Pollard HM: A New Method for the Determination of Glutamic-oxaloacetic and Glutamic Pyruvic Transaminase in Plasma. J Lab Clin Med 46: 785, 1955

52. Hepler OE: Manual of Clinical Laboratory Methods. Springfield, Illinois, Charles C Thomas, Publisher, 1968

53. Hochstrasser H: An Automated Method for the Determination of Carbon Dioxide in Blood Plasma. Am J Clin Pathol 33: 181, 1960

54. Hoffman RG: Establishing Quality Control and Normal Ranges in the Clinical Laboratory. New York, Exposition Press, 1970

55. Huang TC, Chen CP, Wefler V, Raftry A: A Stable Reagent for Liebermann Burchard Reaction. Anal Chem 33: 140, 1961

56. Hyvarinen A, Nikkila EA: Specific Determination of Blood Glucose with O-toluidine. Clin Chim Acta 7: 140, 1962

57. Instrument Laboratory, Inc. Manual: Flame Photometry: Instruction Manual for the Measurement of Sodium and Potassium, 1973

58. Jacobson J, Wennberg RP: Determination of Unbound Bilirubin in the Serum of Newborns. Clin Chem, 2nd Edition, pp. 783–889, 1974

59. Jaffe MZ: Uber den Niederschlag Welchen Pikrinsaure in Normalen Harn Erzeugt, Und Uber Eine Neue Reaction Des Kreatinine. Z Physiol Chem 10: 391, 1886

60. Jendrassik L, Grof P: Vereinfaclite Photomerische Method en Zue Bestinimung des Blut Bilirubins. Biochem Z 297: 81, 1938

61. Jones LM, Shinowara GY, Reinhart HL: The Estimation of Serum Inorganic Phosphate and Acid and Alkaline Phosphatase Activity. J Biol Chem 142: 921, 1942

62. Kalchar HM: Differential Spectrophotometry of Purine Compounds by Means of Specific Enzymes. J. Biol Chem 167: 429, 1947

63. Kaplan A., Szabo L: Clinical Chemistry. Interpretation and Techniques, Philadelphia, Lea and Febiger, 1979

64. Karmen A, Wroblewski F, Ladue JS: Transaminase Activity in Human Blood. J Clin Invest 34: 126, 1955

65. Kessler G: Glutamine Pyruvic Transaminase and Glutamic Oxaloacetic Transaminase. Advances in Automated Analysis, Vol 1: 67, 1971

66. Kessler G, Rush R, Leon L, Delea A, Cupiola R: Automated 340 nm Measurements of SGOT, SGPT and LDH. Advances in Automated Analysis, Technicon International Congress, 1: 67, 1970

67. Kessler G, Wolfman M: An Automated Procedure for the Simultaneous Determination of Calcium and Phosphorus. Clin Chem 10: 686, 1964

68. King J: Practical Clinical Enzymology. Princeton, D. Van Nostrand Co., 1965

69. Klose S, Grief H, Hagen H: Comparison of Two New Developed Enzymatic Cholesterol. Color Tests on Auto-Analyzer Systems with Other Cholesterol Tests. Clin Chem 21: 942, 1975

70. Kohn J: A Cellulose Acetate Supporting Medium for Zone Electrophoresis. Clin Chim Acta 2: 297, 1957

71. Kohn J: Small-scale membrane filter electrophoresis and immuno-electrophoresis. Clin Chim Acta 3: 450, 1958

72. Ladue JS, Wroblewski F, Karmen A: Serum Glutamic-oxaloacetic Transaminase Activity in Human Acute Transmural

Myocardial Infarction. Science 120: 497, 1954

73. Lamela A: Introduction to Medical Laboratory Methods, New York, Harper and Row Publishers, 1971

74. Larsen K: Creatinine Assay by a Reaction Kinetic Principle. Clin Chim Acta 41: 209, 1972

75. Leon LP, Chu DK, Stasiu RO, Snyder LR: New More Specific Methods for the SMA 12/60 Multichannel Biochemical Analyzer, Advances in Automated Analysis, Technicon International Congress, Vol 1, 1976

76. Lumsden JH, Mullen K, McSherry BJ: Canine Hematology and Biochemistry Reference Values. Canad J Comp Med 43: 125, 1979

77. Malloy HT, Evelyn KA: The Determination of Bilirubin with Photoelectronic Colorimeter. J Biol Chem 119: 481, 1937

78. Marbach, EP, Weil MH: Rapid Enzymatic Measurement of Blood Lactate and Pyruvate. Clinical Chem 12: 314, 1967

79. Marsh WH, Fingerhut B and Miller H: Automated Manual Direct Methods for the Determination of Blood Urea Nitrogen. Clin Chem 11: 624, 1965

80. McComb RB, Bowers GN: Study of Optimum Buffer Conditions for Measuring Alkaline Phosphatase Activity in Human Serum. Clin Chem 18: 97, 1972

81. Medway W, Prier JE, Wilkinson JS (Eds): A Textbook of Veterinary Clinical Pathology, Baltimore, Williams and Wilkins, 1969

82. Megraw RE, Hritz AM, Babson AL, Carroll JJ: A Single-Tube Technique For Serum Total Iron and Total Iron Binding Capacity. Clinical Biochemistry 6: 266, 1973

83. Melby EC Jr, Altman NH: Handbook of Laboratory Animal Science. Cleveland, CRC Press, Inc., 1976

84. Menson RC, Narayanswami V, Bussian RW, Adams TH: A Kinetic Method for Determination of CO_2 in Biological Fluids. Clin Chem 20: 872, 1974

85. Michaelsson M: Bilirubin Determination in Serum and Urine. Scand J Clin Lab Invest 13 (Supp) 56:1, 1961

86. Miller SE, Weller JM: Textbook of Clinical Pathology.. Baltimore, Williams and Wilkins, 1971

87. Morgenstern S, Kessler G, Aurbach JA, Flor RV, Klein B: An Automated p-nitrophenol Phosphate Serum Alkaline Phosphatase Procedure for the Autoanalyzer. Clin Chem 1: 876, 1965

88. Musser AW, Ortigoza C: Automated Determination of Uric Acid by the Hydroxylamine Method. Tech Bull Reg Med Tech 36: 21, 1966

89. Oliver IT: A Spectrophotometric Method for the Determination of Creatine Phosphokinase and Myokinase. Biochemical Journal 61: 116, 1955

90. Owen JA, Iggo B, Seandritt FJ, Stewart CF: Determination of Creatinine in Plasma or Serum and in Urine in Critical Examination. Biochem Journal 58: 426, 1954

91. Page CH, Vigoureux P: The International System of Units (SI). Washington D.C., National Bureau of Standards, Special Publication 330, 1972

92. Perkin Elmer Corp Handbook: Perkin Elmer Handbook of Analytical Methods for Atomic Absorption Spectrophotometer, March 1971

93. Pileggi VJ, DiGiorgio J, Wybenga DR: A One Tube Serum Uric Acid Method Using Phosphotungstic Acid on Protein Precipitant and Color Reagent. Clin Chim Acta 37: 141, 1972

94. Rao J, Pelvin MH, Morgenstern S. SMAC High Speed, Continuous Flow, Ion Selective Electrodes for Sodium and Potassium: Theory and Design. "The SMAC SYSTEM" Research and Development Reports, Advances in Automated Analysis, Technicon International Congress, New York, Mediad Inc., 1972

95. Raphael SS: Lynch's Medical Laboratory Technology. Vol 1, Chapter 5, Automation in Clinical Chem, Philadelphia, London, Toronto, W.B. Saunders, 1976

96. Rautela GS, Hall RG Jr, Bekiesz CL, Warmus GR: A Kinetic Method for Rapid Automated Measurement of Triglycerides in Biological Fluids. Clin Chem 20: 857, 1974

97. Rodkey FL: Direct Spectrophotometric Determination of Albumin and Globulin in Human Serum. Clin Chem 11: 478, 1965

98. Rosalki SB: An Improved Procedure for Serum Creatine Phosphokinase Determination. J Lab Clin Med 69: 696, 1967

99. Rosalki SB, Wilkinson JH: Reduction of Alpha-ketobutyrate by Human Serum. Navie 188: 1110, 1960

100. Roy AV, Brower ME, Hayden JE: Sodium Thymolphthalein Monophosphate. A New Acid Phosphatase Substrate with Greater Specificity for Enzyme in Serum. Clin Chem 17: 1093, 1971

101. Sarkar BCR, Chauhan UPS: A New Method for Determining Microquantities of Calcium in Biological Materials. Anal Biochem 20: 155, 1967

102. Schatz N: SMA II Performance Characteristic, Advances in Automated Analysis. Technicon International Congress, Vol 1, 1976

102A. Schmidt EJ: Amylase Determination in Serum. (Klin Wschr, 40: 585. 1961) Translated in English by HU Bergmeyer (ed) Methods of Enzymatic Analysis, 2nd Edition, New York, Academic Press, 1965

103. Schwartz MK, Bethune VG, Fleisher M: Chemical and Clinical Evaluation of the Continuous Flow Analyzer "SMAC" Clin Chem 20: 1062, 1974

104. Segal MA: A rapid Electrotitrimetric Method for Determining CO_2 Combining Power in Plasma or Serum. Am J Clin Pathol 25: 1212, 1955

105. Seligson D, Marino J, Dodson E: Determination of Sulfobromophthalein in Serum. Clin Chem 3: 638, 1957

106. Shihabi ZK, Bishop C: Simplified Turbidimetric Assay for Lipase Activity. Clin Chem 17: 1150, 1971

107. Skeggs LT, Hochstrasser H: Multiple Automatic Sequential Analysis. Clin Chem 10: 918, 1964

108. Slein MW: D-glucose Determination with Hexokinase and Glucose-6-phosphate dehydrogenase. *In,* Methods of Enzymatic Analysis Bergmeyer, HU (ed), New York, Academic Press, 117, 1965

109. Snyder LR, Leon LP: New Chemistry Assays for SMAC, New York, 7th Technicon International Congress, 1976

110. Somogyi M: Modification of Two Methods for the Assay of Amylase. Clin Chem 6: 23, 1960. Modified by Henry RJ, Chiamori N. Clin Chem 6: 434, 1960

111. Stadtman TC: Preparation and Assay of Cholesterol and Ergosterol *In,* SP Colowick and NO Kaplan (eds) Methods in Enzymology. Vol 3, 1957, pp. 392–394

112. Szasz G: A Kinetic Photometric Method for Serum Gamma-Glutamyl-transpeptidase. Clin Chem 15: 124, 1969

113. Talke H, Schubert GE: Enzymatisch Hanstoffbestimung in Blut und Serum im Optischin Nach Warburg. Klin Wschr, 41: 174, 1965

114. Taussky HH: A Microcolorimetric Determination of Creatine in Urine by Jaffe Reaction. J Biol Chem 208: 853, 1954

115. Tietz NW, Friereck EA: A Specific Method for Serum Lipase Determination. Clin Chim Acta 13: 352, 1966

116. Tietz NW (ed): Fundamentals of Clinical Chemistry. Philadelphia, W.B. Saunders, 1970, 1976 (Revised)

117. Trinder P: Determination of Glucose in Blood using Glucose Oxidase with an Alternative Oxygen Acceptor. Am Clin Biochem 6: 24, 1969

118. Trudeau DL, Freier EF: Determination of Calcium in Urine and Serum by Atomic Absorption Spectrophotometry (AAS). Clin Chem 13: 101, 1967

119. Valle BL, Thiers RE: Flame Photometry Treatise on Analytical Chemistry. Kalthoff IM, Elving PJ (eds), New York, Interscience, Vol 6, 1965 p. 3463

120. Van Anken HC, Schiphorst ME: A Kinetic Determination of Ammonia in Plasma. Clin Chim Acta 56: 151, 1974

121. Van den Bergh AAH, Snapper J: Eine Quantitative Bestimnung des Bilirubins im Blutserum. Deut Arch Klin Med 110: 540, 1913

122. Walker WEC, Ulmer DD, Vallee BL: Metal Isoenzymes and Myocardial Infarction, Malic and Lactic Dehydrogenase Activities and Zinc Concentration in Serum. N.Engl. J. Med 255: 449, 1956

123. Wefler V, Raferty A: A Stable Reagent for the Libermann-Burchard Reaction. Clinical Chemistry, Tietz N (ed), Philadelphia, W.B. Saunders, 1976

124. Weichselbaum TE: Accurate and Rapid Method for Determination of Proteins in Small Amounts of Blood Serum and Plasma. Am J Clin Path 16: 40, 1946

125. Werner M: The Concept of Normal Values in Laboratory Medicine. Edited by Race GJ, New York, Harper and Row, 1976

126. Wilkinson JH: An Introduction to Diagnostic Enzymology. Baltimore, Williams and Wilkins, 1962

127. With TK: Bile Pigments: Chemical, Biological and Clinical Aspects, New York, Academic Press, 1968

128. Wolf PL, Williams D, Tsudaka T, Acosta L: Methods and Techniques in Clinical Chemistry. New York, Wiley-Interscience, 1972

129. Wolf PL, Williams D, Von der Muehll E: Practical Clinical Enzymology and Biochemical Profiling. New York, John Wiley and Sons, 1973

130. Wroblewski F, Ladue JS: Serum Glutamic Pyruvic Transaminase (SGPT) in Hepatic Disease. Preliminary Report. Ann Intern Med 45: 801, 1956

131. Wybenga DR, Pileggi VJ, Dirstine PH, Digiorgio J: Direct Manual Determination of Serum Total Cholesterol with a Single Stable Reagent. Clim Chem 16: 980, 1970

132. Wybenga DR, DiGiorgio J, Pileggi VJ: Manual and Automated Methods of Urea Nitrogen Measurement in Whole Serum. Clin Chem 17: 891, 1971

133. Young DS: Standardized Reporting of Laboratory Data: The Desirability of Using SI Units. New Eng J Med 290: 368, 1974

134. Young DS: Normal Laboratory Values (case records of the Massachusetts General Hospital) in SI Units. New Engl J Med 292: 795, 1975

135. Zall DM, Fisher D, Garner MQ: Photometric Determination of Chloride in Water. Anal Chem 28: 1665, 1956

Section VI

1. Ahonen A, Myllyla VV, Hokkanen E: Measurement of Reference Values for Certain Proteins in Cerebrospinal Fluid. Acta Neural Scand 58: 358, 1978

2. Albritton AB: Standard Values in Blood, Philadelphia, W.B. Saunders, 1952

3. Albritton AB: Standard Values in Blood. Philadelphia, W.B. Saunders, 1958

4. Aller JR, Carsters LA: Hematologic Alterations Observed in Newly Acquired Monkeys during their Period of Isolation. Lab Anim Care 15:103, 1965

5. Altman PL, Dittmer DS: Biology Data Book. Washington, D.C., FASEB, 1974

6. Altman PL, Dittmer DS: Blood and Other Body Fluids, Washington, D.C., Fed of Am Soc for Exptl Biol, 1964

7. Altman PL, Dittmer DS: Blood and Other Body Fluids, Washington, D.C., Fed of Am Soc for Exptl Biol, 1968

8. Altman PL, Dittmer DS: Blood and Other Body Fluids, Washington, D.C., Fed of Am Soc for Exptl Biol, 1971

9. Altman PL, Dittmer DS: Blood and Other Body Fluids, Washington, D.C., Fed of Am Soc for Exptl Biol, 1974

10. Altman PL, Katz DD: Human Health and Disease. Federation of American Societies for Experimental Biology, Bethesda, MD, 1977

11. Altshuler HL, Stowell RE, Lowe RT: Normal Serum Biochemical Values of *Macaca Arctoides, Macaca fascicularis* and *Macaca radiata*. Lab Anim Sci 21: 916, 1971

12. Altshuler HL, Stowell RE: Normal Serum Biochemical Values of *Cercopithecus Aethiops, Cercocebus Atys* and *Presbytis Entellus*. Lab Anim Sci 22: 692, 1972

13. Amador E, Bartholomew HP, Massod MF: An Evaluation of the Average of Normals and Related Methods of Quality Control. Am J Clin Pathol 50: 369, 1968

14. Ames AK, Sakanone M, Shinichiso M: Sodium Potassium, Calcium, Magnesium and Chloride in Choroid Plexus Fluid and Cisternal Fluid Compared with Plasma Ultra Filtration. J Neurophysiol 27: 672, 1964

15. Anderson DR: Normal Values for Clinical Blood Chemistry Tests for Macaca Mulatta Monkeys. Am J Vet Res 27: 1484, 1966

16. Anderson JR, Benirschke K: The Armadillo, *Dasypus Novemcinctus*, in Experimental Biology. Lab Anim Care 16: 202, 1966

17. Ayers JL, Ramsey FK, Vreeken E, Jowett D: Serum Glutamic Acid Decarboxylase Activity in Normal Dogs. Am J Vet Res 39: 1802, 1978

18. Baetz AL: Phosphates, Phosphokinase and Transferase Levels in Blood and Tissues of Domestic Animals. Technicon International Congress Vol 3, New York, Mediad Incorp, 1970

19. Barsanti JA, Finco DR: Protein Concentration in Urine of Normal Dogs. Am J Vet Res 40: 1583, 1979

20. Bass AD: Excretion of Uric Acid and Allantoin by Rats Depleted of Liver Xanthene Oxidase. Proc Soc Expt Biol Med, 73: 687, 1950

21. Bauer JD, Ackermann PG, Toro G: Clinical Laboratory Methods, St. Louis, The CV Mosby Company, 1974

22. Becker SV, Schmidt DA, Middleton CC: Selected Biological Values of the African White-tailed Sand Rat. Lab Anim Science 29: 479, 1979

23. Behrens H: Der Liquor Cerebrospinal dis Pferdis Seine/Entnahme Untersuchung und Diagnostiche Bedentung. Proc 15th International Vet Congr 2: 1031, 1953

24. Beland MF, Sehgal PK, Peacock WC: Baseline Blood Chemistry Determinations in the Squirrel Monkey. Lab Anim Science 29: 195, 1979

25. Benirschke K, Garner FM and Jones TC (editors): Pathology of Laboratory Animals, Vols. I and II, New York, Springer-Verlag, 1978

26. Benjamin MM: Outline of Veterinary Clinical Pathology, Second Edition. Ames, Iowa, Iowa State Press, 1961

27. Benjamin MM: Outline of Veterinary Clinical Pathology. Ames, Iowa, The Iowa State University, 1978

28. Bernheim F, Bernheim MLC: Note on the Effect of Caffeine on Ammonia and Urea Excretion in Rabbits. Amer J Physiol 145: 115, 1945

29. Besch EL, Chou BJ: Physiological Responses to Blood Collection Methods in Rats. Proc Soc Exp Biol Med 138: 1019, 1971

30. Best WF, Mason CC, Barron SS, Shephard HG: Automated Twelve Channel Serum Screening. Med Clin N Amer 53: 175, 1969

31. Bito LZ, Davson H: Local Variation in Cerebrospinal Fluid Composition of the Extracellular Fluid of the Cortex. Exptl Neurol 14: 264, 1966

32. Blair CB Jr.: Care and Maintenance of Nine Banded Armadillo, *Dasypus Novem Cinctus Mexicanus,* in a Laboratory Environment. Bull Ora Aca Sci 28: 113, 1970

33. Bloom F: The Blood Chemistry of the Dog and Cat. New York, Gamma Publications, Inc., 1960

34. Blosser TH, Smith VR: Parturient Paresis VI. Some Changes in the Urinary Excretion of Certain Constituents at Parturient and their Possible Association with Changes in the Blood Picture. J. Dairy Sci 33: 329, 1950

35. Blue Spruce Farm Technical Bulletin: Blue Spruce Farms, Inc., New York, Altmont, 1970

36. Bonilla CA: Use of Automated Instrumentation to Establish Normal Serum Parameters in Experimental Animal Models. Adv Automated Anal Technicon Int Congr 7: 85, 1972

37. Boyd JW: The Comparative Activity of Some Enzymes in Sheep, Cattle, Rats Normal Serum and Tissue Levels and Changes During Experimental Liver Necrosis. Res Vet Sci 3: 256, 1962

38. Brunden MN, Clark JJ, Sutter ML: A General Method of Determining Normal Ranges Applied to Blood Values for Dogs. Am J Clin Path 53: 332, 1970

39. Bulgin MS, Shifrine M, Galligar SJ: Electrophoretic Analysis of Normal Beagle Serum Proteins. Lab Anim Sci 32: 275, 1971

40. Burns KF, DeLannoy CE Jr.: Compendium of Normal Blood Values of Laboratory Animals with Indication of Variations. Toxic Appl Pharmacol 8: 429, 1966

41. Burns KR, Ferguson FG, Hampton SH: Compendium of Normal Blood Values for Baboons, Chimpanzees and Marmosets. Am J Clin Path 48: 484, 1967

42. Burns EL, Usher GS, Taylor TW, *et al.:* Individual Biorange Patterns. J Occup Med 13: 138, 1971

43. Butler TM, Wiley GL: Baseline Values for Adult Baboon Cerebrospinal Fluid. Lab Anim Sci 21: 123, 1971

44. Byers SO, Freidman M: Rate of Entrance of Urate and Allantoin into the Cerebrospinal Fluid of the Dalmatian and Non Dalmatian Dog. Amer J Physiol 157: 394, 1949

45. Byrd DF: Union Carbide—Biochemical Data, 1973

46. Caisey JD, King DJ, Clinical Chemical Values for Some Common Laboratory Animals. Clin Chem 26: 1877, 1980

47. Caraway WT: Accuracy in Clinical Chemistry. Clin Chem 17: 63, 1971

48. Cardeilhac JC, Cardeilhac PT: Preliminary Studies on Lactic Dehydrogenase Activity in Blood and Tissues from Normal and Leukemic Cattle. Amer J Vet Res 23: 293, 1962

49. Cardinet GH, Littrel JF, Freeland RA: Comparative Investigation of Serum Creatine Phosphokinase and Glutamic-Oxaloacetic Transaminase Activities in Equine Paralytic Myoglobinuria. Res Vet Sci 8: 219, 1967

50. Carmichael J, Jones ER: The Cerebrospi-

nal Fluid in the Bovine: Its Composition and Properties in Health and Disease with Special Reference to Turning Sickness. J Comp Path (London) 52: 222, 1939

51. Carmichael EB, Petcher PW: Constituents of the Blood of Some Hibernating and Normal Rattle Snake, *Crotalus Horridus.* J Biol Chem 161: 693, 1945

52. Cohen EA: A Comparison of the Total Protein and Albumin Content of the Blood Sera of Some Reptiles. Science 119: 98, 1954

53. Coles EH: Veterinary Clinical Pathology. Philadelphia, W.B. Saunders, 1967

54. Coles EH: Veterinary Clinical Pathology, Philadelphia, London, Toronto, W.B. Saunders Company, 1980

55. Cornelius CE, Bishop J, Swizer J, Rhode EA: Serum and Tissue Transaminase Activities in Domestic Animals. Cornell Vet 49: 116, 1959

56. Crawley GJ, Swenson MJ: Comparison of Two Tests for Serum Transaminase Activities and the Effects of Age of Dog and Method of Handling Samples on Certain Serum Enzymes of Dogs. Am J Vet Res 26: 1468, 1965

57. Criton L, Exley D, Hallpike C: Formation of Circulation and Chemical Properties of Labyrinthine Fluids. J Brit Med Bull 12:101, 1956

58. Criton L, Exley D: Recent Work on the Biochemistry of Labyrinthine Fluids. Proc Royal Soc M (London) 50:697, 1957

59. Davison MM: Stability of Acid Phosphatase in Frozen Serum. Am J Clin Path 23: 411, 1953

60. Davson H: Physiology of the Ocular and Cerebrospinal Fluids, Boston, Little Brown and Co., 1956

61. Dawson CA, Nadel ER, Horvath SM: Arterial Blood and Muscle Lactates during Swimming in the Rat. J Appl Physiol 30: 322, 1971

62. De La Pena A. Goldzieher JW: The Baboon in Medical Research, Vol 2, Edited by Vogtborg H. Austin, University of Texas Press, 1967

63. De La Pena A, Goldzieher JW: Clinical Parameters of the Normal Baboon, in, The Baboon in Medical Research. Edited by Vogtborg H, Vol. 1, Austin, University of Texas Press, 1965

64. De la Pena A, Matthijssen C, Goldzieher JW: Normal Values for Blood Constituents of the Baboon (Papio Species). Lab Anim Care 20: 251, 1970

65. De la Pena, *et al.:* Normal Values for Blood Constituents of the Baboon, Part II. Lab Anim Sci 22: 249, 1972

66. Derwelis SK, Butler TM, Fineg J: Baseline Values of Cerebrospinal Fluid from the Chimpanzee (*Pan Troglodytes*). Lab Anim Care 20: 107, 1970

67. Dessaner HC, Fox W, Gilbert NL: Plasma Calcium, Magnesium, and Protein of Viviparous *Colubrid* Snakes during Estrous Cycle. Proc Soc Exp Biol Med 92: 299, 1956

68. Dimopouleos GT: Plasma Proteins in Clinical Biochemistry of Domestic Animals. Kaneko JJ, Cornelius CE (eds) Vol 1, Second Edition, New York Academic Press, 1972

69. Dinning JS, Day PL: Creatinuria during Recovery from Aminopterin Induced Folic Acid Deficiency in the Monkey. J Biol Chem 181: 897, 1949

70. Dixon E, Bowie WC: Serum and Cerebrospinal Fluid: Their Diagnostic and Prognostic Value. Tuskegee Vet 6: 45, 1962

71. Dumm HC, King JW (editors): Handbook of Clinical Laboratory Data. Cleveland, The Chemical Rubber Co, 1965

72. Dunloo RS: The Use of Animals in Research. Am J Med Tech 35: 423, 1969

73. Earl FL, Melveger BE, Reinwall JE, Wilson RL: Clinical Laboratory Values of Neonatal and Weaning Miniature Pigs. Lab Anim Sci 21: 754, 1971

74. Ellinger P, Kader AMM: Nicotinamide Metabolism in Mammals, Biochem J 44: 77, 1949

75. Emminger A, Reznik G, Reznik-Schuller H, Mohr U: Differences in Blood Values Depending on Age in Laboratory Bred European Hamsters (*Cricetus Cricetus*) Lab Anim 9: 33, 1975

76. Fankhauser R: Der Liquor Cerebrospinalis in der Veterinar Medizin. Zbl Vet Med 1: 136, 1953

77. Fankhauser R, Innes JRM, Saunders LZ: Comparative Neuropathology, New York, Academic Press, 1962

78. Farris EJ, Griffith J Jr (Eds): The Rat in Laboratory Investigation, Second Edition. New York, Hafner Publ Col, 1963

79. Fedetov AI: Cerebrospinal Fluid of Domestic Animals, Translated by Saunders LZ, Distributed by Russian Scientific

Translation Program, Division Gen Med Sci, NIH, 1960

80. Fenn WO: Electrolytes in Muscles. Physical Rev 16: 450, 1936

81. Field JB, Elvelijem CA, Chancey J: A Study of the Blood Constitution of the Carp and Trout. J Biol Chem 148: 261, 1943

82. Finch CE, Foster JR: Hematologic and Serum Electrolyte Values of the C57BL/6J Male Mouse in Maturity and Senescence. Lab Anim Sci 23: 339, 1973

83. Flatt RE, Carpenter AB: Identification of Crystalline Material in Urine of Rabbits. Amer J Vet Res 32: 655, 1971

84. Fletcher WS, Rogers AL, Erikson LF: The Need for a Large Standard Laboratory Dog. Lab Anim Care 16: 1, 1966

85. Foster: Baseline Data on Charles River CD Rats, Charles River Digest, 1973

86. Fox RR, Laird CW, Blau EM, Schultz HS, Mitchell BP: Biochemical Parameters of Clinical Significance in Rabbits, I Strain Variation and II Diurnal Variation. J Heredity 61: 267, 1970

87. Frankel S, Reitman S, Sonnenwirth AC: Gradwohl's Clinical Laboratory Methods and Diagnosis, Second Edition, Volume 2 St. Loius, Mosby, 1970

88. Free AH, Free HM: Urinalysis in Clinical Laboratory Practice. Cleveland CRC Press, Inc. 1976

89. Freedland RA, Hjerpe CA, Cornelius CE: Comparative Studies on Plasma Enzyme Activities in Experimental Necrosis in the Horse. Res Vet Sci 6: 18, 1965

90. Freidman M: Observation Concerning the Effects of (I) Sodium Salicylate and (II) Sodium Salicylate and Glycine upon the production and Excretion of Uric Acid Allantoin in the Rat. Amer J Physiol 152: 302, 1948

91. Friedman SB, Austin WB, Rieselbach RE, Bloc, JB et al.: Effect of Hypochlorinia on Cerebrospinal Fluid Chloride Concentration in a Patient with Anorexia Nervosa and in dogs. Proc Soc Exp Biol Med 114: 801, 1963

92. Galen RS, Gambino SR: Beyond Normality: The Predictive Value and Efficiency of Medical Diagnosis, New York, John Wiley and Sons, 1976

93. Garcia FG, Hunt RD: The Hematogram of the Squirrel Monkey. Lab Anim Care 16: 50, 1966

94. Gaumer AEH, Goodnight CJ: Some Aspects of the Hematology of Turtles as Related to their Activity. Amer Midl Nat 58: 332, 1957

95. Gibbons LV, Kaplan HM: Blood Protein and Mineral Chemistry in Frog Red Leg Disease. Copeia II: 176, 1958

96. Gilman AR: The Blood Sedimentation Rate in the Horse. Amer J Vet Res 13: 77, 1952

97. Griffith JQ: The Rat in Laboratory Investigation. Second Edition, Philadelphia, JB Lippincott, 1949

98. Hahn PF, Baugh P, Heng HC: Alteration of Plasma Electrophoretic Patterns and Anaphylactic Reactions in Protein Depleted Dogs. Proc Soc Exp Biol Med 102: 448, 1956

99. Hamada T, Tabata T, Kusita Y, et al.: Peripheral Blood Analyses of Various Strains of Mice and Rats. Rep Nat Inst Gen (Japan) 10: 31, 1960

100. Hambleton P, Harris-Smith PW, Baskerville A, Bailey NE, Pavey KJ: Normal Values for Some Whole Blood and Serum Components of Grivet Monkeys (Cereopethecus Aethiops). Lab Animals 13: 87, 1979

101. Hansard SL, Comar CL, Plumlee MP: Absorption and Tissue Distribution of Radio Calcium in Cattle. J Anim Sci II: 524, 1952

102. Haris EK, De Mets DL: Estimation of Normal Ranges and Cumulative Proportions by Transforming Observed Distributions to Gaussian Form. Clin Chem 18: 605, 1972

103. Harkness JE, Wagner JE: The Biology and Medicine of Rabbits and Rodents. Philadelphia, Lea and Febiger, 1977

104. Harrison SD Jr, Burdeshaw JA, Crosby RG, Cusic AM, Denise EP: Hematology and Clinical Chemistry Reference Values for Cancer Res 38: 2636, 1978

105. Hartman FA, Lewis LA, Brownell KA, Shelden EF, Walther RF: Some Blood Constituents of the Normal Snake. Physiol Zool 14: 476, 1941

106. Hashimoto K, Nukata S: Plasma Glucose Levels in the Frog. Fol Pharm Jap 47: 9, 1951

107. Hawkins WB, McFarland ML, McHenry EW: Nitrogen Metabolism in Pyridoxine Insufficiency. J Biol Chem 166: 233, 1946

108. Held IR, Freeman S: Binding of Calcium by Human Plasma Proteins Under Simulated Physiologic Conditions. J Appl Physiol 19: 292, 1964

109. Henry JB, Speicher CE: Pitfalls in Interpreting Laboratory Measurements. Postgrad Med 50: 147, 1971

110. Henry RJ: Clinical Chemistry Principles and Techniques. New York, Harper and Row, 1979

111. Henry RJ, Cannon DC, Winkelman JW: Clinical Chemistry: Principles and Technics. Second Edition, Hagerstown, Maryland, Harper and Row, Publishers, 1974

112. Hensley JC, Langham WH: Comparative Fundamental Physiological Parameters of Macaca Mulatta and Macaca Speciasen. Lab Anim Care 14: 105, 1964

113. Herrin RC: The Production and Excretion of Urea and Oxygen Consumption During Intestinal Obstruction in the Rabbit. Amer J Physiol 149: 492, 1947

114. Hill EG: Swine in Biomedical Research. Edited by Bustad LK and McClellan RO, Seattle, Frayn, 1966, pp 163–168

115. Hill BF (Ed): Some Physiological Parameters of Small Animals. Charles River Digest 10, 1971

116. Himwich WA, Himwich HE: Cerebral Circulation, Blood-Brain Barrier and Cerebrospinal Fluid, in, Dukes Physiology of Domestic Animals, Edited by Swenson, MJ, Ithaca, Cornell University Press, 1970

117. Hoffman G(Ed): Abriss der Laboratoriumstierkunde. Jena, G. Fisher, 1961

118. Holmes AN, Passovoy M, Capps RB: Marmosets as Laboratory Animals, III Blood Chemistry of Laboratory-Kept Marmosets with Particular Attention to Liver Function and Structure. Lab Anim Care 17: 41, 1967

119. Houpt RT: Water, Electrolytes, and Acid-Base Balance, in, Dukes Physiology of Domestic Animals, Edited by Swenson MJ, Ithaca, Cornell University Press, 1970

120. Hort I: Paper Electrophoretic Fractionation and Chemical Determination of Horse Serum Proteins and Lipoproteins. Amer J Vet Res 29: 813, 1968

121. Hudgins PC, Cummings MM, Patnode RA: Electrophoretic Distributions of Serum Proteins in the Rabbit, Guinea Pig and Rat Following BCG Administration. Proc Soc Expt Biol 92: 75, 1956

122. Hutton KE: The Blood Chemistry of Terrestrial and Aquatic Snakes. J Cell Comp Physiol 52: 319, 1958

123. Hutton KE, Goodnight CJ: Variations in the Blood Chemistry of Turtles Under Active and Hibernating Conditions. Physiol Zool 30: 198, 1957

124. Jacobi H, Fontaine R: Natrium-Kalinn and Chloridgehalt von Hundensin bei Standard futterung. Z Verusuchstierk 9: 205, 1967

125. Jakobson PE, Monstgard J: Investigation of the Serum Proteins in Pigs from Birth to Maturity. Nord Vet Med 2: 812, 1950

126. Juhn SK, Haugen J, Steven L: Some Chemical Parameters of Serum, Cerebrospinal Fluid, Perilymp and Aqueous Humor of the Chinchilla. Lab Anim Sci 24: 691, 1974

127. Jenning FW, Mulligan W: Levels of Some Chemical Constituents in Normal Horse Serum. J Comp Path Therap 62: 286, 1953

128. Jones RT: Normal Values for Some Biochemical Constituents in Rabbits. Lab Anim 9: 143, 1975

129. Kaneko JJ, Cornelius CE: Clinical Biochemistry of Domestic Animals Vol I, Second Edition, New York, Academic Press, 1972

130. Kaneko JJ, Cornelius CE: Clinical Biochemistry of Domestic Animals Vol II, Second Edition, New York, Academic Press, 1972

132. Kesner L, Muntwyler E: Determination of Non-Nitrogenous Organic Acids in the Urine in Conditions of Altered Acid Base Balance. J Lab Clin Med 61, 604, 1963

133. King TO, Fargus JC: Normal Blood Values of the Adult Female Monkey (M. Mulatta). Lab Anim Care 17: 391, 1967

131. Kaplan HM, Timmons EH: The Rabbit: A Model for the Principles of Mammalian Physiology and Surgery. New York, Academic Press, 1979

134. Kozma CK: Electrophoretic Determinations of Serum Proteins of Laboratory Animals. J Am Vet Med Assoc 151: 865, 1967

135. Kozma C, Weisboth SH, Stratman SL, Conejeros M: Normal Biological Values for Long-Evans Rats. Lab Anim Care 19: 697, 1969

136. Krise GM Jr, Wald N: Normal Blood Values of the Adult Female Monkeys (*Macaca Mulatta* Monkey). J Appl Physiol 12: 482, 1958

137. Kritchevsky D: Cholesterol, New York, John Wiley and Sons, 1958, p. 279

138. Ladenson JN: Non Analytical Sources of

Variation in Clinical Chemistry Results. In, Gradwohl's Clinical Laboratory Methods and Diagnosis. Sonenwirth AC, Jarett L (editors). St Louis. The CV Mosby Company, 1980

139. Laird CW: Clinical Pathology: Blood Chemistry, in Handbook of Laboratory Animal Science, Vol II, Melby EC, Altman NH (Editors), Cleveland, CRC Press, Inc., 1974

140. Laird CW: Representative Values for Animal and Veterinary Populations and Their Clinical Significances. Houston, Hycel, Inc., 1972

141. Lane DR, Robinson R: The Utility of Biochemical Screening in Dogs. Br Vet J 126: 136, 1970

142. Ledoux A: The pH of Labyrinth Fluid (cat) Bull Soc R Sc Liege 4: 254, 1943

143. Lewis JH, Doyle AP: Coagulation, Protein and Cellular Studies on Armadillo Blood. Comp Biol Chem Physiol 12: 61, 1964

144. Lloyd JW, Peterson RA, Collins WE: Effects of an Avian Ultrabronchial Extract in the Domestic Fowl. Poultry Sci 49: 1117, 1970

145. Lossnitzer K, Bajusz E: Water and Electrolyte Alterations During the Life Course of the B10 14-6 Syrian Golden Hamster. A Disease Model of Hereditary Cardiomyopathy. J Mol Cell Cardiol 6: 163, 1974

146. Lumsden JH, Mullen K, McSherry BJ: Canine Hematology and Biochemistry Reference Values. Canad J Comp Med 43: 125, 1979

147. Lumsden JH, Mullen K, Rowe R: Hematology and Biochemistry Reference Values for Female Holstein Cattle. Canad J Comp Med 44: 24, 1980

148. McClellan RO, Vogt GS, Ragan HA: Age-related Changes in Hematological and Serum Biochemical Parameters in Miniature Swine. Swine in Biochemical Research. Seattle, Frayn Publishing Co., 579, 1966

149. McClure HM, Guilland NB, Keeling ME: Clinical Pathology Data for the Chimpanzee and Other Anthropoid Apes, in, The Chimpanzee Vol 6, Basel, Karger, 1973

150. Mc Kelvie DH, Powers S, McKim F: Microanalytical Procedures for Blood Chemistry Long Term Study on Beagles. Am J Vet Res 27: 1405, 1966

151. Majumdar DN, Dasa-Gupta CR: The Blood Picture of the Normal Monkey. Indian J of Med Res 32: 101, 1944

152. Malinow MR, Pope BL, Depaoli BL Jr, Katz S: Laboratory Observations on Living Howlers. Bibl Primate 7: 224, 1968

153. Manning PJ, Lehner NDM, Felder MH, *et al.*: Selected Hematologic, Serum Chemical and Arterial Blood Gas Characteristics of Squirrel Monkeys (*Saimiri Sciureus*). Lab Anim Care 19: 831, 1969

154. Mays A Jr: Baseline Hematological and Blood Biochemical Parameters of the Mongolian Gerbil (*Meriones Unguiculatus*). Lab Anim Care 19: 838, 1969

155. Medway W, Prier JE, Wilkinson JS (Eds.): A Textbook of Veterinary Clinical Pathology. Baltimore, Williams and Wilkins, 1969

156. Medway W: Results of Some Blood Studies on 60 Light Horses and 10 Ponies in Pennsylvania. Technicon International Congress Vol 3, New York, Mediad Incorporated, 1970

157. Melby EC Jr, Altman NH: Handbook of Laboratory Animal Science, Cleveland, CRC Press Inc., 1974

158. Michaelson SM, Scheer K, Gilt S: The Blood of Normal Beagle. J Amer Vet Med Assoc 148: 532, 1966

159. Miller GE, Danzig LS, Talbott JH: Urinary Excretion of Uric Acid in Dalmatian and Non-Dalmatian Dog Following Administration of Diadrast Sodium Salicylate and Mercurial Diuretic. Amer J Physiol 164: 155, 1951

160. Miller ER, Ullrey DW, Ackerman I, Schmidt DA, Hoeffer JA, Luecke RW: Swine Hematology from Birth to Maturity. I. Serum Proteins. J Anim Sci 20:31, 1961a

161. Miller ER, Ullrey DW, Ackerman I, Schmidt DA, Lueke RW, Hoefer JA: Swine Hematology from Birth to Maturity. II. Erythrocyte Population Size and Hemoglobin Concentration. J Anim Sci 20: 89, 1961b

162. Misanik LF: A Multiphase Profile of 500 Physicians, in, Advances in Automated Analysis. Technicon International Congress, Vol 3, New York, Mediad Incorporated, 1970, p. 25

163. Mitruka BM, Rawnsley HM, Vadehra DV: Animals for Medical Research: Models for the Study of Human Disease. New York, John Wiley and Sons, Inc., 1976

164. Mitruka BM, Rawnsley HM: Clinical Biochemical and Hematological Reference Values in Normal Experimental Animals. New York, Masson Publishing USA, Inc., 1977

165. Moore DH: Species Differences in Serum Protein Patterns. J Biol Chem 161: 21, 1945

166. Mooreland AF: (Editor) Biological Values for Various Laboratory Animals. Lab Anim Digest 9: 41, 1974

167. Morgulis S, Spencer HC: Metabolism Studies in Nutritional Muscular Dystrophy. J Nutr 12: 191, 1936

168. Morrow A, Terry M: Enzymes in Blood of Non-human Primates Tabulated from Literature. I. Cholinesterase, Alkaline Phosphates and Other Hydrolyzing Enzymes. Primate Information Centre, Regional Primate Research Center, University of Washington, 1968a

169. Morrow A, Terry M: Enzymes in Blood of Non-human Primates Tabulated from the Literature. II. Serum Glutamic Oxaloacetic Transaminase and Glutamic Pyruvic Transaminase. Primate Information Centre, Regional Primate Research Centre, University of Washington, 1968b

170. Morrow A, Terry M: Enzymes in Blood of Human Primates Tabulated from the Literature. III. Oxido-reductases Including Dehydrogenases. Primate Information Centre, Primate Research Centre, University of Washington, 1968c

171. Munan L, Kelly A, Petitclerc B, Billon B: Atlas Des Normes Sanguines (Atlas of Blood Data) Epidemiology Laboratory, Quebec, Canada, Faculty of Medicine, 1978

172. Neumann GJ: The Determination of Normal Ranges from Routine Laboratory Data. Clin Chem 14: 979, 1968

173. New AE: The Squirrel Monkey. Edited by Rosenblum LA, Cooper RW. New York, Academic Press, 1968

174. Nomura G, Tamura M, Hirohashi K, Nagasi S: Comparison of Constituents of Blood of Several Experimental Animals with Human Blood. Experimental Animals (Tokyo) 22: 321, 1975

175. Ogasawara K: Inorganic Substances in the Animal Serum I. Total and Ionized Calcium and Total and Ionized Magnesium contents in the Serum. Igku to Scibut-Sugaku 29: 250, 1953

176. Ogasawara K: Blood Electrolytes of Fishes I. Magnesium and Inorganic Phosphate in the Carp. Chem Abstr 48: 2797, 1954

177. Oser FO, Lang RE, Vogin EE: Blood Values in Stumptailed Macaques (*M. arctoides*) under Laboratory Conditions. Lab Anim Care 20: 462, 1970

178. Oslen RE, Burns KF: Serum Chemistry Values of Zoo Animals as Determined by Two Different Automated Procedures. Hycel Vet Symp 1:59, 1972

179. Pappenheimer JR, Heisey SR, Jordan EF, Downer JC: Perfusion of the Cerebral Ventricular System in the Unanesthetized Goat. Amer J Physiol 203: 763, 1962

180. Petrey JJ: Ultra-microanalysis of Selected Blood Components of Normal *Macaca Mulatta*. Lab Anim Care 17: 342, 1967

181. Pick JR, Eubanks JW: A Clinicopathologic Study of Heterogenous and Homogenous Dog Population in North Carolina. Lab Anim Care 15: 11, 1965

182. Pierce JR, Laird CW: Normal Blood Chemistry Values of Veterinary Importance—a Preliminary Report. Hycel Vet Symp 1:59, 1972

183. Planas J, Balasch J: Correlation Between Serum Iron and Copper in Different Animals. Rev Exp Fisiol 26: 91, 1970

184. Platner WS: Blood Electrolytes of Fishes. Univ Mich Microfilm Public 1231:289, 1949

185. Pond WG, Banis RJ, Van Vleck LD, Walker EF Jr., Chapmann P: Age Changes in Body Weight and in Several Blood Components of Conventional Versus Miniature Pigs. Proc Soc Exp Biol Med 127: 895, 1968

186. Pridgen WA: Values for Blood Constituents of the African Green Monkey (*Cercopithecus Aethiops*). Lab Anim Care 17: 463, 1967

187. Pryce JD: Interpretation of Laboratory Results. Postgrad Med 36: A56, 1964

188. Ralston Purina Co. Bulletin: Normal Blood Values for Dogs, 1973

189. Rao GN, Shipley EG: Data on Selected Clinical Blood Chemistry Tests of Adult Female Rhesus Monkeys (*M. Mulatta*). Lab Anim Care 20: 226, 1970

190. Rawnsley HM: Clinical Laboratory Data. Dartmouth-Hitchcock Medical Center, Unpublished Reference Value Data, 1981

191. Rawnsley HM: William Pepper Laboratory—Methods in Clinical Chemistry and Hematology. The Hospital of the University of Pennsylvania, 1976

192. Robinson PF: General Aspects of Physiology, in, The Golden Hamster, Its Biology and Use in Medical Research, Hoffman, Robinson and Magalhaes (Eds), Ames, Iowa, Iowa State University Press, 1965

193. Robinson FR, Gisler DB, Dixon DF Jr: Factors Influencing "Normal" SGOT Levels in the Rhesus Monkey. Lab Anim Care 14: 275, 1964

194. Robinson FR, Ziegler RF: Clinical Laboratory Values of Beagle Dogs. Lab Anim Care 18: 39, 1968

195. Roeder F, Rehm O: Die Ceribrospinalflussigkeit Untersucchungsmethoden und Klink for Artze und Tieratze, Berlin, Springer-Verlag, 1942

196. Rollins JB, Hobbs CH, Sperzel RO, Mc Connel S: Hematologic Studies of the Rhesus Monkey (*Macaca Mulatta*). Lab Anim Care 20: 681, 1970

197. Rosenthal ML Cavitz L: Organ, Urine and Feces Vitamin B12 Content of Normal and Starved Rabbits. J Nutr 64: 281, 1958

198. Rothstein R, Hunsaker D II: Baseline Hematology and Blood Chemistry of the South American Wooly Opossum (*Galuromys Derbianus*). Lab Anim Sci 22: 227, 1972

199. Ruppanner R, Norman BB, Adams CJ, Addis DG, Lofgreen GP, Clark JG, Dunbar J: Metabolic and Cellular Profile Testing in Calves Under Feedlot Components—Changes over Time in Feedlot. Am J Vet Res. 39: 845–9, 1978

200. Ruppanner R, Norman BB, Adams CJ, Addis DG, Lofgreen GP, Clark JG, Dunbar J: Metabolic and Cellular Profile Testing in Calves Under Feedlot Conditions: Minerals, Electrolytes and Biochemical Components References Values. Am J Vet Res 39: 841–4, 1978

201. Samadieh B, Bankowski RA, Carroll EJ: Electrophoretic Analysis of Serum Proteins of Chickens, Experimentally Infected with Marek's Disease Agent. Amer J Vet Res 30: 837, 1969

202. Sato T, Oda K, Kubo M: Hematological and Biochemical Values of Thoroughbred Foals in the First Six Month of Life. Cornell Vet 69: 3–19, 1978

203. Scarborough RA; A Compilation of Normal Blood Findings in Laboratory Animals. Thesis, Yale School of Medicine, 1926

204. Schermer S: The Blood Morphology of Laboratory Animals. Philadelphia, F.A. Davis Co., 1967

205. Schnell JV, Siddiquii WA, Geiman QM: Analysis of the Blood of Normal Monkeys and Owl Monkeys Infected with *Plasmodium Falsiparum*. Milit Med 134: 1068, 1969

206. Secord DC, Russell J: A Clinical Laboratory Study of Conditioned Mongrel Dogs and Labrador Retrievers. Lab Anim Sci 23: 567, 1973

207. Sheldon WG, Ross MA: A Generalized Herpes Virus Infection in Owl Monkeys. Report No. 670, Fort Knox, Kentucky, U.S. Army Medical Research Laboratory, 1966

208. Simesen MG: Calcium Inorganic Phosphorus and Magnesium Metabolism in Health and Disease, in, Clinical Biochemistry of Domestic Animals, Kaneko JJ, Cornelius CE (Eds) Vol I. Second Edition, New York, Academic Press, 1972

209. Smith ML, Lee R, Shepparo SJ, Fariss BL: Reference Ovine Serum Chemistry Values. Am J Vet Res 39: 321–2, 1978

210. Soliman MK, Amronsi SE, Youssef LB: Some Studies On-Cerebrospinal Fluid of Healthy Cattle. Zentralbl Veterinaer Med A12: 769, 1965

211. Sonnenwirth AC, Jarett L: Gradwohl's Clinical Laboratory Methods and Diagnosis. Volumes I and II. St Louis, Toronto, London, The CV Mosby Company, 1980

212. Sorenson DK, Kowalczyk T, Hentges JF: Cerebrospinal Fluid Pressure of Normal and Vitamin A Deficient Swine as Determined by a Lumbar Puncture Method. Amer J Vet Res 15: 258, 1954

213. Spector WS: Handbook of Biological Data. Philadelphia, W.B. Saunders Company, 1956

214. Spector WS: Handbook of Biological Data. Philadelphia, W.B. Saunders Company, 1961

215. Spector WS: Handbook of Biological Data. Philadelphia, W.B. Saunders Company, 1972

216. Spiller GA, Spaur CL, Amen RJ: Selected Blood Values for *Macaca Nemestrina* Fed Semipurified Fiber-free Liquid Diets. Lab Anim Sci 25: 341, 1975

217. Steyn DG: Standard Serum Chemical and Haematological Values in Chacma Baboon (*Papio Ursinus*). J of South African Vet Assoc 46: 235, 1975

218. Street RP Jr, Highman B: Blood Chemistry Values in Normal Mystromys Albicaudatus and Osborne-Mendel Rats. Lab Anim Science 21: 394, 1971

219. Strozier LM, Blair CB Jr, Evan BK: Armadillos: Basic Profiles. II. Serum Proteins. Lab Anim Sci 21: 602, 1971

220. Sunderman FW Jr: Current Concepts of "Normal Values," "Reference Values" and "Discrimination Values." Clin Chem 21: 1873, 1975

221. Swenson MJ: Dukes Physiology of Domestic Animals. Eighth Edition, Ithaca, Cornell University Press, 1970

222. Szabuniewica M, McCrady JD: Some Aspects of the Anatomy and Physiology of the Armadillo. Lab Anim Care 19: 843, 1969

223. Talmage RV, Buchanan GD: Armadillo (*Dasypus Novemcinctus*). A Review of Its Natural History, Ecology, Anatomy and Reproductive Physiology. Houston, The Rice Institute Publication, 1954

224. Tasker JB: Fluid and Electrolyte Studies in the Horse I. Blood Values in 100 Normal Horses. Cornell Vet 56: 67, 1966

225. Tasker JB: Reference Values for Clinical Chemistry Using the Coulter Chemistry System. Cornell Vet 68: 460–79, 1978

226. Tegeris AS, Earl FL, Curtis JM: Swine in Biochemical Research. Seattle, Frayn Publishing Co., 1966

227. Tennussen GHG, Verwer MJA: Cerebrospinal Fluids in Dogs. Proc 15th International Vet Congr 2: 1022, 1953

228. Thompson WHS: Determination and Statistical Analysis of the Normal Ranges of Five Serum Enzymes. Clin Chim Acta 21: 469, 1968

229. Thompson JH, Lee YH, Campbell LB: Urinary Electrolyte Excretion in Young Rat. Amer J Vet Res 27: 1093, 1966

230. Tietz NW(Ed): Fundamentals of Clinical Chemistry. Philadelphia, W.B. Saunders, 1976

231. Timmons EH, Marques PA: Blood Chemical and Hematological Studies in the Laboratory Confined Unanesthetised Opossum (*Didelphis Virginiana*). Lab Anim Care 19: 342, 1969

232. Tumbleson ME: Protein-Caloric Undernutrition in Young Sinclair (S-1) Miniature Swine: Serum Biochemic and Hematologic Values. Adv Automated Anal Technicon Int Congr 7: 51, 1972

233. Tumbleson ME, Burks MF, Evans PS, Hutcheson DP: Serum Electrolytes in Undernourished Sinclair (S-1) Miniature Swine. Am J Clin Nutr 25: 476, 1972

234. Tumbleson ME, Badger TM, Baker PC, Hutcheson DP: Systemic Oscillations of Serum Biochemic and Hematologic Parameters in Sinclair (S-1) Miniature Swine. J Anim Sci 35: 48, 1972

235. Tumbleson ME, Hutcheson DP: Serum Proteins and Hematologic Parameters in Mongrel Dogs. Adv Automated Anal Technicon Int Congr 7: 81, 1972

236. Tumbleson ME, Middleton CC, Tinsely OW, *et al.*: Serum Biochemic and Hematologic Parameters of Normal Miniature Swine from 4–9 months of Age. Lab Anim Care 19: 345, 1969

237. Tumbleson ME, Middleton CC, Wallarh JD: Serum Biochemic and Hematologic Parameters of Adult Aoudads (*Ammotragus Lervia*) in Captivity. Lab Anim Care 20: 242, 1970

238. Tumbleson ME, Hutcheson DP: Age-related Serum Cholesterol, Glucose and Total Bilirubin Concentrations of Female Dairy Cattle. Proc Soc Exp Biol Med 138: 1083, 1971

239. Tumbleson ME, Burks MD, Wingfield WE: Serum Protein Concentrations as a Function of Age in Female Dairy Cattle. Cornell Vet 63: 65, 1973

240. Turbyfill CL, Cramer MB, Dewes WA, *et al.*: Serum and Cerebrospinal Fluid Chemistry Values for the Monkey (*Macaca Mulatta*). Armed Forces Radiobiology Research I, November 1968

241. Turbyfill CL, Cramer MB, Dewes WA, Huguley JW: Serum and Cerebrospinal Fluid Chemistry Values for the Monkey (*Macaca Mulatta*). Lab Anim Care 20: 269, 1970

242. Ulrey DE, Miller ER, Brent BE, Bradley BL, Hoefer J: Swine Hematology from Birth to Maturity. IV. Serum Calcium. Magnesium, Sodium, Potassium, Copper, Zinc, and Inorganic Phosphorus. J. Animal Sci 26: 1024, 1967

243. Union Carbide Biochemical Data: 1973

244. Upton PK, Morgan DJ: The Effect of Sampling Technique on Some Blood Parameters in the Rat. Lab Anim Care 9: 85, 1975

245. Van Stewart E, Longwell BB: Normal Clinical Chemical Values for Certain Con-

stituents of Blood of Beagle Dogs 12± 1 months old. Am J Vet Res 30: 907, 1969

246. Vogin EE, Oser F: Comparative Blood Values in Several Species of Non-human Primates. Lab Anim Sci 21: 937, 1971

247. Vondruska JF: Certain Hematologic and Blood Chemical Values in Adult Stump-tailed Macaques (*Macaca Arctoides*). Lab Anim Care 20: 97, 1970

248. Vondruska JF, Greco RA: Certain Hematologic and Blood Chemical Values in Charles River CD Albino Rats. Bull Am Soc Vet Clin Path 2: 3, 1973

249. Vondruska JF, Greco RA: Certain Hematologic and Blood Chemical Values in Juvenile Rhesus Monkeys. Bull Am Soc Vet Clin Pathol 3: 27, 1974

250. Wesibroth SH, Flatt RE, Kraus AL (Eds): Biology of the Laboratory Rabbit. New York, Academic Press Inc., 1974, p 552

251. Weiser M, Stock W, Henk F: Univer die Koneyentration Einiger Melavoliten im Plasma Von Eitagskiiken. Wie Tieraerztl Monatsschr 52: 237, 1965 (Nutr Abstr Rev 35: 996, 1965)

252. Wellde BT, Johnson AJ, Williams JS, Langbehn HR, Sadeen EH: Hematologic, Biochemical and Parasitologic Parameters of the Night Monkey (*Aotus Trivirgatus*). Lab Anim Sci 21: 575, 1971

253. Werner M: The Concept of Normal Values in Laboratory Medicine. New York, Harper and Row, 1976

254. Whelan M: The Effect of Intravenous Injection of Inorganic Chlorides on the Composition of Blood and Urine. J Biol Chem 63: 585, 1925

255. Williams JS, Neroney FC, Hutt G, *et al.*: Serum Chemical Components in Mice Determined by the Use of Ultra Micro Techniques. J Appl Physiol 21: 1026, 1966

256. Wolf HG, Moshe S, Lein AK, *et al.*: Hematological Values for Laboratory Reared Marmosa Mitis. Lab Anim Sci 21: 249, 1971

257. Wolf PL, Williams D, Tsudaka T, Acosta L: Methods and Techniques in Clinical Chemistry. New York, Wiley Inter Science, 1972

258. Wolf PL, Williams D, Von der Muchll E: Practical Clinical Enzymology and Biochemical Profiling. New York, Wiley Interscience, 1973

259. Worden AN, Waterhouse CE, Sellwood EHB: Studies on Composition of Normal Cat Urine. J Small Anim Pract 1:11, 1960

260. Wretlind B, Orstadins K, Lindberg P: Transaminase and Transferase Activities in Blood Plasma and Tissues of Normal Pigs. Zentral Vet Med 6: 963, 1959

261. Yeary RA, Shannon IL, Prigmore JR: Cerebrospinal Fluid Electrolytes in the Dog. Amer J Vet Res 21: 306, 1960

262. Yu L, Pragay DA, Chang D, Wicher K: Biochemical Parameter of Normal Rabbit Serum. Clin Biochem 12: 83, 1979

Section VII

1. Ash KO: Reference Internals (Normal Ranges). A Challenge to Laboratoriano. Am J Med Technol 46: 504, 1980

2. Baselt RC, Wright JA, Cravey RH: Therapeutic and Toxic Concentrations of More than 100 Toxicologically Significant Drugs in Blood, Plasma or Serum. A Tabulation. Clin Chem 21: 44, 1975

3. Benirschke K, Garner FM, Jones TC (Eds): Pathology of Laboratory Animals. Vols I and II, New York, Springer-Verlag, 1978

4. Benjamin MM: Outline of Veterinary Clinical Pathology. Ames, Iowa, The Iowa State University Press, 1978

5. Benjamin MM, McKelvie DM, in, Pathology of Laboratory Animals. Benirschke K, Garner FM, Jones TC (Eds). New York, Springer-Verlag, 1978, p 1749

6. Caraway WT: Accuracy in Clinical Chemistry. Clin Chem 17: 63, 1971

7. Cherian GA, Hill JG: Percentile Estimates of Reference Values for Fourteen Chemical Constituents in Sera of Children and Adolescents. Amer J Clin Path 69: 24, 1978

8. Christian DG: Drug Interference with Laboratory Blood Chemistry Determination. Am J Clin Path 54: 118, 1970

9. Coles EH: Veterinary Clinical Pathology. Philadelphia, London, Toronto, W.B. Saunders Company, 1980.

10. Doaley JF: The Role of Clinical Chemistry in Chemical and Drug Safety Evaluation by Use of Laboratory Animals. Clin Chem 25: 345, 1979

11. Doumas BT, Perry BW, Sasse EA, Stroumfjord JV: Standardization in Bilirubin Assays. Evaluation of Selected Methods

and Stability of Bilirubin Solutions. Clin Chem 19: 984, 1973

12. Elking MP, Kabat HF: Drug Induced Modifications of Laboratory Test Values. Am J Hosp Pharm 25: 485, 1968

13. Fraser C, Peake MJ: Problems Associated with Clinical Chemistry Quality Control Materials Critical Reviews in Clin Lab Sci 12: 59, 1980

14. Friedel R, Mattanheimer H: Release of Metabolic Enzymes from Platelets During Blood Clotting of Man, Dog, Rabbit and Rat. Clin Chim Acta 30: 37, 1970

15. Friedman RE, Entine SM, Hirshberg SB, Anderson RB: Effect of Diseases on Clin Lab Tests. Clin Chem 26 (4 Suppl): 10-476D, 1980

16. Hayashi T, *et al.*: Individual Differences within Normal Values—A Study of the Normal Value of the Individuals. Rinsho Byori (Suppl) 38: 190, 1979

17. Henry RJ, Chiamori N, Goulub OJ, Berkman S: Revised Spectrophotometric Methods for the Determination of Glutamic-oxaloacetic Transaminase and Glutamic Pyruvic Transaminase and Lactic Acid Dehydrogenase. Am J Clin Pathol 34: 381, 1960

18. Henry JB: Todd, Sanford, Davidsohn: Clinical Diagnosis and Management by Laboratory Methods. Philadelphia, London, Toronto, W.B. Saunders Company, 1979

19. Hicks JM, Rouland GL, Buffone GJ: Evaluation of a New Blood Collecting Device ("Microtainer") That Is Suited for Pediatric Use. Clin Chem 22: 2034, 1976

20. Jone JD: Factors that Affect Clinical Laboratory Values. J. Med 22: 316, 1980

21. Laessig RH, Indriksons AA, Hassemer DJ, Paskey TA, Schwartz TH: Changes in Serum Chemical Values as a Result of Prolonged Contact with the Clot. Amer J Clin Path 66: 598, 1976

22. Laessig RH, Hassemer DJ, Passkey TA, Schwartz TH: The Effects of 0.1 and 1 Percent Erythrocytes and Hemolysis on Serum Chemistry Values. Amer J Clin Path. 66: 639, 1976

23. Ladenson JH, Tsai LMB, Michael JM: Serum Versus Heparinized Plasma for Eighteen Common Chemistry Tests. Is Serum the Appropriate Specimen. Amer J Clin Pathol 62: 545, 1974

24. Letllier G and Gagne M: Study of Drug Interferences with Biochemical Tests in an Interlaboratory Quality Control Program.

Seist G, Young DS (eds) Drug Measurement and Drug Effects in Laboratory Health Science, Basel, Karger, 1980, pp 175–179

25. Lubran M: The Effects of Drugs on Laboratory Values. Med Clin N Am 53: 222, 1969

26. Lum G, Gambino SR: A Comparison of Serum Versus Heparinized Plasma for Routine Chemistry Tests. Am J Clin Pathol 61: 108, 1974

27. Lumsden JH, Mullen K: On Establishing Reference Values. Can J Comp Med 42: 293, 1978

28. Lundberg GD: The Reporting of Laboratory Data Interpretations: To Omit or Commit? JAMA 243: 1554, 1980

29. Meites S: Pediatric Clinical Chemistry. Am Assoc of Clin Chem, Washington, 1977

30. Melby EC, Altman NH (eds): Handbook of Laboratory Animal Science. Vol II, Cleveland, CRC Press, Inc., 1976

31. Mia AS, Jacobs AR, Koger HP: Effects of Anticoagulant on Blood Chemistry Tests. The Practicing Veterinarian Autumn: 18-19, 1976

32. Morrison B *et al.*: Intra-individual Variation in Commonly Analyzed Serum Constituents. Clin Chem 28: 1799, 1979

33. Munan L. Kelly A, Petitclerc C: Statistical and Clinical Norms: Variation on a Familiar Theme. Clin Biochem 11: 75, 1978

34. Myers LJ, Pierce KR: The Effect of EDTA Anticoagulant on Enzyme Activity and pH of Blood. The Southwest Veterinarian. Summer: 287–292, 1972

35. O'Kell RT, Knepper DF, Spoon DF, Elliott JR: Influence of Antibiotics on Laboratory Tests. Clin Chem 17: 352, 1971

36. Panek E, Young DS, Bente J: Analytical Interferences of Drugs in Clinical Chemistry. Am J Med Technol 44: 217, 1978

37. Rawnsley HM: Clinical Laboratory Data. Dartmouth Hitchcock Medical Center, Hanover, 1981 (Unpublished)

38. Rosenblatt G, Stokes J, Bassett DR: Whole Blood Viscosity Hematocrit and Serum Lipid Levels in Normal Subjects and Patients with Coronary Heart Disease. J Lab Clin Med 65: 202, 1965

39. Rosalki SB: Creatine Phosphokinase Determination by UV Enzymatic Method. J Lab Clin Med 69: 696, 1967

40. Schlang HA, Kirpatrick CA: The Effect of Physical Exercise on Serum Transaminase. Am J Med Sci 242: 338, 1961

41. Schoning P, *et al.:* Postmortem Biochemical Changes in Canine Blood. J. Forensic Sci 25: 336, 1980

42. Siest G: Reference Values in Human Chemistry. Effects of Analytical and Individual Variations, Food Intake, Drugs and Toxics—Applications in Preventive Medicine Proceedings of the Second International Colloquium Pont-a-Mousson, 1972 S. Karger, Basel, Munichen, Paris, London New York, Sydney 1973

43. Sonnenwirth AC, Jarett L: Gradwohl's Clinical Laboratory Methods and Diagnosis. Vol I and II. St Louis, Toronto, London, The CV Mosby Company, 1980

44. Speicher CE, Smith JW Jr: Interpretive Reporting in Clinical Pathology. JAMA 243: 1556, 1980

45. Street AE, Chesterman H, Smith GKA, Quinton RM: Prolonged Blood Urea Elevation Observed in the Beagle After Feeding. Toxicol Appl Pharmacol 13: 363, 1968

46. Sunderman FW Jr: Drug Interference in Clinical Biochemistry. Crit Rev Clin Lab Sci 1: 427, 1970

47. Szasz G: A Kinetic Photometric Method for Serum Gamma-Glutamyl Transpeptidase. Clin Chem 15: 124, 1969

48. Tallone-Lombardi L, Sverzellate PP, Casiraghi E, Celia MT: Estimation of Normal Ranges for Various Biochemical Parameters. Biochem Exp Biol 14: 319, 1978

49. Tasker JB: Studies on the Use of Coulter Chemistry in the Veterinary Laboratory Day to Day Reproducibility and the Effects of Storage, Hemolysis, Lipemia, Hyperbilirubinemia and Anticoagulants of Test Results. Cornell Vet 68: 480, 1978

50. Thurston GB: Effects of Hematocrit on Blood Viscoelasticity and in Establishing Normal Values. Biorheology 15: 239, 1978

51. Van Peenen MJ, Files JB: The Effect of Medication on Laboratory Test Results. Am J Clin Pathol 52: 666, 1969

52. Vinet B, Letellier G: The in-vitro Effect of Drugs on Biochemical Parameters Determined by a SMAC System. Clin Biochem 10: 47, 1977

53. Wacker, WEC, Ulner, DD, Vallee BL: Metal Isoenzymes and Myocardial Infarction, Malic and Lactic Dehydrogenase Activities and Zinc Concentration in Serum. N Engl J Med 255–449, 1956

54. Winek CL: Tabulation of Therapeutic, Toxic and Lethal Concentrations of Drugs and Chemicals in Blood. Clin Chem 22: 832, 1976

55. Wirth WA, Thompson RL: The Effects of Various Conditions and Substances on the Results of Laboratory Procedures. Am J Clin Pathol 43: 579, 1965

56. Wright LA, Foster MG: Effect of Some Commonly Prescribed Drugs on Certain Chemistry Tests. Clin Biochem 18: 249, 1980

57. Wright LA, Foster MG: Drug Interference in Clinical Chemistry. Clin Biochem 9: 8, 1976

58. Young DS, Thomas DW, Friedman RB, Pestaner LC: Effects of Drugs on Clinical Laboratory Tests. Clin Chem 18: 1041, 1972

59. Young DS, Pestaner LC, Gilbertman V: Effects of Drugs on Clinical Laboratory Tests. Clin Chem 21: 1D-421D, 1975

60. Young DS, Panek E: Effects of Drugs on the Analytical Procedures of a Multitest Analyzer. Siest G, Young DS (eds), Drug Interference and Drug Measurement in Clinical Chemistry. Basel, Karger, 1976